Remote Sensing Based Building Extraction II

Remote Sensing Based Building Extraction II

Editors

Jiaojiao Tian
Qin Yan
Mohammad Awrangjeb
Beril Kallfelz-Sirmacek
Nusret Demir

MDPI • Basel • Beijing • Wuhan • Barcelona • Belgrade • Manchester • Tokyo • Cluj • Tianjin

Editors
Jiaojiao Tian
Remote Sensing Technology Institute, German Aerospace Center (DLR)
Wessling, Germany

Qin Yan
Chinese Academy of Surveying and Mapping
Beijing, China

Mohammad Awrangjeb
Griffith University
Nathan, Australia

Beril Kallfelz-Sirmacek
Independent Researcher,
Overijssel, The Netherlands

Nusret Demir
Akdeniz University,
Antalya, Turkey

Editorial Office
MDPI
St. Alban-Anlage 66
4052 Basel, Switzerland

This is a reprint of articles from the Special Issue published online in the open access journal *Remote Sensing* (ISSN 2072-4292) (available at: https://www.mdpi.com/journal/remotesensing/special_issues/Building_Detection_Volume_2).

For citation purposes, cite each article independently as indicated on the article page online and as indicated below:

LastName, A.A.; LastName, B.B.; LastName, C.C. Article Title. *Journal Name* **Year**, *Volume Number*, Page Range.

ISBN 978-3-0365-7064-8 (Hbk)
ISBN 978-3-0365-7065-5 (PDF)

Contents

About the Editors

Jiaojiao Tian

Jiaojiao Tian (Dr) is a senior research fellow at the Photogrammetry and Image Analysis department of the Remote Sensing Technology Institute, German Aerospace Center, Germany, where she is currently heading the 3D and modelling group. She received her Ph.D. degree in Mathematics and Computer Sciences from Osnabrueck University in 2013. She holds IEEE senior membership and serves as co-chair of the ISPRS Commission WG I/8: Multi-sensor Modelling and Cross-modality Fusion. Her research interests include 3D change detection, building reconstruction, 3D point cloud segmentation, forest monitoring, and DSM-assisted object extraction and classification.

Qin Yan

Qin Yan is the President of the Chinese Academy of Surveying and Mapping. As a professor of remote sensing, her research interests focus on the remote sensing monitoring of natural resources, and high-resolution imagery mapping and interpretation. She has conducted about 20 research projects and gained three national S&T awards. Additionally, she has published more than 50 papers. She is currently editor-in-chief of the International Journal of Image and Data Fusion and the Journal of Surveying and Mapping Science.

Mohammad Awrangjeb

Dr. Mohammad Awrangjeb is a Senior Lecturer at Griffith University, Australia. His research interests include object extraction and modelling from remote sensing data. His research provides solutions to automated 3D city modelling, the automated modelling and monitoring of power line corridors, the automatic solar potential estimation on buildings, forest vegetation modelling, and biomass estimation. He is the co-author of more than 80 research articles in internationally renowned journals and conferences, and a recipient of the Discovery Early Career Researcher Award of the Australian Research Council (www.arc.gov.au) for the period of 2012–2015.

Beril Kallfelz-Sirmacek

Dr. Beril Kallfelz (Sirmacek) is a Dutch scientist and a professional lover of planet Earth. She holds a PhD degree in Electrical and Electronics Engineering. Her research field focuses on developing automated detection and mapping algorithms via computer vision and AI methods using earth observation data from remote sensing satellite images. She received her PhD degree from Istanbul Yeditepe University in Turkey in collaboration with the Technical University of Munich in Germany. Following her PhD studies, in 2009, she began working as a research scientist at the German Aerospace Centre (DLR). In 2011, she moved back to the Netherlands, where she worked on topics pertaining to earth observation at the Technical University of Delft. In the same period, she also pursued a habilitation study at the University of Osnabrueck in Germany. In 2017, she moved to the east Netherlands where she worked at the University of Twente as a postdoctoral researcher. In 2019, she worked as an assistant professor at Jonkoping University in Sweden; however, due to the pandemic, she mostly stayed in the Netherlands and conducted her educational and research activities remotely. Between March 2021 and 2022, she worked at the Saxion University of Applied Sciences in the Netherlands as an associate professor. For more information, visit: www.BerilSirmacek.com.

Nusret Demir

Nusret Demir (Dr) is an Associate Professor and Vice Dean at the Faculty of Science, as well as a member of the Institute of Science and Technology's Remote Sensing division, at the Space Science and Technologies Department of Akdeniz University in Turkey. He earned his Msc in Geodetic and Photogrammetric Engineering from Yıldız Technical University (YTU) and his PhD in Geomatic Engineering from the ETH Zurich Geomatics Engineering Department. He holds two B.S.C. degrees from YTU: one in Industrial Engineering and the other in Geodetic and Photogrammetric Engineering. Additionally, he is the head of the Turkish Surveying Engineers Chamber's Photogrammetry and Remote Sensing Technical Commission. He has a track record of conducting research on building detection, roof modelling, and LIDAR and SAR data processing. He is currently co-chair of ISPRS Working Group I/5: Microwave and InSAR Technology for Earth Observation.

remote sensing

Editorial

Editorial for Special Issue: "Remote Sensing Based Building Extraction II"

Jiaojiao Tian [1,*], Qin Yan [2], Mohammad Awrangjeb [3], Beril Kallfelz (Sirmacek) [4] and Nusret Demir [5]

1 Remote Sensing Technology Institute, German Aerospace Center (DLR), 82234 Wessling, Germany
2 Chinese Academy of Surveying and Mapping, Beijing 100830, China
3 Institute for Integrated and Intelligent Systems, Griffith University, Nathan, QLD 4111, Australia
4 Independent Scientist, 7553 LL Hengelo, The Netherlands
5 Department of Space Science and Technologies, Remote Sensing Division, Akdeniz University, 07058 Antalya, Turkey
* Correspondence: jiaojiao.tian@dlr.de

Citation: Tian, J.; Yan, Q.; Awrangjeb, M.; Kallfelz (Sirmacek), B.; Demir, N. Editorial for Special Issue: "Remote Sensing Based Building Extraction II". *Remote Sens.* **2023**, *15*, 998. https://doi.org/10.3390/rs15040998

Received: 28 November 2022
Accepted: 1 February 2023
Published: 10 February 2023

1. Introduction

Accurate building extraction from remotely sensed images is essential for topographic mapping, urban planning, disaster management, navigation, and many other applications [1]. The easily available very-high resolution 2D/3D dataset and the rapid development of image processing techniques, especially the convolutional neural networks (CNN) and deep learning techniques have further boosted the research on building-extraction-related topics. Especially in recent years, many research institutes and associations have provided open-source datasets and annotated training data to meet the demand for advanced artificial intelligence models, which brings new opportunities to develop advanced approaches for building extraction and monitoring. Hence, there are higher expectations of the efficiency, accuracy, and robustness of building extraction approaches. They should also meet the demand of processing large datasets at the city, national, and global levels. Moreover, challenges remain on transform learning and dealing with imperfect training data, as well as unexpected objects in urban scenes such as trees, clouds, and shadows.

As a follow-on Special Issue of "Remote Sensing based Building Extraction", this Special Issue "Remote Sensing based Building Extraction II" has further collected the cutting-edge approaches for automatic building segmentation [1–4], vectorization [5,6], and regularization [7], dense matching [8], 3D reconstruction [9–11], and road detection [12]. The proposed methods fall into two main categories depending on the use of the input data sources: 2D building extraction and 3D reconstruction/segmentation.

2. 2D Building Extraction

Deep learning (DL) shows remarkable performance in extracting buildings from high-resolution remote sensing images. How to improve the performance of DL methods, especially the perception of spatial information, is worth further study. Paper [2] proposed a building extraction network (B-FGC-Net) with a feature highlighting, global awareness, and cross-level information fusion to achieve improved profitability of accurate extraction and information integration for both small- and large-scale buildings. Focusing on the promotion of the robustness of the interactive segmentation, Shu et al. [1] propose one Progress Guidance Representation Net (PGR-Net) to utilize the distance of newly added clicks to the boundary of the previous segmentation mask as an indication of the interactive segmentation progress, and this information is employed with the previous segmentation mask and positive and negative clicks to form a progress guidance map. This progress guidance map is then fed into a CNN with the original RGB image. Furthermore, they propose an iterative training strategy for the training of the network and adopt an adaptive zoom-in technique during the inference stage for further performance promotion. Farmland constitutes an important resource for human survival and development. With complex ground features and scattered distribution, building extraction from farmland remains a challenging topic. To this end, Paper [3] proposes an attention-enhanced U-Net for building extraction from farmland based on Google and WorldView-2 remote sensing images. First, a Resnet unit is adopted as the infrastructure of the U-Net network encoding part, then the spatial and channel attention mechanism module is introduced between the Resnet unit and the maximum pool and the multi-scale fusion module is added to improve the U-Net network. Second, the buildings are extracted from WorldView-2 and Google images through farmland boundary constraints. Third, boundary optimization and fusion processing are carried out to further refine the building extraction results. In order to investigate the photovoltaic potential of urban buildings, Paper [4] proposed a pseudo-label-guided self-supervised learning (PGSSL) semantic segmentation network structure to extract building information from high-resolution remote sensing images. The pseudo-label-guided learning method allows the feature results extracted by the pretext task to be more applicable to the target task and ultimately improves segmentation accuracy.

To further close the gap between airborne images and vector representation, Van den Broeck and Goedemé [5] propose a fully automated end-to-end workflow for large-scale roof-part polygon extraction from UHR orthoimagery (0.03 m GSD). Their workflow comprised three steps: (1) An multitask fully convolutional network (FCN) was utilized for the semantic segmentation of roof-part objects and edges; (2) A bottom–up clustering algorithm was used, given the predicted roof-part edges, to derive individual roof-part clusters, where the predicted roof-part object area distinguish roof from non-roof; and (3) The roof-part clusters were vectorized and simplified into polygons. The methodology is trained and tested on a challenging dataset comprising of UHR aerial RGB orthoimagery (0.03 m GSD) and LiDAR-derived digital elevation models (DEMs) (0.25 m GSD) of three Belgian urban areas (including the famous touristic city of Bruges). Li et al. [6] explore the idea of combining three deep learning models, each model performing specific tasks, for automated extraction of building footprint polygons from very high-resolution aerial imagery. Their approach uses the U-Net, Cascade R-CNN, and Cascade CNN models to obtain building segmentation maps, building bounding boxes, and building corners, respectively, thus allowing for the direct production of building maps in a vector format. A polygon construction strategy based on Delaunay triangulation is designed to integrate the outputs from the deep learning models effectively, as well as to generate high-quality vector data. To solve the problem of edge discontinuity and incompleteness generated by semantic edge detection, Xia et al. [7] propose a multitask learning Dense D-LinkNet (DDLNet), which adopts full-scale skip connections and edge guidance module to ensure the effective combination of low-level information and high-level information.

3. 3D Reconstruction/Segmentation

The use of 3D building models is essential and provides realistic data for spatial and environmental analysis for various applications such as creating digital, generating simulations to predict and prepare for future scenarios, and creating various urban analytical processes, especially those that consider environmental impact, which is a growing global concern. To obtain a precise 3D model with lower cost, dense stereo matching has been studied persistently in the field of computer vision, remote sensing, and photogrammetry. Along with the development of deep learning, the Guided Aggregation Network (GA-Net) achieves state-of-the-art performance via the proposed Semi-Global Guided Aggregation layers and reduces the use of costly 3D convolutional layers. To solve the problem of GA-Net requiring large GPU memory consumption, Xia et al. [8] propose an efficient end-to-end network GA-Net-Pyramid for dense matching a pyramid architecture to modify the model. Starting from a downsampled stereo input, the disparity is estimated and continuously refined through the pyramid levels. Thus, the disparity search is only applied for a small size of stereo pair and then confined within a short residual range for minor correction, leading to highly reduced memory usage and runtime. Manual modelling of urban buildings is very time-consuming and costly. Due to the complexity of the dense urban regions, research oriented toward the automatic reconstruction of buildings is still an open topic. In the manuscript titled "Parameter-Free Half-Spaces Based 3D Building Reconstruction Using Ground and Segmented Building Points from Airborne LiDAR Data with 2D Outlines" [9], the authors propose a new half-spaces based algorithm for building reconstruction from airborne laser point clouds. In contrast to the related algorithms, which divide 2D outlines of buildings into smaller parts and then process them while taking only convex shapes into account, the proposed algorithm performs reconstruction without division, while also considering concave parts of the rooftops. The method works in two stages, where the input data is processed first to obtain the definition of the base model of each building and the corresponding half-spaces. In the second stage, a building shape is generated by performing 3D Boolean operations over the analysed half-spaces.

A major challenge of large-scale building reconstruction from airborne LiDAR point clouds is the reconstruction of missing vertical walls. Paper [10] provided a fully automatic building reconstruction approach to infer vertical walls based on the connection between planar segments of both roofs and walls. The reconstruction model is obtained by using an extended hypothesis-and-selection-based polygonal surface reconstruction framework. Experimental results demonstrated that the proposed method is superior to the state-of-the-art methods in terms of reconstruction accuracy and robustness. The study also generated a new dataset consisting of the point clouds and 3D models of 20k real-world buildings which can stimulate research in urban reconstruction and the use of 3D city models in urban applications. To further refine the extracted building boundaries, Hui et al. [11] propose a multi-constraints graph segmentation method for building extraction from airborne LiDAR data and achieve satisfactory results. The graph structure is generated using the three-dimensional spatial features of points. To reduce computational cost the point-based building extraction is transformed into an object-based building extraction and geometric morphological features are computed for each segmented object. Finally, a multi-scale progressively growing optimisation method is employed to recover the omitted building parts.

Besides buildings, digital maps of road networks are a vital part of digital cities and intelligent transportation. This study [12] provided a comprehensive review of road extraction based on various remote sensing data sources. It is divided into three parts. Part 1 provides an overview of the existing data acquisition techniques for road extraction, including data acquisition methods, typical sensors, application status, and prospects. Part 2 underlines the main road extraction methods based on four data sources. Part 3 presents the combined application of multisource data for road extraction. It can provide a comprehensive reference for research on existing road extraction technologies.

Acknowledgments: We want to thank the authors who contributed to this Special Issue on "Remote Sensing Based Building Extraction II", as well as the reviewers who provided the authors with comments and very constructive feedback.

Conflicts of Interest: The authors declare no conflict of interest.

References

1. Shu, Z.; Hu, X.; Dai, H. Progress Guidance Representation for Robust Interactive Extraction of Buildings from Remotely Sensed Images. *Remote Sens.* **2021**, *13*, 5111. [CrossRef]
2. Wang, Y.; Zeng, X.; Liao, X.; Zhuang, D. B-FGC-Net: A Building Extraction Network from High Resolution Remote Sensing Imagery. *Remote Sens.* **2022**, *14*, 269. [CrossRef]
3. Li, C.; Fu, L.; Zhu, Q.; Zhu, J.; Fang, Z.; Xie, Y.; Guo, Y.; Gong, Y. Attention Enhanced U-Net for Building Extraction from Farmland Based on Google and WorldView-2 Remote Sensing Images. *Remote Sens.* **2021**, *13*, 4411. [CrossRef]
4. Chen, D.-Y.; Peng, L.; Zhang, W.-Y.; Wang, Y.-D.; Yang, L.-N. Research on Self-Supervised Building Information Extraction with High-Resolution Remote Sensing Images for Photovoltaic Potential Evaluation. *Remote Sens.* **2022**, *14*, 5350. [CrossRef]
5. Van den Broeck, W.A.J.; Goedemé, T. Combining Deep Semantic Edge and Object Segmentation for Large-Scale Roof-Part Polygon Extraction from Ultrahigh-Resolution Aerial Imagery. *Remote Sens.* **2022**, *14*, 4722. [CrossRef]

6. Li, Z.; Xin, Q.; Sun, Y.; Cao, M. A Deep Learning-Based Framework for Automated Extraction of Building Footprint Polygons from Very High-Resolution Aerial Imagery. *Remote Sens.* **2021**, *13*, 3630. [CrossRef]
7. Xia, L.; Zhang, J.; Zhang, X.; Yang, H.; Xu, M. Precise Extraction of Buildings from High-Resolution Remote-Sensing Images Based on Semantic Edges and Segmentation. *Remote Sens.* **2021**, *13*, 3083. [CrossRef]
8. Xia, Y.; D'Angelo, P.; Fraundorfer, F.; Tian, J.; Fuentes Reyes, M.; Reinartz, P. GA-Net-Pyramid: An Efficient End-to-End Network for Dense Matching. *Remote Sens.* **2022**, *14*, 1942. [CrossRef]
9. Bizjak, M.; Žalik, B.; Lukač, N. Parameter-Free Half-Spaces Based 3D Building Reconstruction Using Ground and Segmented Building Points from Airborne LiDAR Data with 2D Outlines. *Remote Sens.* **2021**, *13*, 4430. [CrossRef]
10. Huang, J.; Stoter, J.; Peters, R.; Nan, L. City3D: Large-Scale Building Reconstruction from Airborne LiDAR Point Clouds. *Remote Sens.* **2022**, *14*, 2254. [CrossRef]
11. Hui, Z.; Li, Z.; Cheng, P.; Ziggah, Y.Y.; Fan, J. Building Extraction from Airborne LiDAR Data Based on Multi-Constraints Graph Segmentation. *Remote Sens.* **2021**, *13*, 3766. [CrossRef]
12. Jia, J.; Sun, H.; Jiang, C.; Karila, K.; Karjalainen, M.; Ahokas, E.; Khoramshahi, E.; Hu, P.; Chen, C.; Xue, T.; et al. Review on Active and Passive Remote Sensing Techniques for Road Extraction. *Remote Sens.* **2021**, *13*, 4235. [CrossRef]

remote sensing MDPI

Article

Precise Extraction of Buildings from High-Resolution Remote-Sensing Images Based on Semantic Edges and Segmentation

Liegang Xia *, Junxia Zhang, Xiongbo Zhang, Haiping Yang and Meixia Xu

College of Computer Science and Technology, Zhejiang University of Technology, Hangzhou 310023, China; 2111912201@zjut.edu.cn (J.Z.); 2111812141@zjut.edu.cn (X.Z.); yanghp@zjut.edu.cn (H.Y.); 2111912149@zjut.edu.cn (M.X.)
* Correspondence: xialg@zjut.edu.cn; Tel.: +86-571-8529-0027

Abstract: Building extraction is a basic task in the field of remote sensing, and it has also been a popular research topic in the past decade. However, the shape of the semantic polygon generated by semantic segmentation is irregular and does not match the actual building boundary. The boundary of buildings generated by semantic edge detection has difficulty ensuring continuity and integrity. Due to the aforementioned problems, we cannot directly apply the results in many drawing tasks and engineering applications. In this paper, we propose a novel convolutional neural network (CNN) model based on multitask learning, Dense D-LinkNet (DDLNet), which adopts full-scale skip connections and edge guidance module to ensure the effective combination of low-level information and high-level information. DDLNet has good adaptability to both semantic segmentation tasks and edge detection tasks. Moreover, we propose a universal postprocessing method that integrates semantic edges and semantic polygons. It can solve the aforementioned problems and more accurately locate buildings, especially building boundaries. The experimental results show that DDLNet achieves great improvements compared with other edge detection and semantic segmentation networks. Our postprocessing method is effective and universal.

Keywords: building extraction; high-resolution remote-sensing image; semantic edge detection; semantic segmentation

Citation: Xia, L.; Zhang, J.; Zhang, X.; Yang, H.; Xu, M. Precise Extraction of Buildings from High-Resolution Remote-Sensing Images Based on Semantic Edges and Segmentation. *Remote Sens.* **2021**, *13*, 3083. https://doi.org/10.3390/rs13163083

Academic Editors: Jiaojiao Tian, Qin Yan, Mohammad Awrangjeb, Beril Sirmacek and Nusret Demir

Received: 1 July 2021
Accepted: 3 August 2021
Published: 5 August 2021

1. Introduction

The automatic extraction and analysis of buildings from high-resolution remote-sensing images is an important research topic in the field of remote sensing [1–3]. The results are widely used in urban and rural planning, sociology, change detection [4], natural disaster assessment and other fields and are also important for updating geographic information databases [5]. Compared with natural images, high-resolution remote-sensing images have richer spatial, spectral, and texture feature information. The rapid development of deep learning and computer vision has provided strong technical support for the analysis and use of high-resolution remote-sensing images [6]. Compared with artificial remote-sensing interpretation and vectorization, the CNN model can automatically extract buildings from remote-sensing images, which greatly reduces the consumption of human and material resources. However, due to the complex characteristics and background of geographic objects (geo-objects) [7], cases in which different geo-objects have the same spectrum or the same geo-objects have different spectra are commonly encountered, which makes accurate pixel-level classification a difficult problem. In addition, problems such as cloud, tree and shadow occlusion [4,8], different imaging angles [9], difficulty in drawing labels, and label omissions hinder the accurate estimation of buildings by CNN. Using the characteristics of high-resolution remote-sensing images to perform tasks such as image recognition and detection while avoiding the negative effects of redundant information has become the most challenging frontier issue in the field of remote sensing.

In high-resolution remote-sensing images, buildings are geo-objects with artificial features, rich in types and semantic information, which are great differences in scale, architectural style, and form. It is difficult to ensure the accuracy of building location that depending on high-level semantic features and the accuracy of building boundary that depending on low-level edge features. Currently, most building extraction algorithms are based on semantic segmentation and semantic edge detection.

With the development of deep learning, edge detection algorithm has developed rapidly. HED [10] uses a fully convolutional network with deep supervision to automatically learn multilevel representations and effectively solve the problem of edge ambiguity. However, HED only considers the last convolutional layer information of each stage, while RCF [11] makes full use of multiscale and multilevel information of all convolutional layers. BDCN [12] introduced the scale enhancement module to use multiscale representations to improve the edge detection capabilities. DFF [13] adaptively assigns appropriate fusion weights strategy which helps to produce more accurate and clear edge predictions. CaseNet [14] is an end-to-end semantic edge network based on ResNet [15] that proposes a new skip layer in which the category edge activations of the top layer share and merge the same group features.

The rapid development of deep learning edge detection algorithms provides strong support for building edge detection in high-resolution remote-sensing images. Reda et al. [16] proposed a faster edge region convolutional neural network (FER-CNN). FER-CNN uses the parametric rectified linear unit (PReLu) [17] activation function, which adds only a very small number of parameters to improve the edge detection of buildings. Lu et al. [2] adopted a building edge detection model based on RCF [11], which obtains the edge strength map through the RCF and then refines the edge strength map according to the geometric analysis of the terrain surface. Semantic edge detection which aims at extracting edges as well as semantic information can generate an edge strength map to describe the confidence of the predicted building boundary, but complete edges are difficult to guarantee, and incomplete edges are not sufficient to support the accurate extraction of buildings.

Semantic segmentation is a joint task that requires the positioning and classification of both spatial information and semantic information, paving the way for a complete understanding of the scene. FCN [18] uses the concept of full convolution to perform end-to-end semantic segmentation, and it creatively employs a skip connection that combines high-level information with low-level information to improve segmentation. U-Net [19] modified and expanded the FCN that adds an upsampling stage and a feature channel fuse strategy that uses a connection operation to directly pass high-level information from an encoder to the decoder of the same height. SegNet [20] performs forward evaluation of the fully learned function to obtain smooth predictions and increases the depth of the network so that the network can consider the larger context information. RefineNet [21] is a multipath optimization network that refines low-resolution semantic features in a recursive manner, and proposes a chain residual pool that can capture the context of the background. PSPNet [22] expand the receptive field by dilated convolution to obtain feature maps that can acquire the global scene. PSPNet perform pooling at different levels, and then fuse the local information and global context information.

The powerful feature extraction and interpretation ability of semantic segmentation provides a new method for the interpretation of high-resolution remote-sensing images, which is helpful for the accurate extraction and positioning of buildings and reduces the problems of false extraction and missing detection of buildings. Liu et al. [23] proposed a spatial residual inception network (SRI-Net), in which an SRI module captures and aggregates multiscale contexts for semantic understanding by successively fusing multilevel features. SRI-Net is capable of accurately detecting large buildings while retaining global morphological characteristics and local details. Delassus et al. [24] proposed a fusion strategy based on U-Net [19], which combines the segmented output of the combined model and multiple channels of the input image. Lin et al. [25] proposed an efficient separable factorized network (ESFNet), which uses separable residual blocks and dilated convolution to maintain a small loss of accuracy as well as low computational cost and memory consumption. At the same time, the high precision of semantic segmentation is maintained. Yi et al. [7] proposed a ResUNet based on U-Net and Resnet. It uses a deep residual learning method to promote training and alleviate the problem of model training degradation. Shuang Wang [26] proposed a full convolutional network with dense connections that designed top-down short connections to facilitate the fusion of high and low feature information. However, there are usually irregular boundaries that are difficult to completely match the boundaries of the actual building, and it is impossible to effectively distinguish adjacent buildings [5].

Multitask learning is a learning mechanism inspired by human beings to acquire knowledge of complex tasks by performing different shared subtasks simultaneously [27]. Its aim is to leverage useful information contained in related tasks to help improve the generalization performance of all the tasks [28]. Multitask learning is currently a mainstream direction of deep learning. At present, there are some methods that use multitask learning to integrate edge information and semantic information in CNN to output semantic polygons with precise edges. Even if deep learning methods are widely used in high-resolution remote-sensing building extraction, it is still a challenging task to achieve the precise extraction of building.

Inspired by the multitask learning and the aforementioned problems of semantic edge detection and semantic segmentation in building extraction, we designed a novel CNN model based on multitask learning to achieve accurate extraction of buildings, Dense D-LinkNet (DDLNet), based on D-LinkNet [29] and DenseNet [30]. DDLNet has good adaptability to both semantic segmentation tasks and semantic edge detection tasks. A new universal postprocessing method focuses on the complementarity between edge information and semantic information. The main contributions of this work are as follows.

1. We designed a CNN model named Dense D-LinkNet (DDLNet) to extract buildings from high-resolution remote-sensing images. This model uses full-scale skip connections and edge guidance module to ensure the effective combination of low-level information and high-level information. DDLNet can adapt to both semantic segmentation tasks and edge detection tasks. DDLNet can effectively solve the problem of boundary blur and the problem of edge disconnection.
2. We proposed an effective and universal postprocessing method that can effectively combine edge information and semantic information to improve the final result. Semantic polygon from the semantic segmentation to accurately locate and classify buildings at the pixel level. semantic edges from semantic edge detection to extract precise edges of buildings. This method uses semantic polygons to solve the problem of incompleteness of semantic edges and uses semantic edges to improve the boundary of semantic polygons, realize the accurate extraction of buildings.

2. Materials and Methods

The main purpose of this article is to overcome the incompleteness of edges from semantic edge detection and the problem of boundary blur from semantic segmentation and realize the precise extraction of buildings from remote-sensing images.

In this paper, we designed a novel CNN model Dense D-LinkNet (DDLNet), which can adapt to semantic segmentation tasks and semantic edge detection tasks. In addition, we propose a new postprocessing method to effectively fuse edge and semantic information and achieve the precise extraction of buildings from high-resolution remote-sensing images. The process of extracting buildings can be divided into three stages, as shown in Figure 1. First, the high-resolution remote-sensing images are labeled, and then the vector data are gridded into line label and polygon label, respectively. Second, DDLNet are trained and predicted to generate edge strength maps and semantic polygons, respectively. Then, in the postprocessing stage, edge information is used to improve the boundary of the semantic polygon to achieve more accurate boundary positioning, and the semantic polygon is used to repair discontinuous edges to ensure the continuity and integrity of the edge. Supplemented by the application of the watershed algorithm, semantic information is used as the seed point to select the result of the building to ensure the accuracy of the positioning and topological structure of the building.

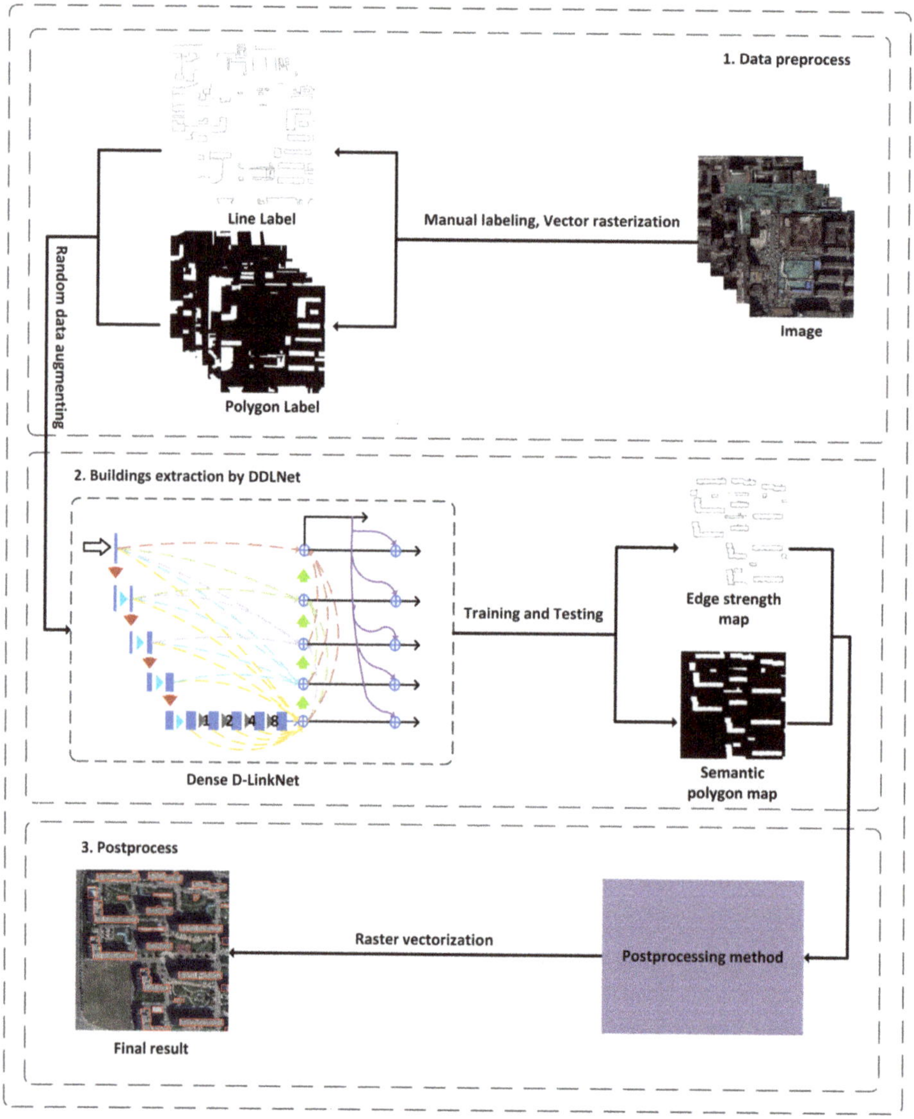

Figure 1. Overall architecture of our method. This architecture can be divided into three stages. The first stage is data preprocessing, the second stage is deep learning training and prediction, which are used for DDLNet to generate edge strength maps and semantic polygons. The last stage is postprocessing, which is used for fusing edge and semantic information to obtain more accurate building results.

2.1. Dense D-LinkNet

Dense D-LinkNet (DDLNet) keeps the D-LinkNet [29] core structure and adds a full-scale skip connection, deep multiscale supervision, edge guidance module on this basis. D-LinkNet is built with the LinkNet [30] for road extraction tasks and adds dilated convolution layers in its center part. The encoder-decoder structure is the core of D-LinkNet. The encoder with pooling layers increases the receptive field but loses low-level details at the same time [5]. It is difficult to recover lost features with only the upsampling operation of the decoder. D-LinkNet only directly maps the features from the encoder to the corresponding decoder. Inspired by DenseNet [31] and U-Net3+ [32], we add the full-scale skip connection which is used to combine all the underlying low-level details with the decoder features to produce more accurate results. What's more, the edge guidance module and deep multiscale supervision are also used to produce more accurate results. The structure of DDLNet is shown in Figure 2.

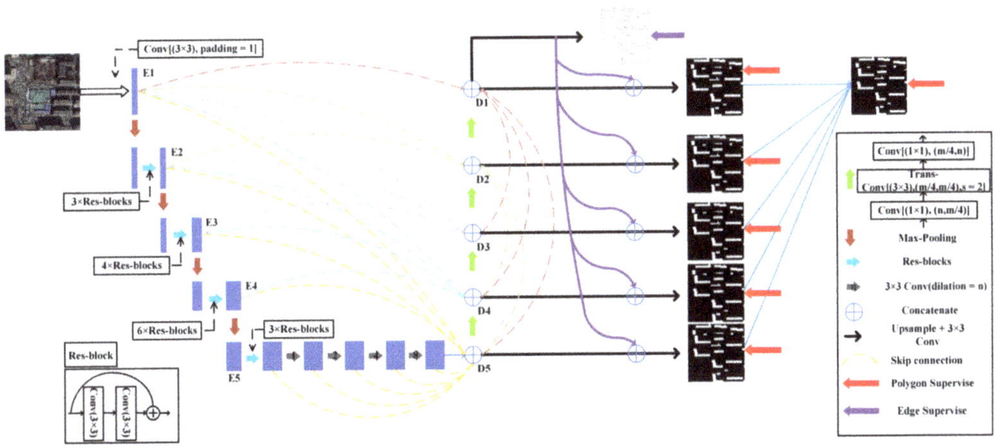

Figure 2. Dense D-LinkNet architecture. Each blue block represents a multichannel feature. Left is the encoder, and the right is the decoder. The dotted curves represent full-scale skip connection. The purple curves represent edge guidance module. Each convolution layer is activated by ReLU, except the last convolution layer, which uses sigmoid activation.

2.1.1. Full-Scale Skip Connections

D-LinkNet just uses the skip connection to combine the same-scale feature from encoder to decoder at the same height. Our full-scale skip connections incorporate low-level details with high-level semantics from feature maps at different scales. Each decoder layer in DDLNet contains larger and same-scale feature maps from the encoder and smaller-scale feature maps from the decoder, which capture fine-grained details and coarse-grained semantics at full scales [32]. We mark the five encoding layers of the encoder as E1, E2, E3, E4, E5 from top to bottom, and mark the five decoding layers of the decoder as D1, D2, D3, D4, D5 from top to bottom, simultaneously. In Figure 2, we have made the corresponding mark. Each decoding layer combines low-level edge features and high-level semantic features through channel concatenate. Therefore, the D1 is the combination of E1, D2, D3, D4, D5. D2 is the combination of E1, E2, D3, D4, D5. D3 is the combination of E1, E2, E3, D4, D5. D4 is the combination of E1, E2, E3, E4, D5. D5 is the combination of E1, E2, E3, E4, and the features of E5 after a series of dilated convolutions.

It is well known that low-level features have richer details, while high-level features have richer semantic information. Therefore, the features after combination have both rich semantic information and spatial details [33]. This means that the full-scale features from the encoder can be used to improve the features in the decoder and the final result which contains high-precision boundary and semantic information.

2.1.2. Deep Multiscale Supervision

To learn hierarchical representations from multiscale feature, deep multiscale supervision is adopted in DDLNet. Multiscale objects are also a challenge for CNN. Currently, the extraction of target feature is conducted on a certain scale because of receptive field of convolution. Different levels have dissimilar high-level information and dissimilar low-level information [4].The feature map from each decoder layer (D1, D2, D3, D4, D5) should be predicted, and loss could be calculated with the groundtruth separately to realize detection at different scales, which is conducive to achieving deep supervision of each layer of the decoder and enhancing the learning ability [33]. Feature fusion is of great help to the promotion of targets at different scales.

The final polygon output is fused with each layer, as shown in Equation (1),

$$\hat{Y}_{final} = \sum_{i=1}^{5} w_i \hat{Y}_i \tag{1}$$

Here, $\hat{Y}$ is the predicted polygon result, subscript i is the number of each scale and subscript final represents the final polygon output, and w is the weight to fuse the polygon output of the layer. We choose w = 0.2 for each polygon output.

2.1.3. Edge Guidance Module

In this module, we aim to extract the precise edge features, then leverage the edge features to guide the polygon features to perform better on both segmentation and boundary. To obtain precise edge features, we decided to perform the edge supervise at the last layer of the decoder. It is well known that low-level features have richer details. However, only low-level information is not enough, while high-level semantic information also needed. The last layer of the decoder (D1) that contains the low-level features from the corresponding encoder and full-scale high-level features from the previous decoder is appropriate and effective for edge detection.

After obtaining the edge feature and polygon features, we aim to leverage the edge features to guide the polygon features. In our module, we propose the one-to-one guidance method. The polygon features from different decoder layer (D2, D3, D4, D5) need to be upsampled to the size of the edge feature and the combination of edge feature and each polygon feature is realized by channel concatenation. By fusing the edge feature into polygon features, the location of high-level predictions is more accurate, and more importantly, the boundary details become better.

2.1.4. Loss

Semantic Edge Loss: In end-to-end training, the loss function is computed over all pixels in a training image X and edge label Y. For a typical high-resolution remote-sensing image, the distribution of edge/nonedge pixels is heavily biased: 90% of the groundtruth is nonedged and 10% is edge [10]. HED introduces a class-balancing weight β to offset this imbalance between edges and nonedges. HED defines the following class-balanced cross-entropy (CBCE) loss function used in Equation (2):

$$Loss_{cbce} = -\beta \sum \log\left(\hat{Y}_j \in |Y-|\right) - (1-\beta) \sum \log\left(\hat{Y}_j \in |Y+|\right) \tag{2}$$

where $\beta = |Y+|/|Y|$ and $1-\beta = |Y-|/|Y|$. $|Y+|$ and $|Y-|$ denote the edge and nonedge, respectively, $|Y| = |Y+| + |Y-|$ denote the number of pixels. $\hat{Y}$ is the prediction map, and subscript $j \in [0, 1, \ldots \text{H} \times \text{W}]$.

Currently, most edge detection networks, such as RCF, BDCN, and DexiNed, use the CBCE loss function to achieve edge detection for natural images. However, for high-resolution remote-sensing images, the CBCE loss will produce fuzzy and rough edges that cannot satisfy the requirement of drawing tasks and engineering applications. Therefore, we use the class-balanced mean square error (CMSE) loss that add the class-balanced parameter based on MSE loss. The CMSE loss can generate thin edge strength maps that are plausible for human eyes. The CMSE loss represents the sum of squares of the differences between the predicted value and the target value and then averages them. Equation (3) for CMSE loss is as follows: m = H × W, $\beta = |Y+|/|Y|$:

$$Loss_{cmse} = \frac{1}{m}\sum(\beta((\hat{Y}_j \in |Y-|) - Y_j)^2 + (1-\beta)((\hat{Y}_j \in |Y+|) - Y_j)^2) \quad (3)$$

Semantic Segmentation Loss: For semantic segmentation, we used binary cross-entropy (BCE) and dice coefficient loss as the loss function. The formula of BCE loss as show in Equation (4).

$$Loss_{bce} = -\sum Y_i \times \log(\hat{Y}_j) - \sum(1 - Y_i) \times \log(1 - \hat{Y}_j) \quad (4)$$

The dice coefficient is a set similarity measure function, which is usually used to calculate the similarity of two samples, and the value range is [0, 1]. The formula of dice loss is shown in Equation (5):

$$Loss_{dice} = 1 - 2 \times \frac{\hat{Y} \cap Y}{\hat{Y} + Y} \quad (5)$$

$\hat{Y} \cap Y$ is the intersection between $\hat{Y}$ and Y. $\hat{Y}$ and Y sub tables represent the number of elements of $\hat{Y}$ and Y, where the molecular coefficient is 2. The total loss can use the following Equation (6) to represent:

$$Loss_{polygon} = Loss_{bce} + Loss_{dice} \quad (6)$$

A major challenge in multitask learning comes from the optimization process itself. In particular, we need to carefully balance the joint training process of all tasks to avoid the situation that one or more tasks have a dominant influence in the network weights. In extreme cases, when the loss of one task is very large and the loss of other tasks is very small, the multitask is almost degenerated into single task goal learning, and the weight of the network is almost completely updated according to the large loss task, gradually losing the advantage of multitask learning. Therefore, we need the weight to balance semantic edge loss and semantic segmentation loss. The formula of final weighted loss is shown in Equation (7):

$$Loss_{all} = w_e Loss_{cmse} + w_p Loss_{polygon} \quad (7)$$

where w_e is the weight of semantic edge loss and w_p is the weight of semantic segmentation loss. We observed the final loss convergence of the DDLNet and determined that w_e = 100, w_p = 6 are suitable parameters.

2.2. Postprocessing

To fully fuse edge information and semantic information, complementary information is used to obtain better prediction results and achieve the precise extraction of buildings. We propose a new postprocessing method.

The semantic edge detection networks and semantic segmentation networks are used to generate the edge strength maps and semantic polygon results, respectively. The prediction image is a grayscale image with pixel values in the range of 0 to 255. The choice of binarization threshold is crucial. Considering that our postprocessing method can remove redundant pixels and more pixels are needed to ensure the results. Thus, we choose the threshold of 100 instead of the threshold of 127 as usual. After binarization, multipixel-width edges cannot represent the building edges. Therefore, we use a skeleton extraction algorithm to refine the edges to a single-pixel width and delete some of the broken lines that exist separately. Thus, we obtained a single-pixel edge map (as shown in Figure 3a) and a binary semantic polygon (as shown in Figure 3b).

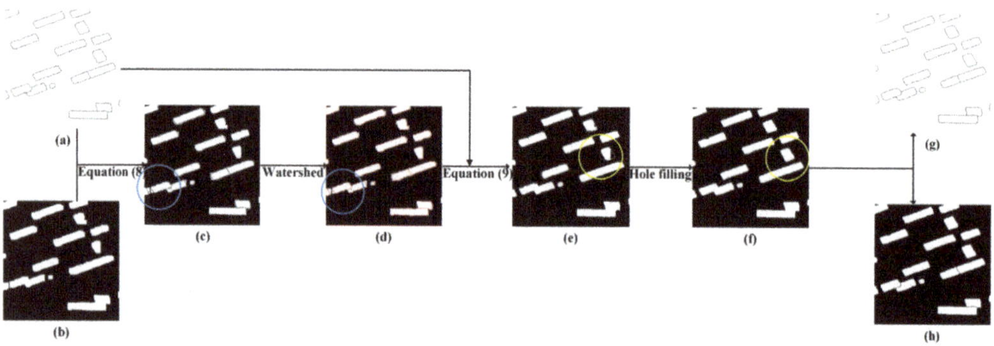

Figure 3. Postprocessing flow chart. (**a**) is a single-pixel edge map, and (**b**) is a semantic polygon map. (**c**) is the result in which Equation (8) is used to overlay edges on the semantic polygon. (**d**) is the result in which the watershed algorithm is used to extract the boundaries. (**e**) is the result in which Equation (9) is used to add edges on the semantic polygon. (**f**) is a complete semantic polygon. (**g**) is the final edge result, and (**h**) is the final semantic segmentation result. The most obvious improvement is indicated by the circle. The blue line represents the effect of Equation (8) and the watershed algorithm, and the yellow line represents the effect of Equation (9) and hole filling.

The problem of edge discontinuity can be solved by the integrity of the semantic polygon. The problem of inaccurate boundaries of semantic polygons can be solved by semantic edges. The improvement in the semantic polygon boundaries is mainly reflected in the following two aspects of our method:

1. If the boundary of the semantic polygon is beyond the edge of a single pixel, it needs to be deleted.
2. If the boundary of the semantic polygon is within the boundary of a single pixel, it needs to be supplemented.

In the first case, we propose Equation (8) to overlay the accurate single-pixel edge to the precise semantic polygon result. This formula can highlight the edge information in the semantic polygon, where the edge will be set to 0. Therefore, we obtain a semantic polygon limited by the edge (as shown in Figure 3c). Equation (8) is given below. The result represents the result image, the edge represents the single-pixel edge image, the polygon represents the semantic polygon image, and (x, y) represents the coordinate of the image.

$$Result(x,y) = \begin{cases} 255, polygon(x,y) > edge(x,y) \\ 0, polygon(x,y) \leq edge(x,y) \end{cases} \tag{8}$$

we use the watershed algorithm [34] to delete the redundant semantic polygon and extract the edges. According to the input seed points, the watershed algorithm delimits the region ownership of each pixel, and the value of the boundary between regions is set to "−1" to distinguish. Accurate semantic polygons are used to mark the seed points inside the restricted semantic polygon. The seed points come from the erosion operation of semantic polygons. To prevent the disappearance of seed points due to excessive erosion operations, we choose to iterate five times to obtain seeds after many experiments. Then, the desired correctly predicted polygon boundary is obtained (as shown in Figure 3d). Then, we use the hole filling algorithm to fill boundaries into semantic polygons. At this time, we can ensure that all the semantic polygons are within the single-pixel edge. The details of the watershed algorithm are shown in Figure 4.

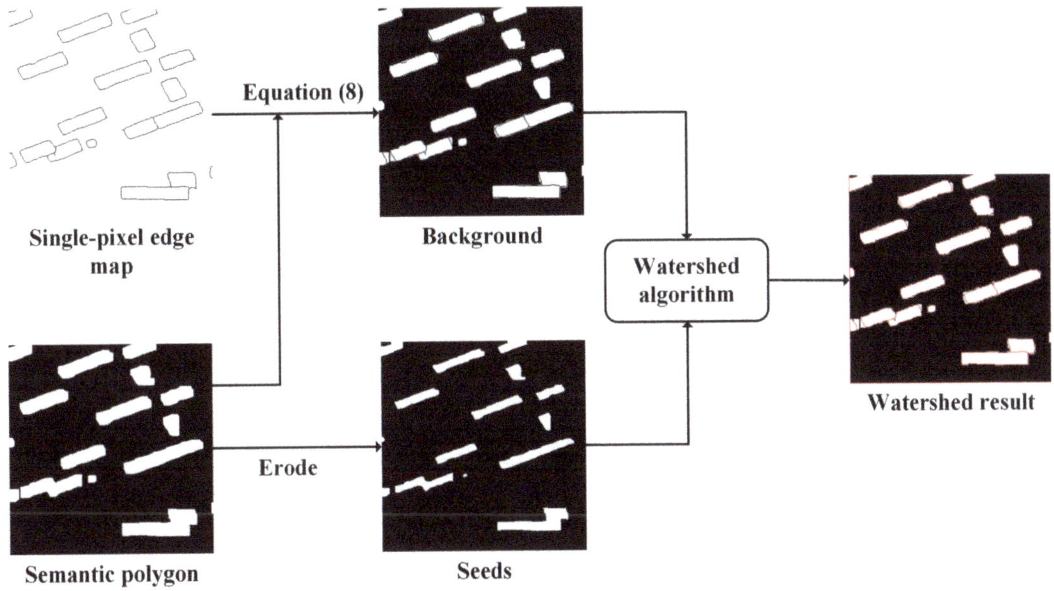

Figure 4. Details of watershed algorithm. The single-pixel edge and semantic polygon use Equation (8) to generate background. The semantic polygon is eroded to generate seeds for the watershed algorithm. The watershed algorithm uses seeds to extract the correct edge (red line in watershed result) in the background.

In the second case, the semantic polygon object and single-pixel edge are fused by Equation (9), which can highlight the edge information outside the semantic polygon. (as shown in Figure 3e), after filling the holes, a complete semantic polygon (as shown in Figure 3f) can be obtained. At this time, the boundary of the obtained semantic polygon results becomes more regular and fits the real building boundary.

$$Result(x,y) = \&\begin{cases} 255, polygon(x,y) \,||\, edge(x,y) = 255 \\ 0, polygon(x,y) \,\&\&\, edge(x,y) = 0 \end{cases} \tag{9}$$

Finally, we solve the problem that adjacent buildings cannot be distinguished in semantic segmentation. The eight-neighborhood algorithm is used to determine whether the edge is the boundary of adjacent buildings and obtain the result of precise semantic segmentation of buildings (as shown in Figure 3h). This operation overcomes the above problems by fusing accurate edge information, and the accurate edge of the building (as shown in Figure 3g) is obtained.

More importantly, through the series of operations mentioned above, the integrity of the semantic polygon is used to repair the broken line in the single-pixel edge map, and the details of repairing a broken line are shown in Figure 5. If the boundary of the semantic polygon cannot realize the connection of the broken line, the broken line will be removed, and the boundary of the semantic polygon will be retained. The final edge result is guaranteed to be complete.

In the postprocessing stage, by fusing the accurate edge information with the accurate semantic segmentation information, our method can make the positioning of high-level prediction more accurate. More importantly, the edge and segmentation details become better, especially the edge of buildings, which cannot be recognized by semantic segmentation, and the problem of broken lines in edge detection.

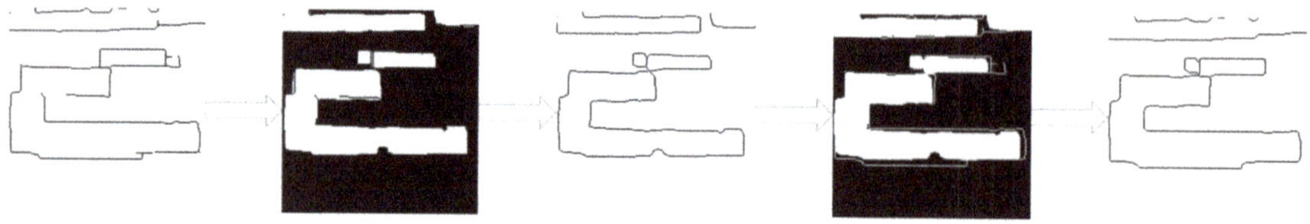

Figure 5. Details of repairing a broken line.

3. Results

3.1. Dataset

We choose two high-resolution remote-sensing image areas of representative buildings as our experimental area to evaluate our method on different datasets. Moreover, one is from Google images, and the other is aerial images.

(1) The first is Beijing, the capital of China. The Beijing scene represents a typical Chinese urban landscape, including different types of buildings, which are difficult to accurately discriminate and extract. We selected four districts in the center of Beijing. The original Google images with a spatial resolution of 0.536 m are Chaoyang District, Haidian District, Dongcheng District, and Xicheng District.

(2) The second is Zhonglu countryside, located in Weixi County, Diqing Prefecture, Yunnan Province. As a typical representative of rural architecture, its buildings have a relatively regular and uniform building shape. The original aerial images with a spatial resolution of 0.075 m.

We select samples with typical architectural features and a certain number of negative samples that do not contain buildings in the abovementioned study area, draw the precise boundary of the building manually, generate corresponding line labels and polygon labels, and randomize divide 80% of the samples are used as the training set, and 20% of the samples are used as the test set. The Beijing area contains 320 training sets of 512 pixel × 512-pixel images and 80 test sets of 512 pixel × 512-pixel images. The Zhonglu area contains 75 training sets of 1024 pixel × 1024-pixel images and 23 test sets of 1024 pixel × 1024-pixel images. The data set is shown in Figure 6.

3.2. Training Details

In our experiments, all deep learning network models are implemented using the PyTorch framework. DDLNet and the comparative experiments are all trained on an NVIDIA RTX TITAN (with 24 G memory) graphics card. We initialize the weights of the DDLNet with the weights of a ResNet34 model pretrained via ImageNet [35]. Some hyperparameters are set as follows: The batchsize on Beijing dataset and Zhonglu dataset are 4 and 2, respectively. The initial learning rate of DDLNet is 2×10^{-4}, and the learning rate is updated for 1/4 of the total epochs. We trained 800 epochs on Beijing dataset and 400 epochs on Zhonglu dataset with DDLNet. Due to the insufficient amount of data, we adopt a data enhancement operation including random cropping, rotation, translation, and horizontal flipping operations after entering the network to expand the dataset and reduce overfitting.

3.3. Evaluation Metrics

We use Intersection over Union (IoU) and F1 score to evaluate the performance of our method. The IoU index represents the overlap ratio between the predicted area and the real area of an image. The higher the overlap rate is, the higher the accuracy of the predicted results. Equation (10) is as follows: A represents the prediction area, and B represents the real label area:

$$\text{Polygon IoU} = \frac{A \cap B}{A \cup B} \tag{10}$$

However, IoU can only evaluate the prediction accuracy of the building polygon result but cannot reflect the prediction accuracy of the building boundary. Therefore, we propose the boundary IoU method to evaluate the accuracy between the predicted building edge and the real building boundary. This method expands the edge by kernel size = 5 pixels and then uses Equation (11) to calculate the accuracy. Exp represents expand and *ks* represents kernel size:

$$\text{Boundary IoU} = \frac{\text{Exp}(A, ks) \cap \text{Exp}(B, ks)}{\text{Exp}(A, ks) \cup \text{Exp}(B, ks)} \tag{11}$$

Polygon IoU can reflect the completeness of the edge, and boundary IoU can reflect the accuracy of the polygon boundary.

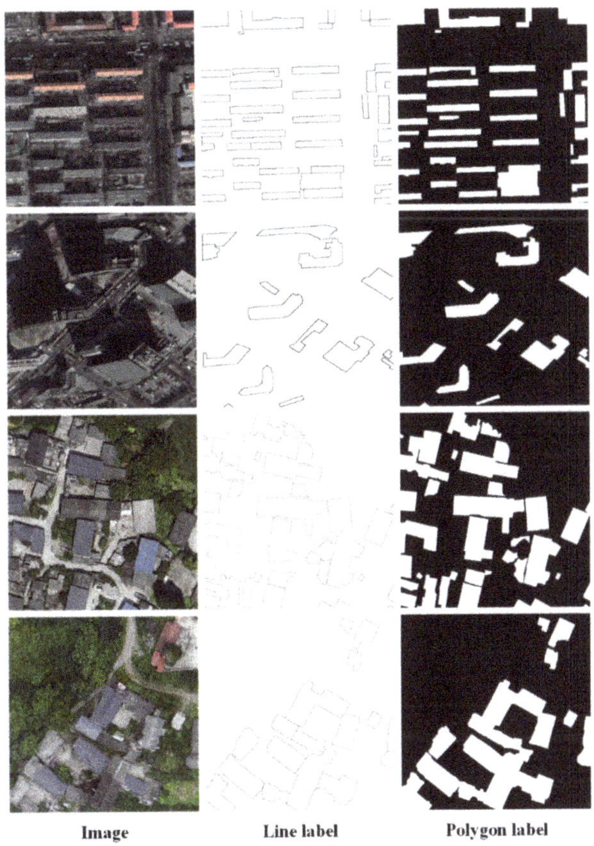

Figure 6. Dataset display. Image represents high-resolution remote-sensing image, line label and polygon label are ground truth of the image, the first two are Beijing dataset, the last two are Zhonglu dataset.

The F1 score is an index used to measure the accuracy of a two-category model. To calculate the F1 score, it is necessary to calculate the precision and recall. In the following formulas, true positives (TP) represent the number of positive pixels belonging to buildings that are correctly identified. True negatives (TN) represent the number of negative pixels belonging to nonbuildings that are correctly identified. False positives (FP) represent the number of negative pixels belonging to nonbuildings that are incorrectly identified as positive pixels belonging to buildings. False negatives (FN) represent the number of positive pixels belonging to buildings that are incorrectly identified as negative pixels belonging to nonbuildings. The explanation of the above TP, TN, FP, FN indicators is shown in Figure 7.

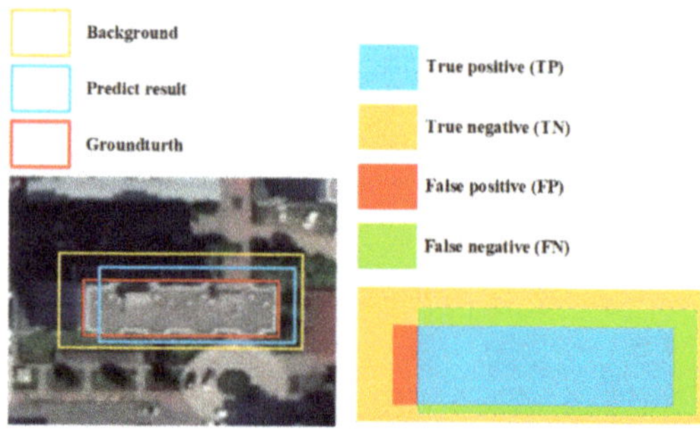

Figure 7. Graphic representation of the TP, TN, FP, FN for the matched ground truth and predicted result.

Precision is the ratio of true positives in the identified positive pixels, and Equation (12) is as follows.

$$\text{Precision} = \frac{\text{TP}}{\text{TP} + \text{FP}} \quad (12)$$

Recall is the proportion of all positive pixels in the test set that are correctly identified as positive pixels, and Equation (13) is as follows.

$$\text{Recall} = \frac{\text{TP}}{\text{TP} + \text{FN}} \quad (13)$$

The F1 score is the harmonic mean value of the precision rate and recall rate, which suggests that the precision rate and recall rate are equally important. The larger the value, the stronger the model's ability, and Equation (14) is as follows.

$$\text{F1 score } = 2 \times \frac{\text{Precision} * \text{Recall}}{\text{Precision} + \text{Recall}} \quad (14)$$

3.4. Results

To further demonstrate the effectiveness of our methods, we select some soft-of-the-art models to compare with our model and the postprocessing method.

First, for semantic edge detection, to assess the quality of our DDLNet model, three models, namely RCF, BDCN, and DexiNed, were selected for comparison with DDLNet on the Beijing and Zhonglu datasets.

In the Beijing dataset, RCF, BDCN and DexiNed achieved boundary IoU of 0.2815, 0.4087 and 0.4503, respectively. DDLNet achieved a boundary IoU of 0.5116, which greatly surpasses the accuracy of the other models. As shown in Table 1, RCF, BDCN, and DexiNed achieved polygon IoU of 0.2751, 0.5110 and 0.1724, respectively. DDLNet achieved a polygon IoU of 0.5295, which is 3.62% more than that of BDCN. The results of those semantic edge detection models on the Beijing test dataset are summarized in Table 1, and their performance are shown in Figure 8.

Table 1. The results of semantic edge detection on the Beijing dataset.

Study Area	Methods	Boundary IoU	Polygon IoU	F1 Score
Beijing	RCF	0.2815	0.2751	0.4300
	BDCN	0.4087	0.5110	0.6341
	DexiNed	0.4503	0.1724	0.6124
	DDLNet	**0.5116**	**0.5295**	**0.7049**

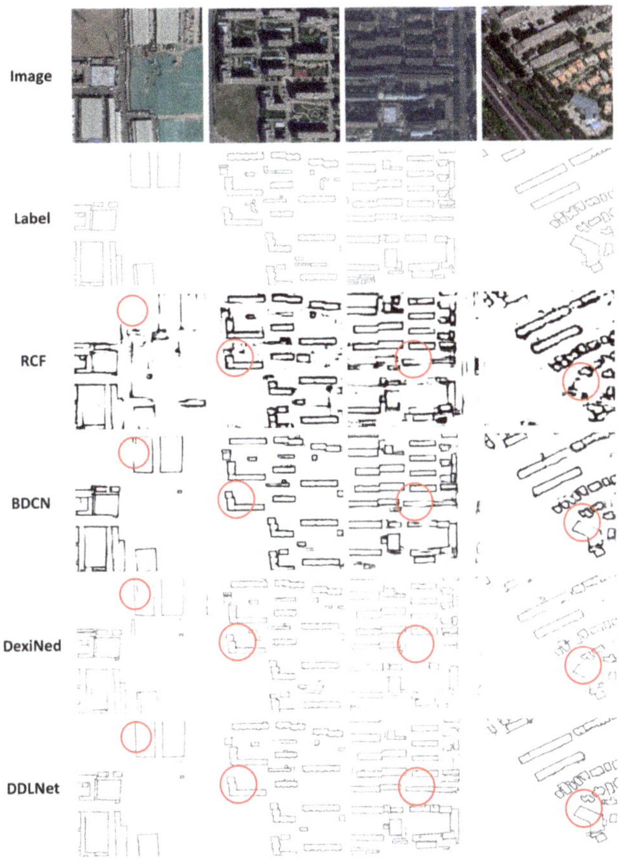

Figure 8. The edge results of RCF, BDCN, DexiNed, and DDLNet on the Beijing dataset.

In the Zhonglu dataset, RCF, BDCN, DexiNed, DDLNet achieved boundary IoU values of 0.4378, 0.7050, 0.6326 and 0.7399 and achieved polygon IoU values of 0.5824, 0.7009, 0.6452 and 0.8719, respectively. The results of those semantic edge detection models on the Zhonglu test dataset are summarized in Table 2, and their performances are shown in Figure 9.

Table 2. The results of semantic edge detection on the Zhonglu dataset.

Study Area	Methods	Boundary IoU	Polygon IoU	F1 Score
Zhonglu	RCF	0.4378	0.5824	0.5677
	BDCN	0.7050	0.7009	0.7182
	DexiNed	0.6326	0.6452	0.6604
	DDLNet	**0.7399**	**0.8719**	**0.7582**

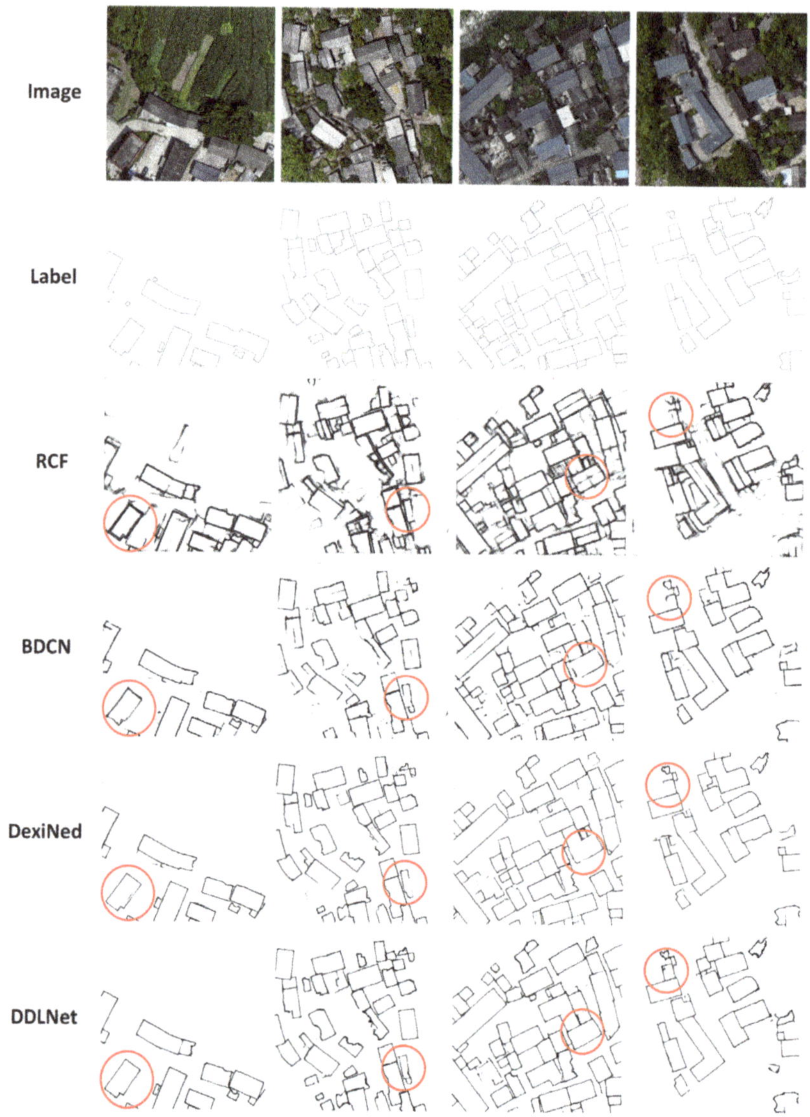

Figure 9. The edge results of RCF, BDCN, DexiNed, and DDLNet on the Zhonglu dataset.

The capability of DDLNet for semantic edge detection tasks is demonstrated on two different data sets. RCF does not perform well on our dataset. BDCN can effectively extract the building boundary and ensure the integrity of the boundary, but its edge is blurred and insufficient in accuracy. DexiNed can produce more accurate and visual edges, but its edge integrity is difficult to guarantee. DDLNet can achieve edge integrity beyond DexiNed and BDCN and effectively extract the boundaries of buildings. It means that it is effective to provide more high-level semantic features for low-level edge features to realize semantic edge extraction.

Second, for semantic segmentation, we also selected three advanced models, namely U-Net, U-Net3+, and D-LinkNet, for the experiment on the Beijing and Zhonglu datasets.

In the Beijing dataset, U-Net, U-Net3+, and D-LinkNet achieved 0.6726, 0.7161, and 0.7212 in polygon IoU, respectively. DDLNet achieved the top performance of 0.7527 of the polygon IoU, which was better than all other models, and even 4.36% more than D-LinkNet. U-Net3+ and DDLNet use the full-scale skip connection to help network learning, and they achieved boundary IoU of 0.4731 and 0.4746, respectively. This greatly surpassed the accuracy of U-Net and D-LinkNet, which achieve boundary IoU of 0.4281 and 0.4438, respectively. The results of those semantic segmentation models on the Beijing test dataset are summarized in Table 3, and their performances are shown in Figure 10.

Table 3. The results of semantic segmentation on the Beijing dataset.

Study Area	Methods	Boundary IoU	Polygon IoU	F1 Score
Beijing	U-Net	0.4281	0.6726	0.8048
	U-Net3+	0.4731	0.7161	0.8352
	D-LinkNet	0.4438	0.7212	0.8398
	DDLNet	**0.4746**	**0.7527**	**0.8607**

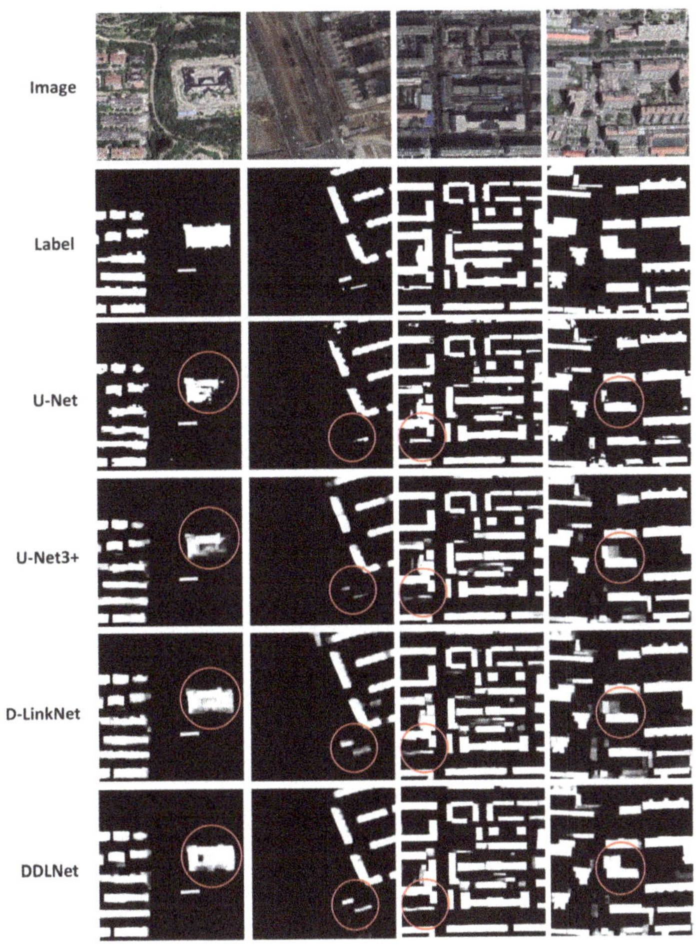

Figure 10. The semantic polygon results of U-Net, U-Net3+, D-LinkNet, and DDLNet on the Beijing dataset.

In the Zhonglu dataset, U-Net, Unet3+, and D-LinkNet achieved 0.7067, 0.8855, and 0.9261 in polygon IoU, respectively. DDLNet achieved the best performance of 0.9364 of the polygon IoU. U-Net3+ and DDLNet achieve boundary IoU of 0.5396 and 0.6905 which greatly surpassed the accuracy of U-Net and D-LinkNet. The results of those semantic segmentation models on the Zhonglu test dataset are summarized in Table 4, and their performances are shown in Figure 11.

Table 4. The results of semantic segmentation on the Zhonglu dataset.

Study Area	Methods	Boundary IoU	Polygon IoU	F1 Score
Zhonglu	U-Net	0.4180	0.7067	0.9004
	U-Net3+	0.5396	0.8855	0.9122
	D-LinkNet	0.6861	0.9261	0.9537
	DDLNet	**0.6905**	**0.9364**	**0.9584**

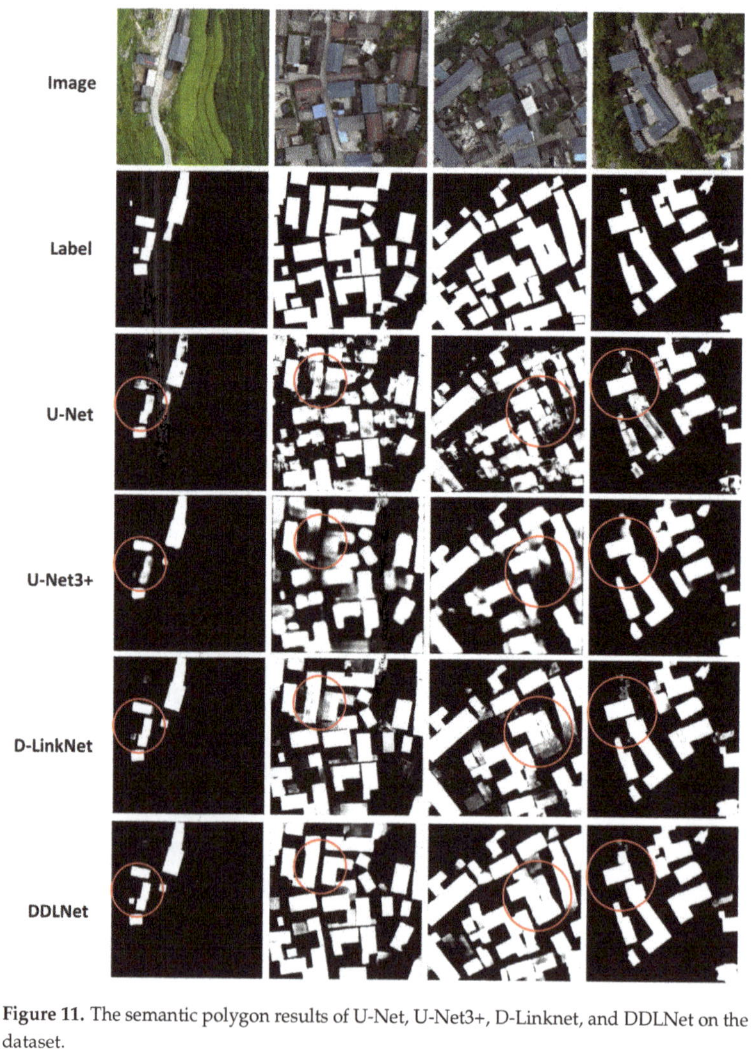

Figure 11. The semantic polygon results of U-Net, U-Net3+, D-Linknet, and DDLNet on the Zhonglu dataset.

The capability of DDLNet for semantic segmentation tasks is demonstrated on two different data sets. U-Net3+ and DDLNet achieve boundary IoU greatly surpassed the accuracy of U-Net and D-LinkNet that proves that the full-scale skip connection is effective in improving the boundary of the polygon from semantic segmentation. The result of DDLNet proves that making full use of low-level edge information proved to be helpful in extracting buildings from high-resolution remote sense images.

Moreover, we evaluated the effectiveness of the postprocessing method. We choose a variety of semantic edge models and semantic segmentation models to verify the effectiveness of our postprocessing scheme. The criteria we chose were that the boundary IoU of the semantic edge model was larger than that of the semantic segmentation model to improve the edge accuracy of the semantic polygon, and the polygon IoU of the semantic segmentation model was larger than that of the semantic edge model to improve the integrity of the semantic edge.

Based on the criteria, in the Beijing test dataset, we choose DDLNet combined with DDLNet, D-LinkNet, U-Net3+, and U-Net. DexiNed combined with D-LinkNet and U-Net. Compared with Tables 1 and 3, the results of postprocessing improve the polygon IoU of semantic edge detection and the boundary IoU of semantic segmentation. In addition, the results are closer to manual vision. The combination and the results of the combination are shown in Table 5 and Figure 12.

Table 5. The results of the postprocessing method on the Beijing dataset.

Study Area	Methods	Boundary IoU	Polygon IoU
Beijing	DDLNet&DDLNet	**0.5227**	**0.7531**
	DDLNet&D-LinkNet	0.5067	0.7297
	DDLNet&U-Net3+	0.5075	0.7217
	DDLNet&U-Net	0.4909	0.7072
	DexiNed&D-LinkNet	0.4775	0.7239
	DexiNed&U-Net	0.4525	0.6784

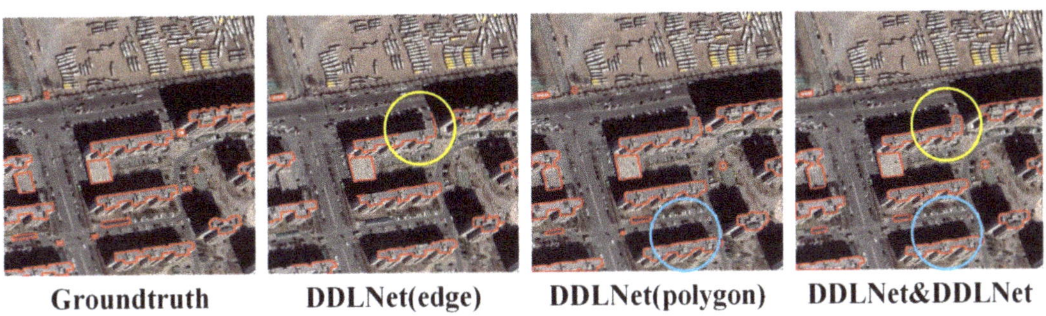

Figure 12. The results of postprocessing method with DDLNet on the Beijing dataset. DDLNet (edge) represent the edge result of DDLNet. In addition, DDLNet (polygon) represent the polygon result of DDLNet. DDLNet&DDLNet represent the result of postprocessing method. The blue circle mark shows the improvement of semantic segmentation where the boundary is closer to the real boundary of building, and the yellow circle mark shows the improvement of semantic edge where the disconnection edge was repaired completely.

Based on the criteria, in the Zhonglu test dataset, we choose DDLNet combined with DDLNet, D-LinkNet, U-Net3+, and U-Net. BDCN combined with DDLNet, D-LinkNet, U-Net3+, U-Net. DexiNed combined with D-LinkNet and U-Net. RCF combined with U-Net. Compared with Tables 2 and 4, the results of postprocessing improve the polygon IoU of semantic edge detection and the boundary IoU of semantic segmentation. The combination and the results of the combination are shown in Table 6 and Figure 13.

Table 6. The results of the postprocessing method on the Zhonglu dataset.

Study Area	Methods	Boundary IoU	Polygon IoU
Zhonglu	DDLNet&DDLNet	**0.7428**	**0.9415**
	DDLNet&D-LinkNet	0.7368	0.9360
	DDLNet&U-Net3+	0.7265	0.9212
	DDLNet&U-Net	0.7256	0.9225
	DexiNed&U-Net3+	0.7085	0.9124
	DexiNed&U-Net	0.7134	0.9131
	BDCN&DDLNet	0.7123	0.9371
	BDCN&D-LinkNet	0.7068	0.9291
	BDCN&U-Net3+	0.6815	0.9038
	BDCN&U-Net	0.6809	0.8906
	RCF&U-Net	0.6270	0.8643

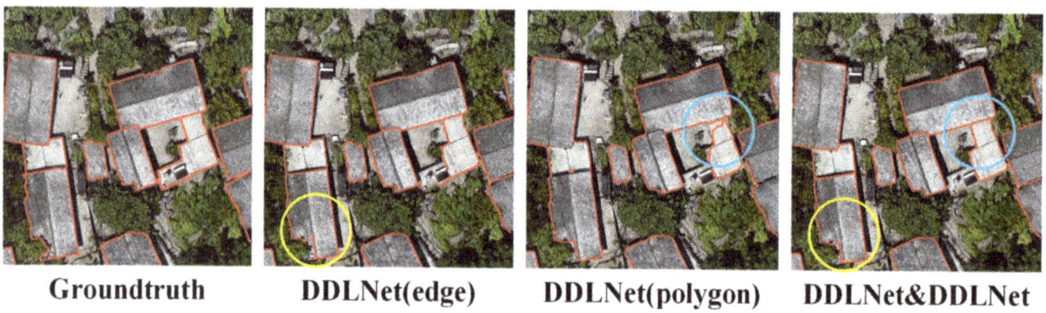

Figure 13. The results of postprocessing method with DDLNet on the Zhonglu dataset. DDLNet (edge) represent the edge result of DDLNet. In addition, DDLNet (polygon) represent the polygon result of DDLNet. DDLNet&DDLNet represent the result of postprocessing method. The blue circle mark shows the improvement of semantic segmentation where the boundary is closer to the real boundary of building, and the yellow circle mark shows the improvement of semantic edge where the disconnection edge was repaired completely.

In summary, we conducted comparative experiments on two different datasets with other SOTA models to verify whether our methods could obtain high-quality results. Experiments confirmed that our model DDLNet had better results than other SOTA models in both semantic edge detection tasks and semantic segmentation tasks and all evaluation metrics, which not only indicated that our models have a good performance in building extraction but also indicates that the edge guidance module and full-scale skip connection are conducive to the automatic extraction of buildings in a network. What's more, our postprocessing method is effective and further improved results of building extraction that helps to improve the vectorization of the result.

4. Discussion

There are certain shortcomings of neural network-based deep learning methods. The edge detection network usually adopts a multiscale fusion strategy to preserve more detailed predictions, resulting in fuzzy and insufficient refinement of the final edge results, as well as difficulty in solving edge disconnection problems. From the experiments and results in the fourth section, DDLNet has better detection accuracy and visual effects than the other edge detection models. The decoder combines the semantic information of the previous layer and the edge information of this layer to improve the accuracy. Through the especially designed CMSE loss, the problem of edge blur, roughness and disconnection is reduced to a certain extent. However, the current loss function design still has problems, and the final edge result still has a disconnection problem.

For the semantic segmentation task, which focuses on the pixel-level classification of targets, the previous semantic segmentation network structure focuses more on the contextual and semantic information of the targets, ignoring the importance of edge information, which leads to difficulty in matching the boundaries of the final semantic polygons with the boundaries of real buildings. Compared with the current semantic segmentation, DDLNet has been improved to a certain extent. From the boundary IoU indicator, we find that the full-scale skip connection and edge guidance module are simple and effective to improve the boundary of buildings.

Convolutional neural networks adopt downsampling pooling operations in encoder and upsampling operations in decoder, and the use of downsampling to compress data is irreversible, resulting in information loss and therefore causing translation invariance and poor results [4]. The loss of information is irreversible, so we design a method to fuse the edge information and semantic information from a postprocessing perspective to achieve precise building extraction. The experimental results show the effectiveness of our postprocessing method, and the final result shows both the edge precision of edge detection and the semantic precision of semantic segmentation. After postprocessing, the boundary IoU and polygon IoU of most final results have improved.

However, some results provide lower-boundary IoU compared with the initial single edge detection in Table 5. As we consider, while the polygon cannot realize the supplement of the disconnection line, we will remove the disconnection line and choose the polygon boundary. Moreover, the polygon may lead to a false detection polygon in the final result. Considering that the polygon boundary may be poor, the final result may have a lower-boundary IoU. At the same time, the hyperparameter setting of the watershed algorithm will also influence the result. This means that the postprocessing process can be further improved.

5. Conclusions

This article focuses on solving the problem of edge discontinuity and incompleteness generated by semantic edge detection, and the polygon shape generated by semantic segmentation is irregular, which does not match the actual building boundary. We propose a novel CNN model named Dense D-LinkNet (DDLNet). DDLNet uses full-scale skip connection, deep multiscale supervision and edge guidance module to overcome the aforementioned problem. The experimental results show that DDLNet is useful and has a certain degree of improvement in the evaluation indicator boundary IoU and polygon IoU in both semantic edge detection tasks and semantic segmentation tasks. Moreover, our postprocessing method is effective and universal and can arbitrarily fuse semantic edge information from edge detection with semantic polygons from semantic segmentation to improve the quality of the final result of buildings.

Author Contributions: L.X. and J.Z. designed and completed the experiments and wrote the article. X.Z., H.Y. and M.X. guided this process and helped with the writing of the paper. All authors have read and agreed to the published version of the manuscript.

Funding: This work was supported in part by the National Key Research and Development Program of China under Grants 2018YFB0505300 and 2017YFB0503600, in part by the National Natural Science Foundation of China under Grant 41701472, in part by the Zhejiang Provincial Natural Science Foundation of China under Grant No. LQ19D010006.

Institutional Review Board Statement: Not applicable.

Informed Consent Statement: Not applicable.

Data Availability Statement: The code of DDLNet is publicly available at https://github.com/Pikachu-zzZ/DDLNet (accessed on 20 June 2021).

Acknowledgments: We would like to acknowledge the Landthink in Suzhou for supporting the Google Earth data and aerial image data.

Conflicts of Interest: No potential conflict of interest was reported by the authors.

References

1. Gilani, S.A.N.; Awrangjeb, M.; Lu, G. Segmentation of airborne point cloud data for automatic building roof extraction. *GISci. Remote Sens.* **2018**, *55*, 63–89. [CrossRef]
2. Lu, T.; Ming, D.; Lin, X.; Hong, Z.; Bai, X.; Fang, J. Detecting building edges from high spatial resolution remote sensing imagery using richer convolution features network. *Remote Sens.* **2018**, *10*, 1496. [CrossRef]
3. Hung, C.-L.J.; James, L.A.; Hodgson, M.E. An automated algorithm for mapping building impervious areas from airborne LiDAR point-cloud data for flood hydrology. *GISci. Remote Sens.* **2018**, *55*, 793–816. [CrossRef]
4. Yang, G.; Zhang, Q.; Zhang, G. EANet: Edge-aware network for the extraction of buildings from aerial images. *Remote Sens.* **2020**, *12*, 2161. [CrossRef]
5. Bischke, B.; Helber, P.; Folz, J.; Borth, D.; Dengel, A. Multi-task learning for segmentation of building footprints with deep neural networks. In Proceedings of the 2019 IEEE International Conference on Image Processing (ICIP), Taipei, Taiwan, 22–25 September 2019; pp. 1480–1484.
6. Yang, H.; Yu, B.; Luo, J.; Chen, F. Semantic segmentation of high spatial resolution images with deep neural networks. *GISci. Remote Sens.* **2019**, *56*, 749–768. [CrossRef]
7. Yi, Y.; Zhang, Z.; Zhang, W.; Zhang, C.; Li, W.; Zhao, T. Semantic segmentation of urban buildings from vhr remote sensing imagery using a deep convolutional neural network. *Remote Sens.* **2019**, *11*, 1774. [CrossRef]
8. Huang, H.; Sun, G.; Rong, J.; Zhang, A.; Ma, P. Multi-feature Combined for Building Shadow detection in GF-2 Images. In Proceedings of the 2018 Fifth International Workshop on Earth Observation and Remote Sensing Applications (EORSA), Xi'an, China, 18–20 June 2018; pp. 1–4.
9. Ji, S.; Wei, S.; Lu, M. Fully convolutional networks for multisource building extraction from an open aerial and satellite imagery data set. *IEEE Trans. Geosci.* **2018**, *57*, 574–586. [CrossRef]
10. Xie, S.; Tu, Z. Holistically-nested edge detection. In Proceedings of the IEEE International Conference on Computer Vision, Santiago, Chile, 7–13 December 2015; pp. 1395–1403.
11. Liu, Y.; Cheng, M.-M.; Hu, X.; Wang, K.; Bai, X. Richer convolutional features for edge detection. In Proceedings of the IEEE Conference on Computer Vision and Pattern Recognition, Honolulu, HI, USA, 21–26 July 2017; pp. 3000–3009.
12. He, J.; Zhang, S.; Yang, M.; Shan, Y.; Huang, T. Bi-directional cascade network for perceptual edge detection. In Proceedings of the IEEE Conference on Computer Vision and Pattern Recognition, Long Beach, CA, USA, 15–20 June 2019; pp. 3828–3837.
13. Hu, Y.; Chen, Y.; Li, X.; Feng, J. Dynamic feature fusion for semantic edge detection. *arXiv* **2019**, arXiv:1902.09104.
14. Yu, Z.; Feng, C.; Liu, M.-Y.; Ramalingam, S. Casenet: Deep category-aware semantic edge detection. In Proceedings of the IEEE Conference on Computer Vision and Pattern Recognition, Honolulu, HI, USA, 21–26 July 2017; pp. 5964–5973.
15. He, K.; Zhang, X.; Ren, S.; Sun, J. Deep residual learning for image recognition. In Proceedings of the IEEE Conference on Computer Vision and Pattern Recognition, Las Vegas, NV, USA, 27–30 June 2016; pp. 770–778.
16. Reda, K.; Kedzierski, M. Detection, Classification and Boundary Regularization of Buildings in Satellite Imagery Using Faster Edge Region Convolutional Neural Networks. *Remote Sens.* **2020**, *12*, 2240. [CrossRef]
17. He, K.; Zhang, X.; Ren, S.; Sun, J. Delving deep into rectifiers: Surpassing human-level performance on imagenet classification. In Proceedings of the IEEE International Conference on Computer Vision, Santiago, Chile, 7–13 December 2015; pp. 1026–1034.
18. Long, J.; Shelhamer, E.; Darrell, T. Fully convolutional networks for semantic segmentation. In Proceedings of the IEEE Conference on Computer Vision and Pattern Recognition, Boston, MA, USA, 8–10 June 2015; pp. 3431–3440.
19. Ronneberger, O.; Fischer, P.; Brox, T. U-net: Convolutional networks for biomedical image segmentation. In Proceedings of the International Conference on Medical Image Computing and Computer-Assisted Intervention, Munich, Germany, 5–9 October 2015; pp. 234–241.
20. Badrinarayanan, V.; Kendall, A.; Cipolla, R. Segnet: A deep convolutional encoder-decoder architecture for image segmentation. *IEEE Trans. Pattern Anal. Mach. Intell.* **2017**, *39*, 2481–2495. [CrossRef] [PubMed]
21. Lin, G.; Milan, A.; Shen, C.; Reid, I. Refinenet: Multi-path refinement networks for high-resolution semantic segmentation. In Proceedings of the IEEE Conference on Computer Vision and Pattern Recognition, Honolulu, HI, USA, 21–26 July 2017; pp. 1925–1934.
22. Zhao, H.; Shi, J.; Qi, X.; Wang, X.; Jia, J. Pyramid scene parsing network. In Proceedings of the IEEE Conference on Computer Vision and Pattern Recognition, Honolulu, HI, USA, 21–26 July 2017; pp. 2881–2890.
23. Liu, P.; Liu, X.; Liu, M.; Shi, Q.; Yang, J.; Xu, X.; Zhang, Y. Building footprint extraction from high-resolution images via spatial residual inception convolutional neural network. *Remote Sens.* **2019**, *11*, 830. [CrossRef]
24. Delassus, R.; Giot, R. CNNs Fusion for Building Detection in Aerial Images for the Building Detection Challenge. In Proceedings of the CVPR Workshops, Salt Lake City, UT, USA, 18–22 June 2018; pp. 242–246.
25. Lin, J.; Jing, W.; Song, H.; Chen, G. ESFNet: Efficient network for building extraction from high-resolution aerial images. *IEEE Access* **2019**, *7*, 54285–54294. [CrossRef]
26. Wang, S.; Zhou, L.; He, P.; Quan, D.; Zhao, Q.; Liang, X.; Hou, B. An Improved Fully Convolutional Network for Learning Rich Building Features. In Proceedings of the IGARSS 2019–2019 IEEE International Geoscience and Remote Sensing Symposium, Yokohama, Japan, 28 July–2 August 2019; pp. 6444–6447.

27. Batra, A.; Singh, S.; Pang, G.; Basu, S.; Jawahar, C.; Paluri, M. Improved road connectivity by joint learning of orientation and segmentation. In Proceedings of the IEEE/CVF Conference on Computer Vision and Pattern Recognition, 15–20 June 2019; pp. 10385–10393.
28. Zhang, Y.; Yang, Q. A survey on multi-task learning. *IEEE Trans. Knowl. Data Eng.* **2021**. [CrossRef]
29. Zhou, L.; Zhang, C.; Wu, M. D-LinkNet: LinkNet With Pretrained Encoder and Dilated Convolution for High Resolution Satellite Imagery Road Extraction. In Proceedings of the CVPR Workshops, Salt Lake City, UT, USA, 18–22 June 2018; pp. 182–186.
30. Chaurasia, A.; Culurciello, E. Linknet: Exploiting encoder representations for efficient semantic segmentation. In Proceedings of the 2017 IEEE Visual Communications and Image Processing (VCIP), Saint Petersburg, FL, USA, 10–13 December 2017; pp. 1–4.
31. Huang, G.; Liu, Z.; Van Der Maaten, L.; Weinberger, K.Q. Densely connected convolutional networks. In Proceedings of the IEEE Conference on Computer Vision and Pattern Recognition, Honolulu, HI, USA, 21–26 July 2017; pp. 4700–4708.
32. Huang, H.; Lin, L.; Tong, R.; Hu, H.; Zhang, Q.; Iwamoto, Y.; Han, X.; Chen, Y.-W.; Wu, J. Unet 3+: A full-scale connected unet for medical image segmentation. In Proceedings of the ICASSP 2020–2020 IEEE International Conference on Acoustics, Speech and Signal Processing (ICASSP), Virtual Conference, Barcelona, Spain, 4–8 May 2020; pp. 1055–1059.
33. He, C.; Li, S.; Xiong, D.; Fang, P.; Liao, M. Remote Sensing Image Semantic Segmentation Based on Edge Information Guidance. *Remote Sens.* **2020**, *12*, 1501. [CrossRef]
34. Vincent, L.; Soille, P. Watersheds in digital spaces: An efficient algorithm based on immersion simulations. *IEEE Trans. Pattern Anal. Mach. Intell.* **1991**, *13*, 583–598. [CrossRef]
35. Deng, J.; Dong, W.; Socher, R.; Li, L.-J.; Li, K.; Fei-Fei, L. Imagenet: A large-scale hierarchical image database. In Proceedings of the 2009 IEEE Conference on Computer Vision and Pattern Recognition, Miami, FL, USA, 20–25 June 2009; pp. 248–255.

remote sensing

MDPI

Article

A Deep Learning-Based Framework for Automated Extraction of Building Footprint Polygons from Very High-Resolution Aerial Imagery

Ziming Li [1], Qinchuan Xin [1,2,*], Ying Sun [1] and Mengying Cao [1]

1 Guangdong Key Laboratory for Urbanization and Geo-Simulation, School of Geography and Planning, Sun Yat-sen University, Guangzhou 510275, China; lizm9@mail2.sysu.edu.cn (Z.L.); sunying23@mail.sysu.edu.cn (Y.S.); caomy7@mail2.sysu.edu.cn (M.C.)

2 State Key Laboratory of Desert and Oasis Ecology, Research Center for Ecology and Environment of Central Asia, Chinese Academy of Sciences, Urumqi 830011, China

* Correspondence: xinqinchuan@mail.sysu.edu.cn

Abstract: Accurate building footprint polygons provide essential data for a wide range of urban applications. While deep learning models have been proposed to extract pixel-based building areas from remote sensing imagery, the direct vectorization of pixel-based building maps often leads to building footprint polygons with irregular shapes that are inconsistent with real building boundaries, making it difficult to use them in geospatial analysis. In this study, we proposed a novel deep learning-based framework for automated extraction of building footprint polygons (DLEBFP) from very high-resolution aerial imagery by combining deep learning models for different tasks. Our approach uses the U-Net, Cascade R-CNN, and Cascade CNN deep learning models to obtain building segmentation maps, building bounding boxes, and building corners, respectively, from very high-resolution remote sensing images. We used Delaunay triangulation to construct building footprint polygons based on the detected building corners with the constraints of building bounding boxes and building segmentation maps. Experiments on the Wuhan University building dataset and ISPRS Vaihingen dataset indicate that DLEBFP can perform well in extracting high-quality building footprint polygons. Compared with the other semantic segmentation models and the vector map generalization method, DLEBFP is able to achieve comparable mapping accuracies with semantic segmentation models on a pixel basis and generate building footprint polygons with concise edges and vertices with regular shapes that are close to the reference data. The promising performance indicates that our method has the potential to extract accurate building footprint polygons from remote sensing images for applications in geospatial analysis.

Keywords: building footprint; map vectorization; convolutional neural network; semantic segmentation

Citation: Li, Z.; Xin, Q.; Sun, Y.; Cao, M. A Deep Learning-Based Framework for Automated Extraction of Building Footprint Polygons from Very High-Resolution Aerial Imagery. *Remote Sens.* **2021**, *13*, 3630. https://doi.org/10.3390/rs13183630

Academic Editors: Mohammad Awrangjeb, Qin Yan, Beril Sirmacek, Jiaojiao Tian and Nusret Demir

Received: 13 July 2021
Accepted: 9 September 2021
Published: 11 September 2021

1. Introduction

Information on the spatial distribution and changes of buildings has a wide range of applications in urban studies, such as urban planning, disaster management, population estimation, and map updating [1,2]. Local bureaus of urban planning and natural resource management used to expend high levels of manpower and material resources to obtain an accurate raster (i.e., map features described by a matrix of pixels, where each pixel contains an associated value) and vector (i.e., map features delineated by discrete vertices where each vertex defines the coordinates of the spatial objects) data of buildings [3]. The spaceborne and airborne technology provides abundant remote sensing images that have become increasingly important for extracting building information [4]. In early studies, pixel mixture is one important factor that influences building extraction when only fine- to coarse-resolution satellite images were available. Nowadays, advanced remote sensing technology offers very high-resolution images at sub-meter spatial resolution, making them attractive to extract accurate building footprint polygons. In the very high-resolution images, the influence of mixed pixels is minor, and the scene complexity becomes a new challenge for building footprint extraction. Currently, many government departments and industrial companies adopt manual methods to delineate the vector data of building footprints from high-resolution remote sensing images so as to obtain the vector maps that meet the accuracy requirements of surveying and mapping. As manual annotation is time-consuming and requires expertise [5], there is a need to develop an efficient and robust scheme for automated extraction of building footprint polygons from remote sensing images.

Classifying high-resolution remote sensing images to obtain the raster data of buildings has been extensively studied for decades [6,7]. Traditional methods include discovering and designing handcrafted features, such as spectral features, texture features, morphological features, and boundary features, as empirical indicators to distinguish buildings from other land surface objects in remote sensing images [8,9]. These methods often use empirical thresholds for classifying building objects and have drawbacks, such as insufficient uses of spectrum or spatial information and a high sensitivity to the choice of parameters or thresholds [10,11]. Machine learning models can overcome the shortcomings of the empirical methods and have been successfully applied for classifying buildings in remote sensing images [12]. Machine learning models, such as support vector machine, random forests, and artificial neural networks, use computer algorithms to establish the nonlinear relationships between inputs and targets through the training and learning processes based on many known samples. Machine learning methods have a high degree of automation and can efficiently reduce the confusion between buildings and other man-made features, but the input features to the machine learning models are often manually designed, such as low-level features or enhanced semantic features [13,14]. In recent years, with the development of computer technology, deep learning methods can automatically extract the features of different levels, such as low-level, middle-level, and high-level features from images [15], and have been widely used in the fields of computer vision and remote sensing image processing [16]. Earlier studies used patch-based convolutional neural networks (CNNs) to process image blocks to train and predict the class of the center pixel, but CNNs have difficulties in providing full-resolution classification maps [17,18]. Many studies on classifying buildings from remote sensing images now favor fully convolutional networks (FCNs), which implement an encoder–decoder framework and normally include convolution, down-sampling, and up-sampling layers for pixel-wise prediction or sematic segmentation [19,20]. For example, Maggiori et al. [21] utilized a fully convolutional architecture and designed a multiscale neuron module to produce dense predictions of building maps by alleviating the trade-off between recognition and localization. Ji et al. [22] proposed a Siamese U-Net that improves the classification performance of buildings, particularly large buildings. Huang et al. [23] designed an end-to-end trainable gated residual refinement network that has a competitive performance in extracting building raster data across urban and suburban scenes based on tests in a publicly available dataset. The above studies provide feasible methods for generating the classification maps of buildings using high-resolution remote sensing images. Some studies utilized the light detection and ranging (LiDAR) point cloud data to carry out building reconstruction given that the LiDAR data contain three-dimensional information related to buildings [24–26]. Gilani et al. [27] defined three general steps, including feature preservation, surface growing, and false plane elimination, to achieve automatic detection of building roof planes from the LiDAR point cloud. Although the LiDAR data can provide supplemental geometry features, the data acquisition of LiDAR is also more costly and complicated than those of optical sensors [28].

Extracting the vector data of building footprints from high-resolution remote sensing images is much more challenging than image classification. Traditional methods that directly extract the vector data of building footprints often manually design features or indexes that allow for generating the vector data of building footprints, but these methods are normally empirical and have difficulties for uses across images and scenes. Wang et al. [29] proposed a method to automatically extract rectangular buildings of different sizes and directions by combining image segmentation, scale-invariant feature transformation, and adaptive window Hough transform. Qin et al. [30] adopted straight line detection and graph loop searching to extract building boundaries with different shapes, sizes, and densities. Recently, some researchers have attempted using the deep learning methods to improve the extraction of building footprint polygons. Girard and Tarabalka [31] developed a deep learning-based model that could directly generate the vector maps, but the developed model has limitations in terms of extracting buildings with different shapes.

An indirect method that extracts building footprint polygons is to first classify buildings in remote sensing images, and then convert the raster format of buildings into a vector format. Buildings often have regular boundaries that consist of straight lines, while the pixel-wise semantic segmentation methods are often ineffective in describing the details of building footprint boundaries even if the produced classification map is highly accurate. It is, therefore, necessary to optimize the building footprint boundaries when converting them from raster data to vector data. One optimization approach is to improve the classification results of building boundaries in remote sensing images. Wu et al. [32] proposed a boundary-regulated network that consists of a modified U-Net and a multi-tasking framework to generate segmentation maps and building outlines by accounting for boundary regulation. To resolve the problem of blurry object boundaries, Marmanis et al. [33] proposed a DCNN models for semantic segmentation of high-resolution aerial images, which explicitly accounts for the boundaries of classes in the segmentation process. Taking the contours into consideration, Liao et al. [34] proposed a boundary-preserved building extraction approach. By embedding the contour information in the labels, the proposed approach can enhance the representation of building boundaries and improve the performance on boundaries of adjacent buildings. Another optimization approach is to improve the building boundary vertices when converting data formats. Some classic vector data processing algorithms, such as the Douglas–Peucker [35], Wang–Müller [36] and Zhou–Jones [37] algorithms, have been widely used by researchers. Maggiori et al. [38] proposed a new algorithm that uses a labeled triangular mesh to approximate the classification maps. Although many methods can improve the generated vector data of buildings, the obtained polygons generally do not match the building footprints well. One reason is that it is difficult to distinguish between adjacent buildings and deal with misclassified small patches during the map vectorization. Existing methods that extract building footprint polygons often have limitations; for example, they are only applicable for generating polygons of individual buildings with regular shapes such as rectangles.

Compared with pixel classification or segmentation, polygon extraction is a different and challenging task because it involves the transition of data representation. Specifically, pixel classification or segmentation of images transforms raster data into raster data, while the process that extract polygons from images transforms raster data into vector data. In the task of pixel-wise classification, the deep learning model only needs to discriminate the object pixels from the background pixels. In the task of polygon extraction, it needs to obtain at least the positions, the belongings, and the connections of the key vertices. Only a few studies have applied the deep learning models to detect the key points of geographical objects from remote sensing images. There are two main difficulties when using deep learning to directly detect a key point. First, compared to segmentation, the imbalance between positive and negative samples is severe. The annotated key points are much fewer than the background samples, and as a result, training and converging the deep learning network is difficult if we treat this task as a classification problem. Second, the number of key points that exist in an image is unknown. It is difficult to handle the outputs with irregular length using a convolutional neural network directly. Song et al. [39] proposed an FCN-based approach to detect building corners in aerial images. The corners are extracted by the contours of the predicted building footprints, which means that the performance of the corner detector largely relies on the accuracy of semantic segmentation. A deep learning-based model that treats the geolocation of building corners as direct objectives of optimization is not yet available.

One main problem is that existing segmentation-based methods extract building footprints with irregular outlines and blurred boundaries, which results in vectorized polygons with irregular shapes and redundant vertices. Given that existing deep learning-based studies mainly focus on improving the map classification or semantic segmentation of buildings, we aim to explore the ability of deep learning methods to learn and extract the vector data of buildings. The goal of this study is to develop a framework that synthesizes deep learning models to extract building footprint polygons from the remote sensing images and allow for the direct production of building maps in a vector format. In the developed framework, we use three deep learning models to perform different image processing tasks, including semantic segmentation, bounding box detection, and key point detection. A polygon construction strategy based on Delaunay triangulation was also designed to integrate these outputs effectively and thus generate high-quality building polygon data. The proposed framework is able to achieve comparable mapping accuracies with semantic segmentation models on a pixel basis and generate building footprint polygons with concise edges and vertices with regular shapes that are close to the reference data. Our method has the potential to extract accurate building footprint polygons from remote sensing images for applications in geospatial analysis.

2. Methodology

2.1. Overview

Our goal is to detect and generate the building footprint polygons that accurately describe the outline of individual buildings in a vector format from very high-resolution remote sensing images. Compared with the pixel-based raster data, the vector data contains not only the geometric information but also topological information, which provides another efficient way to represent real-world features in a geographic information system. Geometric information includes the positions of objects and their components in Euclidean space. Topological information includes the number, relationship, connection, and types of topological elements, such as vertices. Both geometric information and topological information are needed to represent objects correctly. To construct an individual building vector, we need both the accurate position of the vertices and the spatial relationship among vertices, which are able to generate polylines and polygons for the maps. The vector data use interconnected vertices to represent the object shapes, and each vertex describes its position using geographic coordinates in a spatial reference frame. As mentioned earlier, extracting building footprint polygons from aerial images involves tasks of different purposes, so it is difficult to optimize an individual deep learning network for different tasks. Hence, our approach is to split the main task of polygon extraction into several sub-tasks and utilize an appropriate deep learning model for each subtask. By integrating several methods in one framework, we are able to extract building footprint polygons from aerial images automatically and efficiently. As the vertices of building footprint polygons are often the corners of the building rooftop and the connections between building footprint corners are often straight lines, our strategy is to extract building polygons from remote sensing images by detecting the vertices of building footprint polygons first, and then finding the correct connections among them. Figure 1a illustrates the schematic workflow of the proposed building extraction method, which consists of four main steps: corner detection, semantic segmentation, object detection, and polygon construction. The framework takes very high-resolution aerial remote sensing images with RGB bands as inputs and extracts key information related to the building footprints from them. It adopts U-Net to produce the classification maps of buildings, Cascade R-CNN to detect building objects in the images, and Cascade CNN to detect building footprint corners. The features extracted using deep learning approaches from the remote sensing images are combined to produce building footprint polygons. The proposed deep learning-based framework for automated extraction of building footprint polygons (DLEBFP) is described in detail in the following sections.

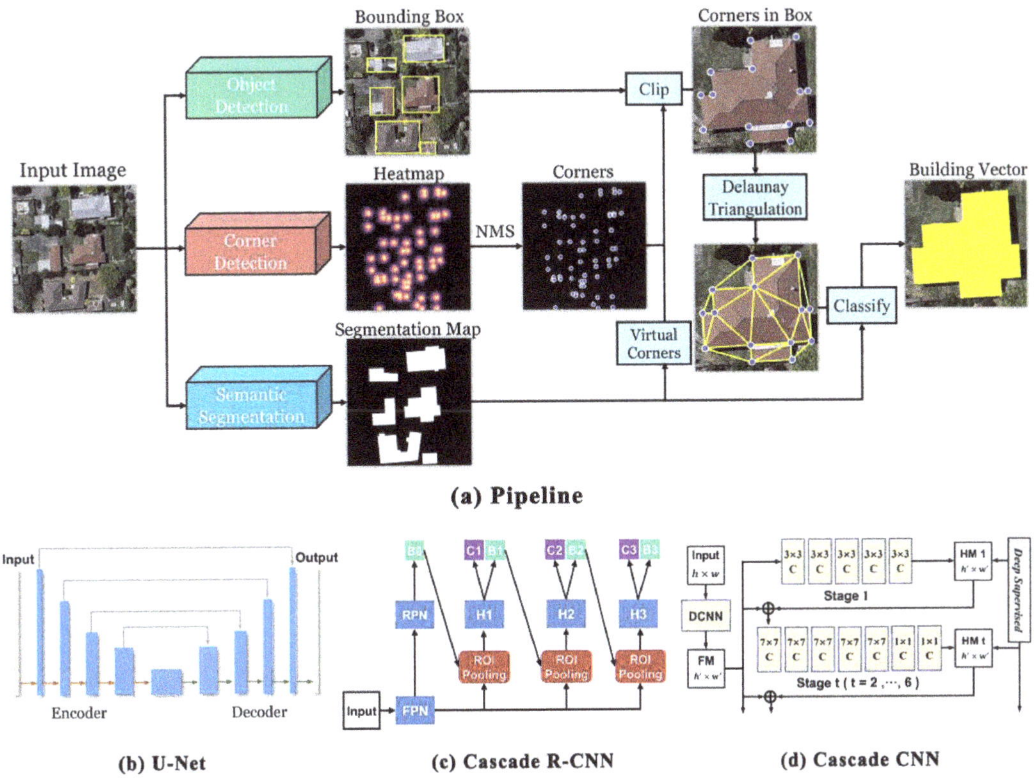

Figure 1. The diagram (**a**) shows the workflow that extracts building footprint polygons from very high-resolution remote sensing images using the deep learning models, where the subplots illustrate the architectures of (**b**) U-Net, (**c**) Cascade R-CNN, and (**d**) Cascade CNN. In (**b**), the red arrows denote the operations of convolution and down-sampling, and the green arrows denote the operations of convolution and up-sampling. In (**c**), FPN denotes feature pyramid network; RPN denotes the region proposal network; H1, H2, and H3 denote the network of detection head at different stages; B0, B1, B2, and B3 denote the results of the bounding boxes predicted at different stages; and C1, C2, and C3 denote the classification results predicted at different stages. In (**d**), DCNN represents the deep convolutional neural network; FM denotes the feature maps generated by DCNN; and HMt (t = 1, 2, . . . , 6) indicates the predicted heat maps at different stages, respectively.

2.2. Building Segmentation

U-Net, originally proposed for biomedical image processing and implemented in Caffe [40] was found to be effective in the semantic segmentation of remote sensing images [41,42]. U-Net used skip connections between the encoder and the decoder such that the decoder can receive low-level features containing abundant spatial and geometric information from the encoder, and thus it can generate precise classification maps. U-Net is widely used to extract various natural and man-made objects from remote sensing images. We use U-Net as a deep learning model for the semantic segmentation of the buildings. Figure 1b illustrates that U-Net utilizes the encoder–decoder architecture to make dense pixel-wise predictions. In the original implementation of U-Net, there are five convolutional blocks in the encoder and four upsampling blocks in the decoder. A convolutional block consists of two 3×3 unpadded convolutions, each followed by a rectified linear unit (ReLU). At the end of the block, a 2 × 2 max pooling operation with stride of 2 is appended for downsampling. After the process of each convolutional block, the number of feature channels doubled. In the decoder, the upsampling block consists of an upsampling operation of the feature map followed by a 2 × 2 convolution, which halves the number of channels. Two 3 × 3 convolutions followed by ReLU are used to process the concatenation of low-level features and features from the previous block. For the last layer in the decoder, a convolutional function followed by a sigmoid function is applied to map the output. In total, the U-Net has 23 convolutional layers, including 9 convolutional layers in the encoder and 14 in the decoder.

In this work, we modified the original U-Net. The modified U-Net shared similar architecture with original U-Net, and we added several modules to improve the performance of semantic segmentation. We adopted the encoder part based on ResNet34 [43] and used a residual unit consisting of multiple combinations of convolution layers, batch normalization (BN), and rectified linear unit (ReLU) activation. ResNet34 contains five convolutional blocks but more convolutional layers in each block, and the total number of convolutional layers was 34. Since many studies have suggested that a deeper network would produce more discriminative features and achieve better performance, the use of ResNet34 helps improve the segmentation results of UNet. Instead of directly passing through the convolution layers, the residual units utilize a shortcut connection and element-wise addition to transfer the input features directly to the output. It was found that the identity and projection shortcuts in ResNet could address the degradation problem during model training and also introduce neither extra parameter nor computation complexity [41]. The ReLU activation function is defined as $f(x) = \max(x, 0)$ which reduces the possibility of vanishing gradient and accelerates the convergence of network during training [44,45]. BN performs the normalization for each training mini-batch and it allows for high learning rates and addresses internal covariate shift [46]. During the training stage, the mean and variance of features in a batch are first calculated and then used to normalize the features. In addition, two learnable parameters γ and β control the rescaling and shifting of the normalized values. As mini-batches are not used during inference, the moving average of the training set mean and variance are computed and used to normalize the features during the test procedure. For the decoder, compared to the blocks used in original U-Net, the BN layer is also inserted between the convolutional layer and the ReLU activation, such that we can accelerate the network convergence and improve the model performance. It should be noted that the cropping operation was removed from our network because the encoder used convolutional operation with the same padding rather than unpadded convolutions utilized in original U-Net.

Since the number of nonbuilding pixels is much higher than the number of the building pixels in most scenes, the effect of class imbalance could cause the learning process to be trapped in a local minimal of the loss function, making the classifiers strongly biased towards the background class [23,47]. To address the issue of class imbalance, we combined DiceLoss with binary cross-entropy loss as the objective function. The total loss calculation can be written as follows:

$$L_{seg} = L_{Dice} + L_{BCE} \tag{1}$$

$$L_{Dice} = 1 - \frac{2 \times \sum_{i=1}^{H} \sum_{j=1}^{W} y_{ij} \times \hat{y}_{ij} + \varepsilon}{\sum_{i=1}^{H} \sum_{j=1}^{W} y_{ij} + \sum_{i=1}^{H} \sum_{j=1}^{W} \hat{y}_{ij} + \varepsilon} \tag{2}$$

$$L_{BCE} = -\frac{1}{H \times W} \sum_{i=1}^{H} \sum_{j=1}^{W} \left[y_{ij} \times \log \hat{y}_{ij} + (1 - y_{ij}) \times \log(1 - \hat{y}_{ij}) \right] \tag{3}$$

where H is the height of the input image, W is the width of the input image, y_{ij} denotes the binary label of the pixel in the image (0 represents nonbuilding and 1 represents building), $\hat{y}_{ij}$ represents the predicted probability of pixel ranging from 0 to 1, and ε denotes a factor used to smooth the loss and the gradient.

2.3. Building Object Localization

We use Cascade R-CNN to identify building objects from the remote sensing images. Note that many object detection models use intersection over union (IoU) to determine positive or negative samples; the use of a prescribed IoU significantly influences the model performance, because a low IoU easily leads to noisy detections and a higher IoU results in low detection accuracy because of model overfitting and sample mismatching. Cascade R-CNN can effectively address the abovementioned problems and improve object detection by adopting a multistage strategy for model training with an increasing IoU [48]. At the beginning of the model run, the results generated by the region proposal network are heavily tilted towards low quality, and thus the model uses a low IoU. Following the first stage of image classification and bounding box regression, the obtained bounding boxes are resampled using a higher IoU to provide samples with a higher quality. These processes go iteratively to improve the model performance for detecting objects from images.

Figure 1c illustrates the architecture of Cascade R-CNN model for building object detection. We use the feature pyramid network [49], which can detect multiscale objects by combining high-resolution low-level features and low-resolution high-level features to extract the feature maps at different scales. Lin, Dollar, Girshick, He, Hariharan, Belongie, and Ieee [49] demonstrated that using more than three stages in Cascade R-CNN would lead to a decreased performance. Thus, we chose the number of stages to be three and set the IoU thresholds of the detector in different stages as 0.5, 0.6, and 0.7, respectively. We used the same loss function in the feature pyramid network and the detection head, binary cross-entropy loss in the task of image classification, and smoothL1 loss in the task of bounding box regression:

$$L_{SmoothL1} = \begin{cases} 0.5 \times (y - \hat{y})^2 & if\ |y - \hat{y}| < 1 \\ |y - \hat{y}| - 0.5 & otherwise \end{cases} \tag{4}$$

where y denotes the reference values, and $\hat{y}$ denotes the predicted values obtained from the models.

2.4. Corner Detection

Cascade CNN, initially proposed for multi-person pose estimation [50], is a key point detection network. We used Cascade CNN to detect building footprint corners. We removed the branch of part affinity field that was used to determine the overall relationship in the models because our network only focuses on predicting building footprint corners. Cascade CNN adopts a multistage strategy. In the first stage, the backbone network extracts high-level features from the input image, and convolutional layers with different kernel sizes are used to learn semantic information and produce a confidence map of the target key point. The model structure in the second stage is similar to that in the first stage, but the network concatenates the feature maps extracted by the backbone network and the confidence map generated from the previous stage as model inputs. The processes in the other stages are similar to those in the second stage.

Training a deep learning model to extract the geolocation of key points in an image is prone to model overfitting when directly using the coordinates as ground truth. We used heat maps as the ground truth for model developments, such that the network predicts the confidence map instead of the direct geographic coordinates of key points. The heat map is a two-dimensional map that represents the possibilities of the occurrence of the key points at each pixel. In a heat map, the pixel value of the possibilities ranges from 0 to 1. If one pixel is close to the annotated key point, the possibility value at the given pixel is close to 1. If multiple key points occur within one pixel, there is a peak corresponding to each key point. Using a heat map is advantageous, as it is possible to visualize the deep learning processes, given that the outputs can be multimodal [51]. We generate the heat map S_k^* for each key point k by placing a Gaussian function with fixed variance at the ground truth position of the building corners. The heat map S_k^* at location p is defined as follows:

$$S_k^*(p) = exp\left(-\frac{\|p - x_k\|_2^2}{2\sigma^2}\right) \tag{5}$$

where p denotes the pixel coordinate in the image, x_k denotes the coordinate of the key point, $\|p - x_k\|_2^2$ is the squared Euclidean distance from the given pixel p to the key point x_k, and σ is a constant that controls the spread of the peak.

As there could be many key points in a single image, a max operator is used to aggregate individual confidence maps of each corner and thus generate the complete confidence map used for training the models. The operator is defined as follows:

$$S^*(p) = \max_k S_k^*(p) \tag{6}$$

The architecture of Cascade CNN is shown in Figure 1d. We utilized VGG19 [52] as the backbone network in the model. The size of the feature map is one eighth of the size of the input image and the number of stages in Cascade CNN is 6. The loss function used in both the intermediate layers and the output layers is a mean square error (MSE) loss, which penalizes the squared pixel-wise differences between the predicted confidence map and the synthesized heat map. The loss function of the predicted confidence map is defined as follows:

$$L_{confmap} = \frac{1}{H \times W}\sum_{i=1}^{H}\sum_{j=1}^{W}(y_{ij} - \hat{y}_{ij})^2 \tag{7}$$

where H denotes the height of image, W denotes the width of the image, y_{ij} is the value of the pixel (i, j) in the ground reference map, ranging from 0–1, and $\hat{y}$ is the value of the pixel (i, j) in the predicted confidence map obtained from Cascade CNN.

Cascade CNN uses the inputs of the aerial images to generate confidence maps of key points. We extracted the locations of the building footprint corners by performing a nonmaximum suppression (NMS) on the heat maps [50,53]. Specifically, we used a max filter with the size of 3×3 slides along the heat map to extract the local maximum pixels as the pixels of key points. Note that we calculate the losses for the confidence maps in each stage during the training processes and avoid the problems of vanishing gradient or exploiting gradient when training the network.

2.5. Polygon Construction

To construct the structure of the building footprint polygons, we used two-dimensional Delaunay triangulation to transform the detected key points into the candidate polygons. Delaunay triangulation, as proposed by Boris Delaunay in 1934, has been widely used in computer graphics and geographical studies [54,55]. Given a set of discrete points, Delaunay triangulation generates a triangulated irregular network, where no points are located inside the circumcircle of any triangle in the network and the minimum angle of all triangles is the largest among all possible networks. We used the bounding boxes obtained by Cascade R-CNN to constrain the key points and constructed a triangulated irregular network for each individual building. Owing to the limitations in computational resources, there is a need to crop large remote sensing images, resulting in buildings being truncated at the borders of the cropped images. We defined the intersection points between the segmentation mask and the border of the cropped image as virtual corners (VC) and used them together with the key points detected by Cascade CNN for constructing a triangulated irregular network. The use of VC when constructing building polygons based on Delaunay triangulation could largely reduce erroneous triangles. For each triangle in the triangulated irregular networks generated by Delaunay triangulation, we calculated the ratio of the building areas obtained from the segmentation map generated using U-Net to the triangle areas and applied an individual threshold to classify the triangles as either building or nonbuilding triangles. All building triangles were merged to produce the building footprint polygons across the entire region.

3. Experiment Setup

3.1. Dataset

The performance of the deep learning model relies on a dataset with high-quality samples. An aerial imagery dataset from the Wuhan University (WHU) building dataset was used (http://gpcv.whu.edu.cn/data/ (accessed on 10 November 2020). This dataset consists of more than 84,000 independent buildings labeled in the vector format and the aerial images at a 0.075 m spatial resolution covering an area of 78 km^2 in Christchurch, New Zealand. This area contains buildings of various architectural types with varied colors, sizes, shapes and usages, making it ideal to evaluate the deep learning models. The original aerial images are open-source data provided by the Land Information New Zealand (LINZ) Data Service (https://data.linz.govt.nz/layer/53451-christchurch-0075m-urban-aerial-photos-2015-2016/ (accessed on 29 Aug 2021)). The photographs were taken around 2015 and 2016, and the images were ortho-rectified digital orthophoto maps (DOMs) with RGB channels in New Zealand Transverse Mercator (NZTM) map projection [22,56]. The spatial accuracy is 0.2 m with 90% confidence level. As shown in Figure 2, the area was divided into three sub-regions for model training, validation, and testing. The main reason for using the WHU dataset for model tests is that the vector data provided by the land information service of New Zealand have been carefully checked and corrected by cartography experts [22]. The high-quality vector data of building footprints can be easily transformed into raster building maps, bounding boxes, and heat maps with vertex coordinates for training and testing the deep learning models.

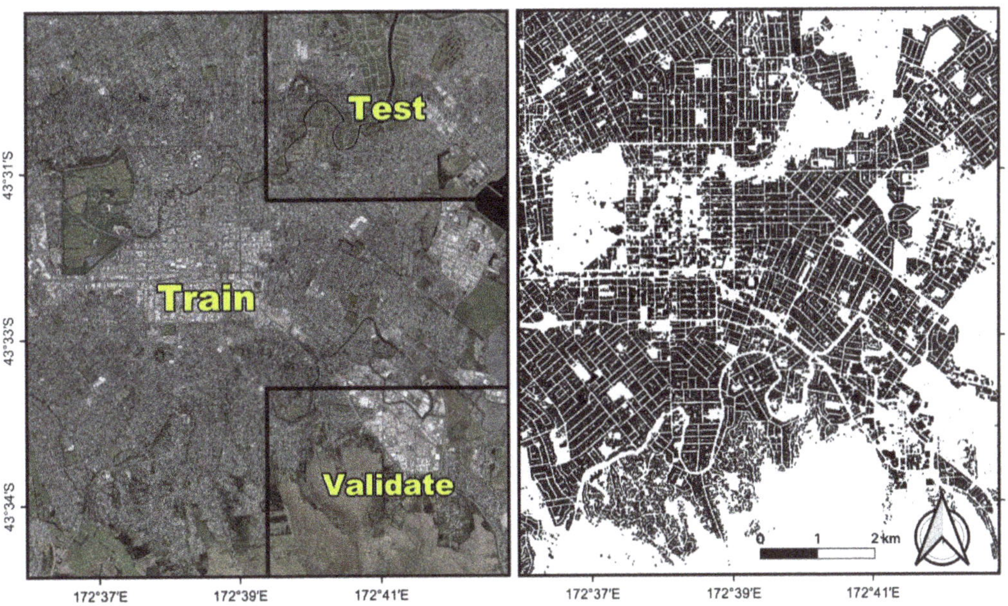

Figure 2. The aerial image (**left**) and the vector map of building footprint polygons (**right**) are shown for the study area in Christchurch, New Zealand.

Additionally, we used the publicly available benchmark dataset, ISPRS Vaihingen semantic labeling dataset (https://www2.isprs.org/commissions/comm2/wg4/benchmark/2d-sem-label-vaihingen/ (accessed on 20 January 2021)), to test the robustness of the proposed framework. The Vaihingen dataset consists of pairs of images and labels at the spatial resolution of 9 cm. The dataset contains 33 true orthophotos with near infrared (NIR), red (R), and green (G) bands, which are beneficial for roof contour detection. These tiles have different image sizes with an average size of 2494 × 2064 pixels. The pixel-based labels have six categories, including impervious surfaces, buildings, low vegetation, trees, cars, and clutter. We manually delineated building footprint polygons based on both images and pixel-based labels for studies. We used six in the 33 tiles for testing and the others for model training and validation.

3.2. Implementation Details

All the deep learning models and experiments were implemented on the Ubuntu 18.04 system equipped with a single NVIDIA RTX 2080Ti GPU with 11 GB of memory under CUDA10.0 and cuDNN7.5. U-Net was implemented based on the open-source machine learning library of PyTorch (https://pytorch.org/ (accessed on 10 November 2020)). The encoder part in the network was initialized using a pretrained ResNet-34 model, and the network was trained with a batch size of four. We only used the feature maps that were generated from the first four stages to make predictions. An Adam optimizer was used to optimize the parameters with a learning rate of 0.0001. Data augmentation, including rotation, flipping, and random manipulation of brightness and contrast, was applied to the images at the training stage. The network was trained for 40 epochs using the training set, and the model that performed well in the validation set was stored. Cascade R-CNN was implemented using MMDetection, an object detection and instance segmentation code based on PyTorch. The backbone was initialized using a pretrained ResNeXt101 model. The basic module of the ResNeXt101 model was different from that of ResNet-34. It used a bottleneck design to decrease the parameters in the network and make it efficient. In ResNeXt101, group convolution was introduced to improve the model accuracy without increasing complexity. Feature maps obtained at all stages were used in the feature pyramid network to detect objects with different scales. The network was trained using the stochastic gradient descent optimizer with a batch size of four, an initial learning rate of 0.02, a momentum of 0.9, and a weight decay of 0.0001. We used the learning rate warmup with a ratio of 1/3 in the first 500 iterations. The total number of epochs was set to 100, and the learning rate was reduced to one-tenth of the current learning rate every 30 epochs. Cascade CNN was implemented based on PyTorch. When generating the heat maps, we set σ in the Gaussian function as 12. Data augmentation was also applied to the training data. The backbone was initialized using the pretrained VGG-19 model. Only the feature maps in the third stage were used for prediction. The entire network was trained using the stochastic gradient descent optimizer with a batch size of four, an initial learning rate of 0.02, a momentum of 0.9, and a weight decay of 0.0001. The network was trained for 50 epochs. All the model configurations and hyper-parameters were chosen according to parallel experiments.

Note that all model configurations and hyper-parameters were carefully designed according to parallel experiments. Hundreds of experiments were conducted to test the performances of possible configurations and hyper-parameters, and we chose the one that outperformed any other settings, namely the architecture and specific values presented above.

3.3. Comparative Methods

To understand the model performance, we compared DLEBFP with three different deep learning models for the semantic segmentation results in the raster format and with a popular approach for generating vector results on the WHU building dataset.

We applied the deep learning methods of U-Net, FCN, and SegNet [57] to produce semantic segmentation maps for model comparisons. FCN alters the original CNN structure to enable dense prediction. FCN uses transposed convolution to upsample feature maps to match the sizes of the images and exploit the skip layer strategy. FCN has proven its performance in terms of model accuracy and computational efficiency across several benchmark datasets. SegNet has an encoder–decoder architecture and is often used for evaluating the performance of semantic segmentation models. In SegNet, the pooling indices computed in the max-pooling step in the encoder are reused in the corresponding decoder to perform nonlinear upsampling. Normally, the memory required in SegNet is much less than FCN for the same task of semantic segmentation.

Additionally, we vectorized the semantic segmentation maps produced by U-Net and applied the Douglas–Peucker algorithm to generalize the vector maps. The Douglas–Peucker algorithm is widely used for processing vector graphics and cartographic generalization [35,58]. Many building extraction researches also chose the Douglas–Peucker algorithm for building vector simplification due to its simplicity and efficiency [59,60]. Additionally, its performance has been proven superior to other classic simplification algorithms, such as the Reumann–Witkam algorithm and the Visvalingam–Whyatt algorithm [58]. The Douglas–Peucker algorithm simplifies a curve that is composed of line segments to a similar curve with fewer points by accounting for the maximum distance between the original curve and the simplified curve. At each step, the Douglas–Peucker algorithm attempts to find a point that is the farthest from the line segment with the first and the last points as end points. If the distance between the farthest point and the line segment is smaller than a prescribed threshold, it decimates all points between the first and the last points; otherwise, it keeps the farthest point and recursively calls itself with the first point and the farthest point and then with the last point and the farthest point. The Douglas–Peucker algorithm can produce objective quality approximations [61]. We applied different thresholds for the maximum distance in the Douglas–Peucker algorithm (i.e., 0.1, 0.5, and 1.0 m) to produce building footprint polygons based on the classification maps derived from U-Net for five sub-regions and the entire study region. By comparing the results simplified by different thresholds, we are able to analyze the performance variance with thresholds. Moreover, the results of simplification by Douglas–Peucker are also used to compared with the vector generated by our methods DLEBFP to verify the superiority of our approach.

To test the robustness of the proposed method, we also conducted experiments in the ISPRS Vaihingen dataset and compared the comparative methods with DLEBFP. Similar to the experiments on the WHU dataset, we trained and analyzed the model performance on the Vaihingen dataset.

3.4. Ablation Studies

Ablation studies aim on investigating how individual or combination of features affect the model performance by gradually removing some features of the model. Ablation studies have been widely adopted in the field of remote sensing and computer science [62–64]. Here, we conduct experiments to investigate the ablated features of both the virtual corner and bounding boxes. In DLEBFP, virtual corners generated from the semantic segmentation maps and bounding boxes detected by Cascade R-CNN are both used to improve the model performance in producing the building vector data. We can still extract the building footprint polygons by methods without these components.

To evaluate the uses of virtual corners and bounding boxes, we set four kinds of experiments on the WHU dataset in the ablation studies, including a baseline method. The baseline method (hereinafter referred to as Baseline) only uses two deep learning models (i.e., Cascade CNN and U-Net) for feature extraction and Delaunay triangulation for polygon construction. In Baseline, we constructed the triangulation network based on the building corners detected by Cascade CNN and classified the constructed triangles into building or nonbuilding triangles based on the segmentation map extracted by U-Net. We merged all building triangles to produce building footprint polygons. In the other three experiments, we tested the virtual corners and bounding boxes detected by Cascade R-CNN and extracted the building footprint polygons by (1) the baseline method with bounding boxes (hereinafter referred to as Baseline + BB), (2) the baseline method with virtual corners (hereinafter referred to as Baseline + VC), and (3) the baseline method with both bounding boxes and virtual corners (hereinafter referred to as Baseline + BB + VC).

3.5. Evaluation Metrics

We evaluated the model performance based on the raster format and used the metrics, including precision, recall, and IoU, which have been widely used in the assessment of building extraction results [65,66]. For the raster map, precision is the ratio of the true positive pixels to all detected building pixels, recall is the ratio of the true positive pixels to the reference building pixels, and IoU is the ratio of the true positive pixels to the total number of true positive, false positive, and false negative pixels. IoU is extensively used in evaluating model performance for image classification, as it provides a measure that penalizes false positive pixels. The abovementioned metrics are defined as follows:

$$\text{Precision} = \frac{\text{TP}}{\text{TP} + \text{FP}} \tag{8}$$

$$\text{Recall} = \frac{\text{TP}}{\text{TP} + \text{FN}} \tag{9}$$

$$\text{IoU} = \frac{\text{TP}}{\text{TP} + \text{FP} + \text{FN}} \tag{10}$$

where true positive (TP) denotes the number of building pixels correctly classified as buildings, false positive (FP) denotes the number of nonbuilding pixels misclassified as buildings, and false negative (FN) denotes the number of building pixels that are not detected.

In addition to assessments based on the pixel-wise metrics, we computed the vertex-based F1-score (VertexF) as proposed by Chen, Wang, Waslander, and Liu [60] to evaluate the performance of the generated building footprint polygons. To derive VertexF, the extracted polygons and the reference polygons were interpreted as two different sets of vertices. We set a buffer distance for every ground truth vertex and then classified all the vertices of the extracted building polygon as true positive (TP), false positive (FP), and false negative (FN). VertexF is calculated as:

$$\text{VertexF}_{\text{s}} = \frac{2\text{TP}_{\text{s}}}{2\text{TP}_{\text{s}} + \text{FN}_{\text{s}} + \text{FP}_{\text{s}}} \tag{11}$$

where the subscript s denotes the buffer distance, TP_s is the number of true positive vertices, FN_s is number of false negative vertices, and FP_s is the number of false positive vertices. We tested the buffer distances at 0.5 m and 1.0 m, respectively.

Evaluating the mapping accuracies of vertices is particularly meaningful for the vector data, because a simple and accurate representation is crucial to the map production. Moreover, the mapping accuracies of vertices better reflect the required manual editing workload when converting the extraction results to real map products [60].

4. Results

4.1. Results on WHU Dataset

Figure 3 shows the rasterized results of the extracted building footprints using different methods in the test area. Visually, the four methods have their own pros and cons. DLEBFP (Figure 3c) has fewer FPs than the three semantic segmentation models but contains more FNs. One reason is that the object detection method used in our approach is functionally similar to a filter that only screens pixels with high confidence of building footprints, such that it would improve the precision but impair the recall of the model. Among the three segmentation-based methods, FCN (Figure 3d) produces results similar to those of our method. FPs produced by FCN are less than those derived from either SegNet (Figure 3e) or U-Net (Figure 3f). Compared with the other methods, SegNet and U-Net generated more misclassified building pixels and generally performed well in extracting relatively large buildings. U-Net occasionally misclassified water pixels as buildings, for example, the pixels that are located in the river in the test area.

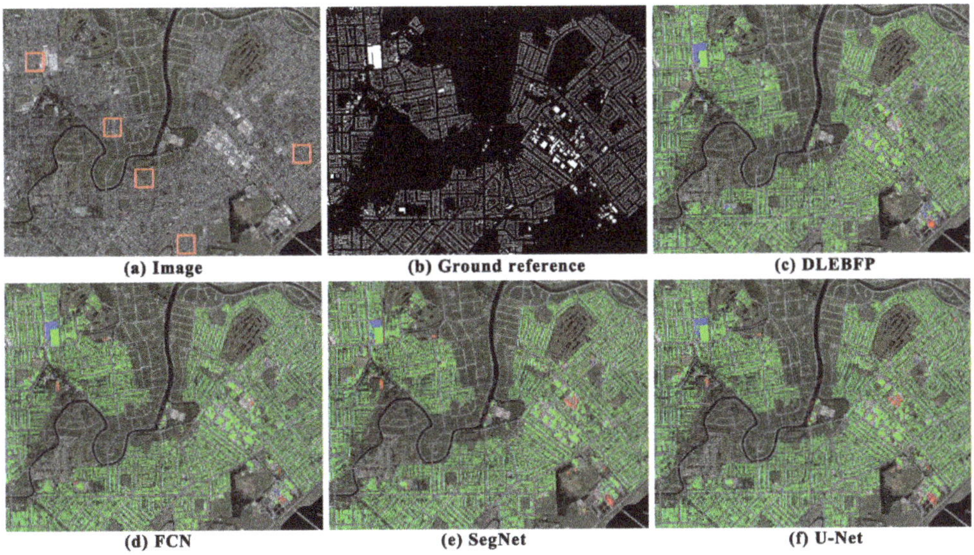

Figure 3. Comparisons of different methods for producing the raster classification maps on the WHU dataset are shown for (**a**) the true-color composite image, (**b**) the binary ground reference image where the white color denotes building areas and the black color denotes nonbuilding areas, (**c**) the building classification map derived from DLEBFP, (**d**) the building classification map derived from FCN, (**e**) the building classification map derived from SegNet, and (**f**) the building classification map derived from U-Net. In (**c**–**f**), the true positives (TPs), false positives (FPs), and false negatives (FNs) are marked in green, red, and blue colors, respectively.

Figure 4 shows five scenes with the window size of 3000 × 3000 pixels, where the extent and location of buildings are marked in red rectangles in Figure 3a, for intuitive comparisons of the mapping results. All methods were able to identify most of the buildings correctly, demonstrating the power of deep learning approaches for the semantic segmentation of remote sensing images. Compared with the semantic segmentation models, our method generates fewer FPs. For example, in the first scene, three semantic–segmentation-based methods misclassified the road pixels as the building pixels (Figure 4c–e), and our method avoided misclassification because we used Cascade CNN to detect building footprint corners such that the falsely classified roads were screened out. In the second scene, many pixels surrounding a large building are also classified as buildings by both SegNet (Figure 4d) and U-Net (Figure 4e), whereas FCN (Figure 4c) produces a relatively lesser misclassification of pixels in the surroundings of the same building. In the same scene, our method distinguishes buildings from other objects and preserves the geometric details of building footprint boundaries. Although DLEBFP can extract most buildings accurately, there are some omission errors in buildings, resulting in higher FNs than the semantic segmentation methods. In general, among the three semantic segmentation methods, FCN did not extract the shape of buildings accurately and lost some details along the building boundaries, and U-Net and SegNet had similar performance across five scenes.

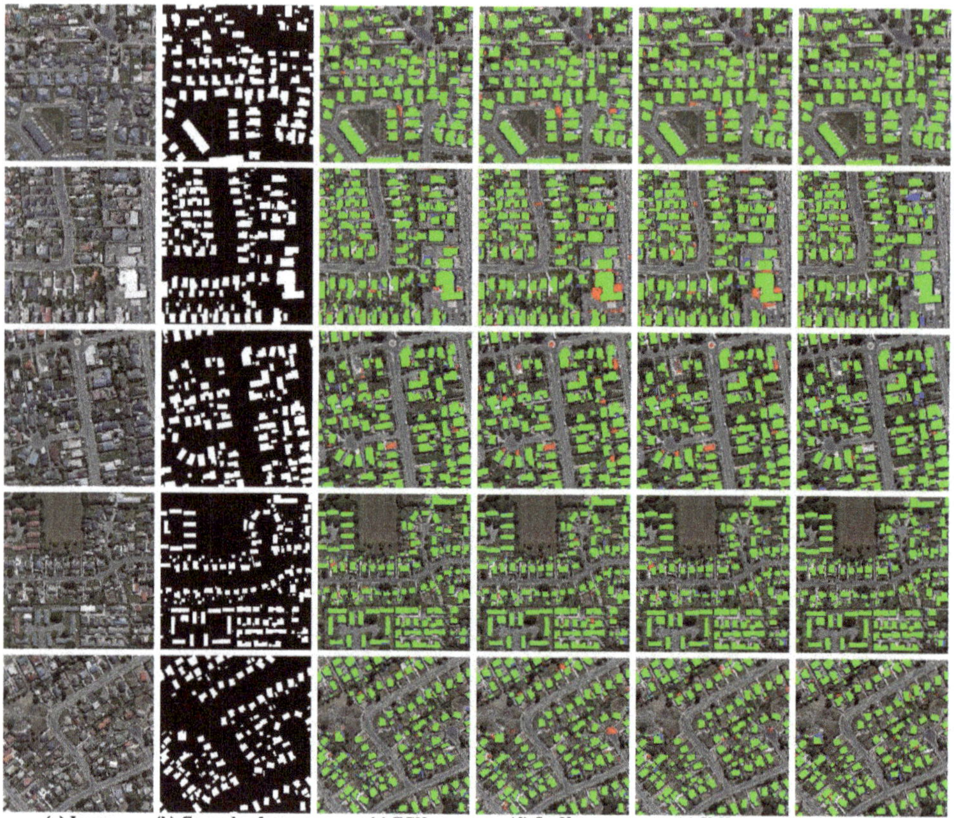

Figure 4. The true-color-composite images and the extracted classification maps of buildings using different approaches for five close-up scenes in the WHU dataset are shown for (**a**) the true-color composite images, (**b**) the binary ground reference images where the white color denotes building areas and the black color denotes nonbuilding areas, (**c**) the building classification maps derived from FCN, (**d**) the building classification maps derived from SegNet, (**e**) the building classification maps derived from U-Net, and (**f**) the building classification maps derived from DLEBFP. The true positives (TPs), false positives (FPs), and false negatives (FNs) of buildings are marked in green, red, and blue colors, respectively.

Table 1 summarizes the quantitative evaluation results of our method and the three deep learning models in the close-up scenes and the entire testing dataset at the pixel level. The results of quantitative comparisons are in line with the visual examination results. Our method achieved the best results in Scene 1, Scene 2, and Scene 3, with the IoU values of 0.932, 0.886, and 0.895, respectively. U-Net achieves the best performance in Scene 4 and Scene 5 with the IoU values of 0.902 and 0.893, respectively, which are both 0.006 higher than our method. SegNet only performs better than our method in Scene 4, with an IoU of 0.898. FCN performs worse than the other methods across all scenes. For the entire testing data, DLEBFP outperforms FCN and SegNet and obtains a precision of 0.926, the recall of 0.914, and the IoU of 0.851. The IoU obtained by DLEBFP is lower than that obtained by U-Net. The high IoU of U-Net is due to more correctly classified building pixels, while the other methods omit these pixels, and thus U-Net has the highest recall values. DLEBFP has the highest precision values among all methods, because it combines the results of three deep learning models and effectively refines the building footprint detection on a step-by-step basis.

Table 1. The statistical results obtained by different models in the close-up scenes and the entire test dataset of the WHU dataset.

Model	Scene 1			Scene 2			Scene 3			Scene 4			Scene 5			Test Dataset		
	Precision	Recall	IoU	Precision	Recall	IoU	Precision	Recall	IoU	Precision	Recall	IoU	Precision	Recall	IoU	Precision	Recall	IoU
FCN	0.941	0.941	0.889	0.925	0.921	0.857	0.918	0.942	0.868	0.942	0.917	0.868	0.938	0.909	0.858	0.932	0.897	0.841
SegNet	0.941	0.959	0.905	0.887	0.957	0.853	0.898	0.957	0.864	0.951	0.941	0.898	0.922	0.940	0.870	0.911	0.925	0.848
U-Net	0.945	0.970	0.917	0.908	0.956	0.872	0.907	0.964	0.879	0.947	0.949	0.902	0.933	0.954	0.893	0.917	0.933	0.861
DLEBFP	0.959	0.970	0.932	0.943	0.939	0.886	0.935	0.954	0.895	0.950	0.941	0.896	0.932	0.949	0.887	0.926	0.914	0.851

Figure 5 displays the vector maps of individual building examples obtained using different methods. Note that the building examples vary considerably in terms of colors, shapes, and surroundings. All methods are able to capture building footprint boundaries in general but with different accuracies. There are large differences among the methods in terms of the vertex number of the constructed polygons. As shown in Figure 5c, if we directly transform the semantic segmentation maps produced by U-Net into building footprint polygons without further processing, dense vertices are located near the building footprint boundaries, which are not useful for survey applications and not appropriate for data storage and transmission. Map simplification using the Douglas–Peucker algorithm with a 0.1 m threshold of the maximum distance (Figure 5d) results in fewer vertices than before, but many redundant vertices still exist when compared with the reference data. The number of vertices decreases as the threshold of the maximum distance in the Douglas–Peucker algorithm increases (Figure 5e,f), but the details of the building footprint boundaries are missing, resulting in inconsistency between the obtained building footprint polygons and the reference data in terms of building shapes. Our method (Figure 5g) performed the best among these methods in depicting the boundaries of buildings with different shapes and sizes. Note that the proposed method uses concise vertices to generate fine-grained building footprint boundaries and preserve geometric details as well as the shapes and structures of buildings. As seen from the second and seventh buildings, our method can accurately detect the buildings obstructed by trees. The multistage prediction in Cascade CNN can infer invisible building corners using the locations of the other corners and the extracted high-level features. In comparison, U-Net does not recover the buildings obstructed by trees and underperforms in comparison to our method in terms of the constructed building footprint polygons.

Table 2 lists the quantitative results obtained using different approaches. The vector results of U-Net have an IoU value of 0.858, which is slightly lower than that of the raster result. The results of our methods in the vector format have an IoU value of 0.850, which is lower than that of U-Net. In terms of the number of extracted buildings, the results of our method are closer to the reference data. DLEBFP generates 14,687 building footprint polygons, only 988 (approximately 6.3%) less than the reference data. One reason is that our method generates a few adjacent polygons that share building footprint corners with the other polygons, and the adjacent polygons could be recognized as a single polygon when conducting statistical analysis. The vector results of U-Net give 18,302 building polygons, 2627 (roughly 16.8%) more than the reference data, because there are many fragmented patches in the results. As mentioned earlier, the number of vertices is important for the management and application of the vector data. The vector results of U-Net have nearly eight million vertices, much higher than those of the ground reference. When generalizing the vector data using the Douglas–Peucker algorithm with the maximum distance thresholds of 0.1 m, 0.5 m, and 1.0 m, the number of vertices decreases to approximately 1 million, 278,541, and 250,132, respectively, and IoU decreases to 0.858, 0.851, and 0.840, respectively. Our method generates building footprint polygons with 135,623 vertices, only 1377 more than the reference data, and obtains an IoU value of 0.850. A vertex-based metric was calculated based on the extraction of different methods. As displayed in Table 2, our method outperforms other methods with $VertexF_{0.5}$ of 0.668 and $VertexF_{1.0}$ of 0.744, which are much higher than those derived from the other methods.

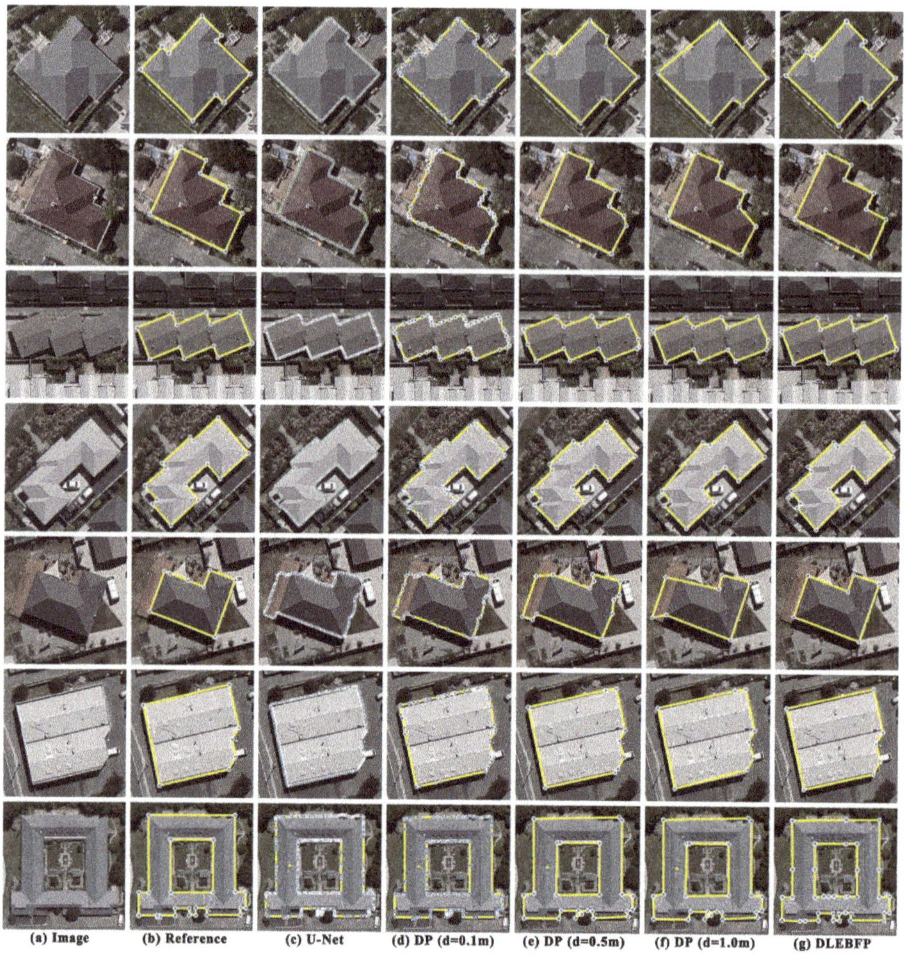

Figure 5. Examples of the true-color-composite images and extracted building footprint polygons using different methods on the WHU dataset are shown for (**a**) the true-color composite images, (**b**) the ground reference images, (**c**) the polygons derived from U-Net, (**d–f**) the polygons derived from U-Net and simplified by Douglas Peucker algorithms with 0.1, 0.5 and 1.0 m threshold, (**g**) the polygons derived from DLEBFP. The building footprint boundaries and vertices are marked by yellow lines and blue dots, respectively.

Table 2. Statistical results obtained using different models after vectorization on the WHU dataset.

Method	IoU Based on the Raster Data	IoU Based on the Vector Data	Changing Rate	Number of Reference Buildings	Number of Extracted Buildings	Number of Reference Vertices	Number of Extracted Vertices	$VertexF_{0.5}$	$VertexF_{1.0}$
DLEBFP	0.851	0.850	0.1%	15,675	14,687	134,246	135,623	0.668	0.744
U-Net	0.861	0.858	0.3%	15,675	18,302	134,246	7,990,152	0.022	0.025
U-Net + Douglas–Peucker (d = 0.1 m)	0.861	0.858	0.3%	15,675	18,302	134,246	1,138,969	0.114	0.134
U-Net + Douglas–Peucker (d = 0.5 m)	0.861	0.851	1.0%	15,675	18,302	134,246	278,541	0.229	0.309
U-Net + Douglas–Peucker (d = 1.0 m)	0.861	0.840	2.1%	15,675	18,302	134,246	250,132	0.216	0.295

4.2. Results on Vaihingen Dataset

In order to test the robustness of the proposed method, we conduct extra experiments on the ISPRS Vaihingen dataset. Figure 6 displays the examples for the extraction results using different methods. Our method can obtain more accurate building footprints with distinctive boundaries and regular shapes. There are less FPs in the extraction results of our proposed method. For example, in the first and third images, there are many FPs located around the buildings in the results obtained using FCN, SegNet, and U-Net, and our method can discriminate it accurately. As shown in the second scene, both FCN and SegNet produce considerable FNs inside a large building, and both DLEBFP and U-Net can extract the large building accurately.

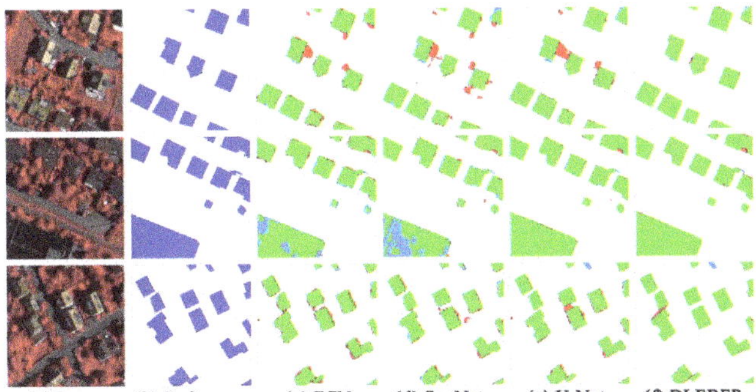

Figure 6. Building extraction results of different models on the ISPRS Vaihingen dataset are shown for (**a**) the false-color composite images (NIR-R-G), (**b**) the binary ground reference images where the blue color denotes building areas and the white color denotes nonbuilding areas, (**c**) the building classification map derived from FCN, (**d**) the building classification map derived from SegNet, (**e**) the building classification map derived from U-Net, and (**f**) the building classification map derived from DLEBFP. In **c**–**f**, the true positives (TPs), false positives (FPs), and false negatives (FNs) are marked in green, red, and blue colors, respectively.

Table 3 lists the statistical results for model comparisons on the ISPRS Vaihingen dataset. DLEBFP outperforms the other models in the metrics of precision, $VertexF_{0.5}$, and $VertexF_{1.0}$. Among all the tested methods, FCN produced the worst results with the IoU of 0.844. For the entire dataset, DLEBFP outperforms both FCN and SegNet and achieves a precision of 0.947, a recall of 0.922, and an IoU of 0.876. The IoU of DLEBFP is slightly lower than that of U-Net. The $VertexF_{1.0}$ of polygons obtained by three semantic segmentation models are less than 0.05, indicating that all these models are not directly suitable for practical applications, and further manual editing is required. We also compared the simplified results of U-Net with the results obtained by our proposed method. Simplification using the Douglas–Peucker algorithm with a 0.1 m threshold of maximum distance results in higher $VertexF_{0.5}$ and $VertexF_{1.0}$ than before, but the metric is still less than one third of that obtained by DLEBFP. As the threshold of maximum distance increases, the IoU of U-Net becomes lower than that of our method, but both Vertex-$F_{1.0}$ and Vertex-$F_{0.5}$ were still lower than that of DLEBFP. Taking both pixel-based and vertex-based evaluation results into consideration, DLEBFP can generate building footprint polygons better than the comparative methods. Overall, the performance of DLEBFP on the Vaihingen dataset are better than on the WHU dataset as indicated by both the pixel-based and vertex-based metrics, probably because the ISPRS Vaihingen datasets have high quality images and accurate image registrations.

Table 3. Statistical results obtained by different models on ISPRS Vaihingen dataset.

Model	Precision	Recall	IoU	$\text{VertexF}_{0.5}$	$\text{VertexF}_{1.0}$
FCN	0.883	0.950	0.844	0.024	0.040
SegNet	0.910	0.932	0.854	0.023	0.030
U-Net	0.932	0.942	0.881	0.037	0.043
U-Net + DP0.1	0.932	0.941	0.879	0.198	0.245
U-Net + DP0.5	0.936	0.929	0.874	0.404	0.555
U-Net + DP1.0	0.938	0.917	0.865	0.417	0.562
DLEBFP	0.947	0.922	0.876	0.731	0.782

4.3. Ablation Studies

Figure 7 displays the results of the ablation experiments in the test area. The results of Baseline (Figure 7d) and Baseline + BB (Figure 7e) are similar, and both have considerable FNs and large omission errors in building detection. The extracted results using the Baseline + VC approach are generally comparable to those of Baseline + BB + VC. Visual comparisons suggest that both methods can extract most of the buildings correctly, but the Baseline + BB + VC method produces more FNs.

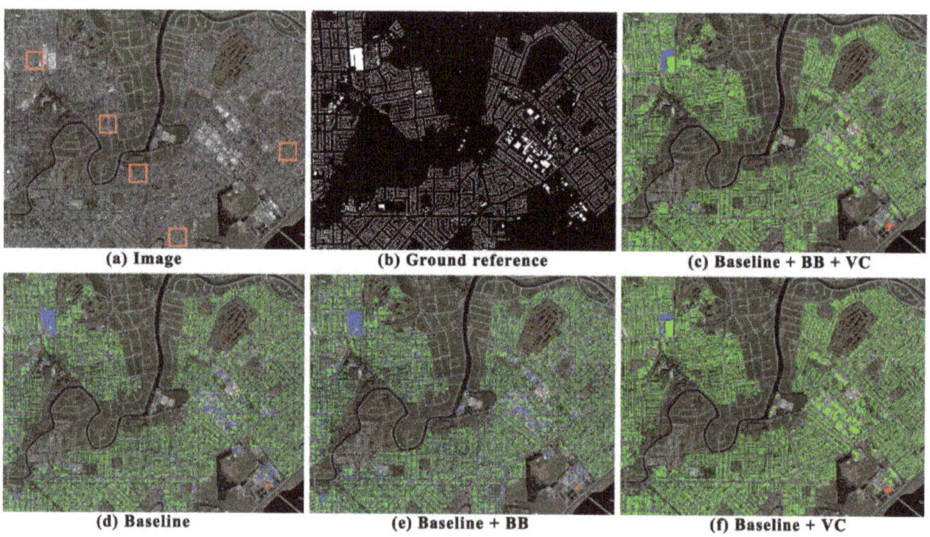

Figure 7. Comparisons of different models with different improved strategy for producing the rasterized classification maps on the WHU dataset are shown for (**a**) the true-color composite image, (**b**) the binary ground reference image where the white color denotes building areas, and the black color denotes nonbuilding areas. In (**c**–**f**), the true positives (TPs), false positives (FPs), and false negatives (FNs) are marked in green, red, and blue colors, respectively.

Figure 8 shows the results of the ablation experiments in the selected scenes of the test areas as marked in red rectangles in Figure 7a. As shown in Figure 8c, the Baseline method produces many FPs in the gaps among buildings and does not extract buildings accurately. Because the Delaunay triangulation generates triangles of buildings that share one or more corners with the other buildings, it could result in the adhesion of buildings. By applying the constraint of bounding boxes, the results of Baseline + BB (Figure 8d) show a reduced number of FPs among buildings but have many FNs because of missing buildings at the edge of the cropped images in the deep learning models. Baseline + VC (Figure 8e) achieves larger improvements over the Baseline method by reducing FNs considerably and detecting most of the erroneously mapped buildings correctly, but still results in considerable FPs among buildings. Baseline + BB + VC (Figure 8f), the complete method, integrates the advantages of the earlier two strategies and considerably reduces both FPs and FNs of buildings.

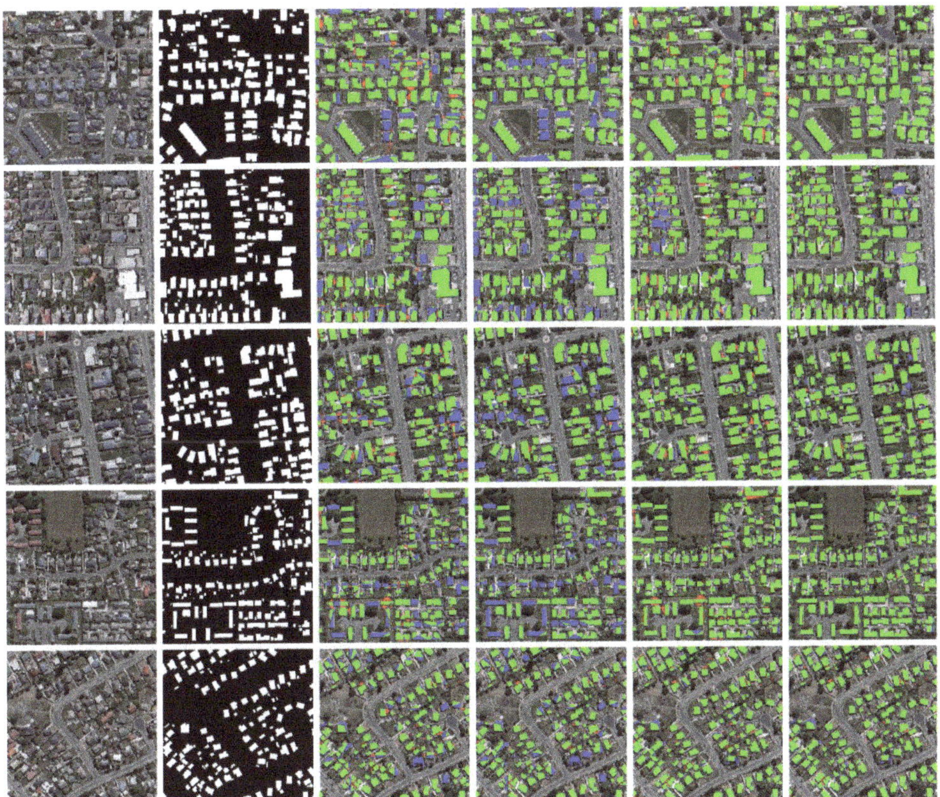

Figure 8. Close-up images and classification results obtained by models with different strategy combinations across five scenes are shown for (**a**) the true-color composite images, (**b**) the binary ground reference image where the white color denotes building areas, and the black color denotes nonbuilding areas, (**c**) the building classification map derived from Baseline, (**d**) the building classification map derived from Baseline + BB, (**e**) the building classification map derived from Baseline + VC, and (**f**) the building classification map derived from Baseline + BB + VC. The true positives (TPs), false positives (FPs), and false negatives (FNs) of buildings are marked in green, red, and blue colors, respectively.

Table 4 lists the quantitative results in the ablation experiments. For all five scenes, Baseline + BB could achieve a higher precision but a lower recall than Baseline. Compared with the Baseline method, the Baseline + BB method has a lower IoU in Scenes 3, 4, and 5, and a higher IoU in Scenes 1 and 2. Applying the constraint of bounding boxes did not enhance the model performance effectively. Compared with Baseline, the Baseline + VC method increases both recall and IoU and achieves the IoU values higher than 0.82 in five scenes, indicating that adding virtual corners helps reduce omission caused by image clipping. By integrating two improvement strategies, the Baseline + BB + VC method achieves the best performance in five different scenes. Compared with the Baseline + VC method, adding the constraints of bounding boxes could increase the values of precision, recall, and IoU. The impact is different from that when only adding the constraints of bounding boxes to Baseline, implying that the strategy of using bounding boxes only improves the model accuracy when buildings can be extracted more accurately. The performance of these methods is consistent in the entire test dataset with that on the five scenes. Compared with Baseline, Baseline + BB reduces IoU by 0.6% and Baseline + VC increases IoU by 16.2%. The Baseline + BB + VC method outperforms the other methods and achieves the best performance with a precision of 0.926, a recall of 0.914, and an IoU of 0.851. In comparison with Baseline, the precision, recall, and IoU of our method increased by 4.4%, 17.8%, and 18.1%, respectively.

Table 4. Statistical results obtained in the ablation experiments.

Model	Scene 1			Scene 2			Scene 3			Scene 4			Scene 5			Test Dataset		
	Precision	Recall	IoU	Precision	Recall	IoU	Precision	Recall	IoU	Precision	Recall	IoU	Precision	Recall	IoU	Precision	Recall	IoU
Baseline	0.885	0.776	0.705	0.891	0.771	0.704	0.875	0.792	0.712	0.869	0.702	0.635	0.906	0.817	0.753	0.882	0.736	0.670
Baseline + BB	0.958	0.737	0.714	0.943	0.743	0.711	0.926	0.736	0.694	0.941	0.653	0.627	0.927	0.790	0.743	0.918	0.706	0.664
Baseline + VC	0.914	0.954	0.877	0.911	0.890	0.820	0.909	0.954	0.871	0.896	0.931	0.840	0.925	0.946	0.879	0.905	0.911	0.832
Baseline + BB + VC	0.959	0.970	0.932	0.943	0.939	0.886	0.935	0.954	0.895	0.950	0.941	0.896	0.931	0.949	0.887	0.926	0.914	0.851

5. Discussion

By combining the deep learning models for different tasks, our method can accurately extract building footprint polygons from very high-resolution aerial images. One advantage of our method is that it can extract building footprints with sharp boundaries in a vector format and use concise vertices to represent building footprint boundaries that are close to the ground reference. Another advantage of our method is that it can be executed on a regional scale instead of on an individual building. The reasons our method performs well are as follows. First, we used a deep learning model to detect the building footprint corners, which are used to guide the construction of building footprint polygons. Compared to the key point detection methods in the traditional methods, such as Harris and scale-invariant feature transform, the deep learning model can detect building footprint corners accurately. The traditional methods likely detect erroneous key points belonging to other man-made objects, such as roads. Second, our method integrates the multi-task results generated by the deep learning models based on the Delaunay triangulation. The ensemble framework enhances the model performance and makes it possible to automatically extract building footprint polygons from remote sensing images on a regional scale.

The accuracy of the corner detection influences the performance of the entire framework and Figure 9 exhibits the corner detection results on the two studied datasets. As shown in the figure, most of the corners can be detected by Cascade CNN despite of errors. First, our method cannot detect the corners that are severely obscured by trees. Although Cascade CNN can infer the existence of corners sheltered by trees in some cases, the predicted location of corners often deviates from their real locations. Second, a few redundant points that did not exist in the ground truth data were detected because these points have representation similar to building corners. The redundant detected points have limited impacts on the overall performance of our results because such points are normally removed when we classified the triangles based on the semantic segmentation maps. In addition, our method performs weakly on the corners that are closely located. It is difficult to accurately and completely distinguish adjacent corners. In this case, an incomplete set of corners causes a loss of geometric details and decreases the accuracy of the extracted building footprint polygons. When comparing with the results of corner detections on the WHU dataset, Cascade CNN performs better in the Vaihingen dataset. Visually, the results in the Vaihingen dataset have less omitted corners than in the WHU dataset. The statistical results associated with vertex-based evaluation metrics illustrate that the model performance in the Vaihingen dataset is better than in the WHU dataset. The differences in the model performance are likely due to different standards of classification labels and image quality. In the Vaihingen dataset, fewer buildings are obscured by trees or the other objects, and the sheltered buildings are not annotated as the building pixels. By comparison, the WHU dataset has more sheltered buildings that are still labeled as the building pixels in the ground truth maps. It is, therefore, challenging to detect building corners in the WHU dataset. DLEBFP achieves higher IoU in the Vaihingen dataset than in the WHU dataset because better corner detection results were obtained in the Vaihingen dataset when using Cascade CNN.

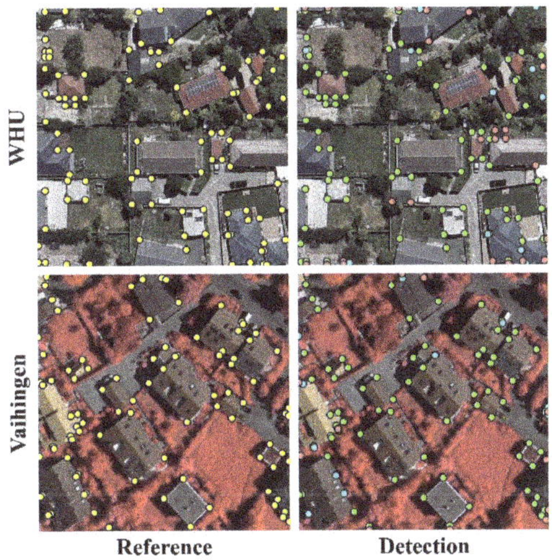

Figure 9. Examples of corner detections using Cascade CNN and reference maps on both the WHU dataset and the Vaihingen dataset. Ground truths are marked in yellow. The true positives (TPs), false positives (FPs), and false negatives (FNs) of building corners are marked in green, red and blue colors, respectively.

Our framework may extract inaccurate building footprint polygons and Figure 10 exhibits four typical types of inaccurate detections. In the first case (Figure 10a), one of the corners obscured by the other objects could not be detected by our model and thus we obtained an incomplete building polygon with FNs. The second error type, as shown in Figure 10b, is mainly caused by the omission of the corner detection model, which results in FNs. The third type, as demonstrated in Figure 10c, is also mainly caused by the occasional omission of the corner detection model but leads to FPs instead of FNs. Figure 10d shows the fourth type of errors, in which two separated buildings were merged into one building, and there are FNs located in the gaps between the two buildings. Although we utilized the bounding box detected by Cascade R-CNN to constrain the extent of polygon construction processes, there are difficulties in a few cases when some buildings are very close to one another.

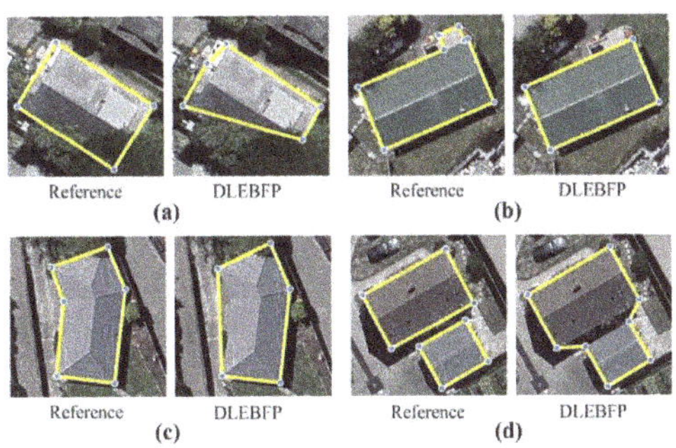

Figure 10. Four examples of inaccurate building footprint polygons extracted by DLEBFP and the corresponding reference polygons are shown in (**a–d**), which is caused by four typical causes.

Model efficiency is also an important indicator when evaluating a method. As shown in Table 5, we compare our method with the others in terms of the time cost of inferencing deep learning models and post-processing, and the storage size of files in the shapefile format. As for the inference time of the deep learning models, the computational cost of our method is approximately four times that of the other methods because we use three deep neural networks for different tasks. As for the post-processing time, the time needed for directly converting the raster data to vector data is acceptable. For example, converting the raster map produced by U-Net to the vector data of building footprints needs 37.94 ms. When applying a vector generalization method of the Douglas–Peucker algorithm, the post-processing time increases by approximately 100 ms. Our method costs the processing time approximately three times more than the method that uses U-Net with vector generalization. Note that the vector file obtained using our method only has a storage size of 3.3 MB, which is less than one third the size of the files obtained using the other methods. It is important to generate files with smaller sizes and maintain both high pixel-based and vertex-based accuracies in the field of surveying and mapping. A large file size indicates that the vector file likely contains redundant vertices and does not meet the requirement of vector map production. In addition, practical applications including data storage, management, transmission, and spatial analysis prefer accurate and concise vector data. As the comparative methods tested here generated data with much redundant information, it is worthwhile producing the vector data with a smaller file size.

Table 5. Comparisons of model efficiency and file sizes.

Model	Inference Time (ms)	Post-Processing Time (ms)	File Size (MB)
FCN	153.11	11.29	95.6
SegNet	131.43	21.22	129.7
U-Net	185.64	37.94	141.8
U-Net + DP0.1	185.64	141.37	26.4
U-Net + DP0.5	185.64	135.95	10.9
U-Net + DP1.0	185.64	135.05	9.9
DLEBFP	523.49	351.52	3.3

The calculation of both inference time and post-processing time is based on the average of the test dataset including patches with the size of 1024 × 1024 pixels. The inference time indicates the time cost of the inference of different deep learning models. The post-processing time denotes the time cost of the vectorization process. File size denotes the storage size of the output vector files in a shapefile format.

6. Conclusions

In this work, we proposed a novel framework that combines three deep learning models for different tasks to directly extract building footprint polygons from very high-resolution aerial images. The framework uses U-Net, Cascade R-CNN, and Cascade CNN to provide semantic segmentation maps, bounding boxes, and corners of building, respectively. Furthermore, a robust polygon construction strategy was devised to integrate three types of results and then generate the building polygon with high accuracy. In the strategy, Delaunay triangulation utilizes the detected building footprint corners to generate the polygons of individual building footprints, which are further refined using the maps of bounding boxes and semantic segmentation. The experiments on a very high-resolution aerial image dataset covering 78 km^2 and containing 84,000 buildings suggest that our method can extract the building polygons accurately and completely. Our method achieves a precision of 0.926, recall of 0.914, and the IoU of 0.851 in the test dataset of the WHU building dataset. The proposed method was compared with benchmark segmentation models and classic map generalization methods. Qualitative and quantitative analyses indicate that our methods can generate a comparable accuracy with fewer redundant vertices and provide high-quality building footprint polygons. These promising results suggest that the developed method could potentially be applied in the mapping of buildings polygon across large areas.

Author Contributions: Conceptualization, Z.L. and Q.X.; methodology, Z.L. and Q.X.; software, Z.L. and M.C.; validation, Z.L.; formal analysis, Z.L.; investigation, Z.L.; resources, Q.X.; data curation, Z.L.; writing—original draft preparation, Z.L. and Q.X.; writing—review and editing, Q.X. and Y.S.; visualization, Z.L. and M.C.; supervision, Q.X. and Y.S.; project administration, Z.L.; funding acquisition, Q.X. and Y.S. All authors have read and agreed to the published version of the manuscript.

Funding: This research was funded by National Natural Science Foundation of China (grant nos. 41875122 and 41801351), National Key R&D Program of China (grant nos. 2017YFA0604300 and 2017YFA0604400), Western Talents (grant no. 2018XBYJRC004), Guangdong Top Young Talents (grant no. 2017TQ04Z359).

Institutional Review Board Statement: Not applicable.

Informed Consent Statement: Not applicable.

Data Availability Statement: The datasets used in this study are openly available. These datasets can be found in http://gpcv.whu.edu.cn/data/ (accessed on 10 November 2020) for WHU dataset and https://www2.isprs.org/commissions/comm2/wg4/benchmark/2d-sem-label-vaihingen/ (accessed on 20 January 2021) for Vaihingen dataset.

Acknowledgments: We would like to acknowledge the Group of Photogrammetry and Computer Vision (GPCV) at Wuhan University for providing the WHU building dataset. We are also grateful to ISPRS for providing the Vaihingen dataset.

Conflicts of Interest: The authors declare no conflict of interest.

References

1. Tong, X.; Lin, X.; Feng, T.; Xie, H.; Liu, S.; Hong, Z.; Chen, P. Use of shadows for detection of earthquake-induced collapsed buildings in high-resolution satellite imagery. *ISPRS J. Photogramm. Remote Sens.* **2013**, *79*, 53–67. [CrossRef]
2. Jensen, J.R.; Cowen, D.C. Remote sensing of urban/suburban infrastructure and socio-economic attributes. *Photogramm. Eng. Remote Sens.* **1999**, *65*, 611–622.
3. Turker, M.; Koc-San, D. Building extraction from high-resolution optical spaceborne images using the integration of support vector machine (SVM) classification, Hough transformation and perceptual grouping. *Int. J. Appl. Earth Obs. Geoinf.* **2015**, *34*, 58–69. [CrossRef]
4. Ma, L.; Li, M.; Ma, X.; Cheng, L.; Du, P.; Liu, Y. A review of supervised object-based land-cover image classification. *ISPRS J. Photogramm. Remote Sens.* **2017**, *130*, 277–293. [CrossRef]
5. Liasis, G.; Stavrou, S. Building extraction in satellite images using active contours and colour features. *Int. J. Remote Sens.* **2016**, *37*, 1127–1153. [CrossRef]
6. Rottensteiner, F.; Trinder, J.; Clode, S.; Kubik, K. Building detection by fusion of airborne laser scanner data and multi-spectral images: Performance evaluation and sensitivity analysis. *ISPRS J. Photogramm. Remote Sens.* **2007**, *62*, 135–149. [CrossRef]
7. Shi, Y.; Li, Q.; Zhu, X.X. Building footprint generation using improved generative adversarial networks. *IEEE Geosci. Remote Sens. Lett.* **2018**, *16*, 603–607. [CrossRef]
8. Huang, X.; Zhang, L. A Multidirectional and Multiscale Morphological Index for Automatic Building Extraction from Multispectral GeoEye-1 Imagery. *Photogramm. Eng. Remote Sens.* **2011**, *77*, 721–732. [CrossRef]
9. Ok, A.O.; Senaras, C.; Yuksel, B. Automated Detection of Arbitrarily Shaped Buildings in Complex Environments From Monocular VHR Optical Satellite Imagery. *IEEE Trans. Geosci. Remote Sens.* **2013**, *51*, 1701–1717. [CrossRef]
10. Ji, S.; Wei, S.; Lu, M. A scale robust convolutional neural network for automatic building extraction from aerial and satellite imagery. *Int. J. Remote Sens.* **2018**, *40*, 3308–3322. [CrossRef]
11. Yuan, J. Learning Building Extraction in Aerial Scenes with Convolutional Networks. *IEEE Trans. Pattern Anal. Mach. Intell.* **2018**, *40*, 2793–2798. [CrossRef]
12. Du, S.; Zhang, F.; Zhang, X. Semantic classification of urban buildings combining VHR image and GIS data: An improved random forest approach. *ISPRS J. Photogramm. Remote Sens.* **2015**, *105*, 107–119. [CrossRef]
13. Wu, G.; Shao, X.; Guo, Z.; Chen, Q.; Yuan, W.; Shi, X.; Xu, Y.; Shibasaki, R. Automatic Building Segmentation of Aerial Imagery Using Multi-Constraint Fully Convolutional Networks. *Remote Sens.* **2018**, *10*, 407. [CrossRef]
14. Liu, W.; Yang, M.; Xie, M.; Guo, Z.; Li, E.; Zhang, L.; Pei, T.; Wang, D. Accurate Building Extraction from Fused DSM and UAV Images Using a Chain Fully Convolutional Neural Network. *Remote Sens.* **2019**, *11*, 2912. [CrossRef]

15. LeCun, Y.; Bengio, Y. Convolutional networks for images, speech, and time series. *Handb. Brain Theory Neural Netw.* **1995**, *3361*, 1995.
16. Zhu, X.X.; Tuia, D.; Mou, L.; Xia, G.-S.; Zhang, L.; Xu, F.; Fraundorfer, F. Deep learning in remote sensing: A comprehensive review and list of resources. *IEEE Geosci. Remote Sens. Mag.* **2017**, *5*, 8–36. [CrossRef]
17. Zhao, W.; Du, S. Learning multiscale and deep representations for classifying remotely sensed imagery. *ISPRS J. Photogramm. Remote Sens.* **2016**, *113*, 155–165. [CrossRef]
18. Ye, Z.; Fu, Y.; Gan, M.; Deng, J.; Comber, A.; Wang, K. Building Extraction from Very High Resolution Aerial Imagery Using Joint Attention Deep Neural Network. *Remote Sens.* **2019**, *11*, 2970. [CrossRef]
19. Marmanis, D.; Wegner, J.D.; Galliani, S.; Schindler, K.; Datcu, M.; Stilla, U. Semantic segmentation of aerial images with an ensemble of CNSS. *ISPRS Ann. Photogramm. Remote Sens. Spat. Inf. Sci.* **2016**, *3*, 473–480. [CrossRef]
20. Fu, G.; Liu, C.; Zhou, R.; Sun, T.; Zhang, Q. Classification for high resolution remote sensing imagery using a fully convolutional network. *Remote Sens.* **2017**, *9*, 498. [CrossRef]
21. Maggiori, E.; Tarabalka, Y.; Charpiat, G.; Alliez, P. Convolutional Neural Networks for Large-Scale Remote-Sensing Image Classification. *IEEE Trans. Geosci. Remote Sens.* **2017**, *55*, 645–657. [CrossRef]
22. Ji, S.; Wei, S.; Lu, M. Fully Convolutional Networks for Multisource Building Extraction From an Open Aerial and Satellite Imagery Data Set. *IEEE Trans. Geosci. Remote Sens.* **2019**, *57*, 574–586. [CrossRef]
23. Huang, J.; Zhang, X.; Xin, Q.; Sun, Y.; Zhang, P. Automatic building extraction from high-resolution aerial images and LiDAR data using gated residual refinement network. *ISPRS J. Photogramm. Remote Sens.* **2019**, *151*, 91–105. [CrossRef]
24. Rottensteiner, F.; Sohn, G.; Gerke, M.; Wegner, J.D.; Breitkopf, U.; Jung, J. Results of the ISPRS benchmark on urban object detection and 3D building reconstruction. *ISPRS J. Photogramm. Remote Sens.* **2014**, *93*, 256–271. [CrossRef]
25. Dey, E.K.; Awrangjeb, M.; Stantic, B. Outlier detection and robust plane fitting for building roof extraction from LiDAR data. *Int. J. Remote Sens.* **2020**, *41*, 6325–6354. [CrossRef]
26. Awrangjeb, M.; Gilani, S.A.N.; Siddiqui, F.U. An effective data-driven method for 3-d building roof reconstruction and robust change detection. *Remote Sens.* **2018**, *10*, 1512. [CrossRef]
27. Gilani, S.A.N.; Awrangjeb, M.; Lu, G. Segmentation of Airborne Point Cloud Data for Automatic Building Roof Extraction. *GIScience Remote Sens.* **2018**, *55*, 63–89. [CrossRef]
28. Mahmud, J.; Price, T.; Bapat, A.; Frahm, J.-M. Boundary-aware 3D building reconstruction from a single overhead image. In Proceedings of the IEEE/CVF Conference on Computer Vision and Pattern Recognition, Seattle, WA, USA, 13–19 June 2020; pp. 441–451.
29. Wang, M.; Yuan, S.; Pan, J. Building detection in high resolution satellite urban image using segmentation, corner detection combined with adaptive windowed hough transform. In Proceedings of the 2013 IEEE International Geoscience and Remote Sensing Symposium-IGARSS, Melbourne, Australia, 21–26 July 2013; pp. 508–511.
30. Qin, X.; He, S.; Yang, X.; Dehghan, M.; Qin, Q.; Martin, J. Accurate Outline Extraction of Individual Building From Very High-Resolution Optical Images. *IEEE Geosci. Remote Sens. Lett.* **2018**, *15*, 1775–1779. [CrossRef]
31. Girard, N.; Tarabalka, Y. End-to-end learning of polygons for remote sensing image classification. In Proceedings of the IGARSS 2018-2018 IEEE International Geoscience and Remote Sensing Symposium, Valencia, Spain, 22–27 July 2018; pp. 2083–2086.
32. Wu, G.; Guo, Z.; Shi, X.; Chen, Q.; Xu, Y.; Shibasaki, R.; Shao, X. A Boundary Regulated Network for Accurate Roof Segmentation and Outline Extraction. *Remote Sens.* **2018**, *10*, 1195. [CrossRef]
33. Marmanis, D.; Schindler, K.; Wegner, J.D.; Galliani, S.; Datcu, M.; Stilla, U. Classification with an edge: Improving semantic image segmentation with boundary detection. *ISPRS J. Photogramm. Remote Sens.* **2018**, *135*, 158–172. [CrossRef]
34. Liao, C.; Hu, H.; Li, H.; Ge, X.; Chen, M.; Li, C.; Zhu, Q. Joint Learning of Contour and Structure for Boundary-Preserved Building Extraction. *Remote Sens.* **2021**, *13*, 1049. [CrossRef]
35. Douglas, D.H.; Peucker, T.K. Algorithms for the reduction of the number of points required to represent a digitized line or its caricature. *Cartogr. Int. J. Geogr. Inf. Geovis.* **1973**, *10*, 112–122. [CrossRef]
36. Wang, Z.; Müller, J.-C. Line generalization based on analysis of shape characteristics. *Cartogr. Geogr. Inf. Syst.* **1998**, *25*, 3–15. [CrossRef]
37. Zhou, S.; Jones, C.B. Shape-aware line generalisation with weighted effective area. In *Developments in Spatial Data Handling*; Springer: Berlin/Heidelberg, Germany, 2005; pp. 369–380.
38. Maggiori, E.; Tarabalka, Y.; Charpiat, G.; Alliez, P. Polygonization of remote sensing classification maps by mesh approximation. In Proceedings of the 2017 IEEE International Conference on Image Processing (ICIP), Beijing, China, 17–20 September 2017; pp. 560–564.
39. Song, W.; Zhong, B.; Sun, X. Building corner detection in aerial images with fully convolutional networks. *Sensors* **2019**, *19*, 1915. [CrossRef] [PubMed]
40. Ronneberger, O.; Fischer, P.; Brox, T. U-Net: Convolutional Networks for Biomedical Image Segmentation. In *Medical Image Computing and Computer-Assisted Intervention, Pt Iii*; Navab, N., Hornegger, J., Wells, W.M., Frangi, A.F., Eds.; Lecture Notes in Computer Science: Munich, Germany, 2015; Volume 9351, pp. 234–241.
41. Zhang, Z.; Liu, Q.; Wang, Y. Road Extraction by Deep Residual U-Net. *IEEE Geosci. Remote Sens. Lett.* **2018**, *15*, 749–753. [CrossRef]
42. Jaturapitpornchai, R.; Matsuoka, M.; Kanemoto, N.; Kuzuoka, S.; Ito, R.; Nakamura, R. Newly built construction detection in sar images using deep learning. *Remote Sens.* **2019**, *11*, 1444. [CrossRef]

43. He, K.; Zhang, X.; Ren, S.; Sun, J. Deep Residual Learning for Image Recognition. In Proceedings of the IEEE Conference on Computer Vision and Pattern Recognition, Las Vegas, NV, USA, 27–30 June 2016; pp. 770–778.
44. Glorot, X.; Bordes, A.; Bengio, Y. Deep sparse rectifier neural networks. In Proceedings of the Fourteenth International Conference on Artificial Intelligence and Statistics, Ft. Lauderdale, FL, USA, 11–13 April 2011; pp. 315–323.
45. Krizhevsky, A.; Sutskever, I.; Hinton, G.E. Imagenet classification with deep convolutional neural networks. *Adv. Neural Inf. Process. Syst.* **2012**, *25*, 1097–1105. [CrossRef]
46. Ioffe, S.; Szegedy, C. Batch normalization: Accelerating deep network training by reducing internal covariate shift. In Proceedings of the International Conference on Machine Learning, Lille, France, 7–9 July 2015; pp. 448–456.
47. Milletari, F.; Navab, N.; Ahmadi, S.-A. V-Net: Fully Convolutional Neural Networks for Volumetric Medical Image Segmentation. In Proceedings of the 2016 Fourth International Conference on 3D Vision (3DV), Stanford, CA, USA, 25–28 October 2016; pp. 565–571.
48. Cai, Z.; Vasconcelos, N. Cascade R-CNN: Delving into High Quality Object Detection. In Proceedings of the 2018 IEEE/CVF Conference on Computer Vision and Pattern Recognition, Salt Lake City, UT, USA, 18–23 June 2018; pp. 6154–6162.
49. Lin, T.-Y.; Dollar, P.; Girshick, R.; He, K.; Hariharan, B.; Belongie, S. Feature Pyramid Networks for Object Detection. In Proceedings of the IEEE Conference on Computer Vision and Pattern Recognition, Honolulu, HI, USA, 21–26 July 2017; pp. 936–944.
50. Cao, Z.; Simon, T.; Wei, S.-E.; Sheikh, Y. Realtime Multi-Person 2D Pose Estimation using Part Affinity Fields. In Proceedings of the IEEE Conference on Computer Vision and Pattern Recognition, Honolulu, HI, USA, 21–26 July 2017; pp. 1302–1310.
51. Pfister, T.; Charles, J.; Zisserman, A. Flowing ConvNets for Human Pose Estimation in Videos. In Proceedings of the IEEE International Conference on Computer Vision, Santiago, Chile, 11–18 December 2015; pp. 1913–1921.
52. Simonyan, K.; Zisserman, A. Very deep convolutional networks for large-scale image recognition. *Arxiv Prepr.* **2014**, arXiv:1409.1556. Available online: https://arxiv.org/abs/1409.1556 (accessed on 1 February 2021).
53. Li, J.; Su, W.; Wang, Z. Simple pose: Rethinking and improving a bottom-up approach for multi-person pose estimation. In Proceedings of the AAAI Conference on Artificial Intelligence, New York, NY, USA, 7–12 February 2020; pp. 11354–11361.
54. Deng, M.; Liu, Q.; Cheng, T.; Shi, Y. An adaptive spatial clustering algorithm based on Delaunay triangulation. *Comput. Environ. Urban Syst.* **2011**, *35*, 320–332. [CrossRef]
55. He, X.; Zhang, X.; Xin, Q. Recognition of building group patterns in topographic maps based on graph partitioning and random forest. *ISPRS J. Photogramm. Remote Sens.* **2018**, *136*, 26–40. [CrossRef]
56. Chen, Q.; Wang, L.; Wu, Y.; Wu, G.; Guo, Z.; Waslander, S.L. Aerial imagery for roof segmentation: A large-scale dataset towards automatic mapping of buildings. *ISPRS J. Photogramm. Remote Sens.* **2019**, *147*, 42–55. [CrossRef]
57. Badrinarayanan, V.; Kendall, A.; Cipolla, R. SegNet: A Deep Convolutional Encoder-Decoder Architecture for Image Segmentation. *IEEE Trans. Pattern Anal. Mach. Intell.* **2017**, *39*, 2481–2495. [CrossRef] [PubMed]
58. Shi, W.; Cheung, C. Performance evaluation of line simplification algorithms for vector generalization. *Cartogr. J.* **2006**, *43*, 27–44. [CrossRef]
59. He, H.; Zhou, J.; Chen, M.; Chen, T.; Li, D.; Cheng, P. Building extraction from UAV images jointly using 6D-SLIC and multiscale Siamese convolutional networks. *Remote Sens.* **2019**, *11*, 1040. [CrossRef]
60. Chen, Q.; Wang, L.; Waslander, S.L.; Liu, X. An end-to-end shape modeling framework for vectorized building outline generation from aerial images. *ISPRS J. Photogramm. Remote Sens.* **2020**, *170*, 114–126. [CrossRef]
61. Heckbert, P.S.; Garland, M. *Survey of Polygonal Surface Simplification Algorithms*; Carnegie-Mellon Univ Pittsburgh PA School of Computer Science: Pittsburgh, PA, USA, 1 May 1997.
62. Liu, Y.; Fan, B.; Wang, L.; Bai, J.; Xiang, S.; Pan, C. Semantic labeling in very high resolution images via a self-cascaded convolutional neural network. *ISPRS J. Photogramm. Remote Sens.* **2018**, *145*, 78–95. [CrossRef]
63. Ienco, D.; Interdonato, R.; Gaetano, R.; Minh, D.H.T. Combining Sentinel-1 and Sentinel-2 Satellite Image Time Series for land cover mapping via a multi-source deep learning architecture. *ISPRS J. Photogramm. Remote Sens.* **2019**, *158*, 11–22. [CrossRef]
64. Woo, S.; Park, J.; Lee, J.-Y.; Kweon, I.S. Cbam: Convolutional block attention module. In Proceedings of the European conference on computer vision (ECCV), Munich, Germany, 8–14 September 2018; pp. 3–19.
65. Yang, H.L.; Yuan, J.; Lunga, D.; Laverdiere, M.; Rose, A.; Bhaduri, B. Building Extraction at Scale Using Convolutional Neural Network: Mapping of the United States. *IEEE J. Sel. Top. Appl. Earth Obs. Remote Sens.* **2018**, *11*, 2600–2614. [CrossRef]
66. Li, Q.; Shi, Y.; Huang, X.; Zhu, X.X. Building footprint generation by integrating convolution neural network with feature pairwise conditional random field (FPCRF). *IEEE Trans. Geosci. Remote Sens.* **2020**, *58*, 7502–7519. [CrossRef]

remote sensing

MDPI

Article

Building Extraction from Airborne LiDAR Data Based on Multi-Constraints Graph Segmentation

Zhenyang Hui [1], Zhuoxuan Li [1], Penggen Cheng [1,*], Yao Yevenyo Ziggah [2] and JunLin Fan [3]

1 Faculty of Geomatics, East China University of Technology, Nanchang 330013, China; huizhenyang2008@ecut.edu.cn (Z.H.); 2020110363@ecut.edu.cn (Z.L.)
2 Faculty of Geosciences and Environmental Studies, University of Mines and Technology, Tarkwa 999064, Ghana; yyziggah@umat.edu.gh
3 Jiangxi Nuclear industry Surveying and Mapping Institute Group Co., Ltd., Nanchang 330038, China; fanjunlin2000@163.com
* Correspondence: pgcheng@ecut.edu.cn

Abstract: Building extraction from airborne Light Detection and Ranging (LiDAR) point clouds is a significant step in the process of digital urban construction. Although the existing building extraction methods perform well in simple urban environments, when encountering complicated city environments with irregular building shapes or varying building sizes, these methods cannot achieve satisfactory building extraction results. To address these challenges, a building extraction method from airborne LiDAR data based on multi-constraints graph segmentation was proposed in this paper. The proposed method mainly converted point-based building extraction into object-based building extraction through multi-constraints graph segmentation. The initial extracted building points were derived according to the spatial geometric features of different object primitives. Finally, a multi-scale progressive growth optimization method was proposed to recover some omitted building points and improve the completeness of building extraction. The proposed method was tested and validated using three datasets provided by the International Society for Photogrammetry and Remote Sensing (ISPRS). Experimental results show that the proposed method can achieve the best building extraction results. It was also found that no matter the average quality or the average F1 score, the proposed method outperformed ten other investigated building extraction methods.

Keywords: airborne LiDAR; building extraction; graph segmentation; object primitive; geometric feature

Citation: Hui, Z.; Li, Z.; Cheng, P.; Ziggah, Y.Y.; Fan, J. Building Extraction from Airborne LiDAR Data Based on Multi-Constraints Graph Segmentation. *Remote Sens.* **2021**, *13*, 3766. https://doi.org/10.3390/rs13183766

Academic Editors: Jiaojiao Tian, Qin Yan, Mohammad Awrangjeb, Beril Sirmacek and Nusret Demir

Received: 5 August 2021
Accepted: 16 September 2021
Published: 20 September 2021

1. Introduction

A building is an essential component of urban construction. The extraction and reconstruction of buildings has been a critical step for many applications, such as urban planning, disaster assessment, cadastral management and so on [1,2]. Airborne LiDAR is an active remote sensing technology which has developed very rapidly in the recent years [3–6]. This technology has become an attractive choice for building extraction [7,8], due to its high efficiency and measuring accuracy, less interference by external environment and strong initiative [9,10].

1.1. Related Works

Currently, building extraction methods can be classified into two categories, including the machine learning methods and the classic methods. In the machine learning methods, the acquired raw data are first transformed into a multidimensional feature space. Then, an optimal feature space is estimated by a learning classifier to map the features into desired outputs [11]. Ni et al. [12] first proposed a stepwise point cloud segmentation method to extract three kinds of segments, including planar, smooth and rough surfaces. Random Forest (RF) was then employed to select features and classify the afore-mentioned segments. Finally, semantic rules were used to optimize the classification result. The experimental results showed that this method was effective for small scale targets. Nahhas et al. [13] proposed a deep learning method to detect buildings by fusing the LiDAR data and orthophotos. In this method, features were first extracted by object-based analysis. The low-level features were transformed into compressed features with a feature-level fusion and an autoencoder-based dimensionality. After that, the compressed features were transformed into the high-level features using a Convolutional Neural Network (CNN) to classify the objects into buildings and background. Maltezos et al. [11] also adopted a CNN model in their method. Firstly, a multi-dimensional feature vector was created using the raw LiDAR data and seven additional features. Experimental results showed that this algorithm could extract buildings with 85% completeness, and the correctness reached 93%, at per-area level. Huang et al. [14] fused high-resolution aerial images and LiDAR points for building extraction with an end-to-end trainable gated residual refinement network. The modified residual learning network was applied as the encoder part of network to learn multi-level features from the fusion data. A gated feature labeling unit was introduced to reduce unnecessary feature transmission and refine classification results. Zhang et al. [7] developed a hybrid attention-aware fusion network based on a novel hybrid fusion architecture to extract the buildings from high-resolution imagery and LiDAR data. Li et al. [15] first split the raw preprocessed LiDAR data into numerous samples to fed into CNNs directly. Then, the graph geometric moments CNNs were proposed to train and recognize building points. Finally, the test scenes were fed into the framework to extract the building points. Wen et al. [6] first proposed a graph attention convolution module. This module could examine spatial relationship among all points and determine the convolution weights. Then, a global-local graph attention CNN was designed to classify the airborne point clouds using multiscale features of the point clouds. Yuan et al. [16] proposed a fully CNN based on the residual network. The training pattern for multi-modal data was provided by combining the advantage of high-resolution aerial images and LiDAR data. Obviously, for the CNN models, the precise labelled point clouds are generally necessary. Zolanvari et al. [17] provided a publicly available annotated benchmark dataset for training and testing. Moreover, they also tested three well-known CNN models, including PointNet, PointNet++ and So-Net on the benchmark. In addition to the geometric features built for the machine learning techniques, some researchers try to calculate the statistical characteristics of the point cloud distribution to realize unsupervised segmentation [18]. For instance, Crosilla et al. [19] proposed a filtering and classification technique for LiDAR points by calculating the statistical features, including skewness and kurtosis iteratively. Specifically, the skewness and kurtosis are third- and fourth-order moments about the mean.

Although the machine learning methods can realize building extraction, these methods unavoidably involve some limitations. For instance, high-quality training data are required, and the differences in the data distribution between the training data and the experiment scenes lead to low accuracy [20]. To avoid these problems, some authors work on the classic methods. In terms of input data, methods can be further classified as two categories, including building extraction based on LiDAR points and based on multi-source data fusion [15]. In the first type of methods, LiDAR points are the only data source for building extraction. In these methods, building points are extracted by relying mainly on their geometric morphological features that are different from other objects. Dorninger et al. [21] proposed a building extraction method based on plane segmentation detection. The hierarchical clustering algorithm was used to obtain the initial candidate seed points. Then, they extracted the buildings with the iterative region growing algorithm. Poullis et al. [22] used the object-based region growing algorithm to detect the building points and refined the boundary with a polygon Boolean operation to achieve better building extraction results. In Sun et al. [23], point cloud data were divided into ground and non-ground points with a graph cut-based method where a novel hierarchical Euclidean clustering method was used to extract rooftop patches. Finally, a region growing-based segmentation method was presented to detect all building points with each rooftop patch. Awrangjeb and Fraser [24] separated the raw LiDAR points into two groups and formed a "building mask" with the group containing the ground points. The group that contains non-ground points was segmented into individual building or tree objects using the building mask. During the segmentation, the planar roof segments were extracted and refined based on the rules, such as the coplanarity, locality and height difference of points. The experimental results showed that this method offered a high successful rate for building detection and roof plane extraction. Fan et al. [25] realized the building extraction based on the fact that each roof can be composed of gabled roofs and single flats. The algorithm Random Sample Consensus (RANSAC) was then used to detect roof ridges. Then, the points on the two roof flats along a roof ridge were identified based on their connectivity and coplanarity. Sampath et al. [26] first analyzed point characteristics to exclude the nonplanar points. The authors used the fuzzy k-means algorithm to cluster the planar points. To extract complete buildings, the clustered points were merged into the integrated rooftop according to the breaklines. Zou et al. [27] proposed a strip strategy-based method to filter the building points and extract the edge point set from LiDAR data. The point cloud was segmented into several data strips. After that, building points were filtered from the data strips with an adaptive-weight polynomial. Finally, building edges were extracted by a modified scanline method. This method is usually suitable for urban areas where buildings were densely distributed. Cai et al. [28] introduced a coarse-to-fine building detection method that was based on semi-suppressed fuzzy C-means and restricted region growing. After a minimum bounding rectangle was adopted to refine the detection results, the method could offer an excellent performance for building detection with over 89.5% completeness and a minimum 91% correctness. In Wang et al. [29], a semantic-based method was employed to extract building points with contexts. A Markov random field optimization model was constructed for postprocessing and segmentation results refinement.

Although the building extraction methods using only LiDAR data can achieve good extraction results, these methods unavoidably have some practical limitations. For instance, the laser pulses emitted by airborne LiDAR system often have a certain tilt angle, resulting in the absence of points in some areas [30]. These data gaps will result in incomplete building extraction. Although LiDAR data provide accurate three-dimensional coordinates, they lack texture information that is very important for identifying buildings with special shapes [31]. To overcome these enumerated limitations, some authors integrate multi-source data to achieve high accuracy. Authors such as Awrangjeb et al. [32] proposed an automatic three-dimensional roof extraction method by combining LiDAR data and multi-spectral orthoimages. In their study, the Normalized Difference Vegetation Index (NDVI) from multi-spectral orthoimages and the entropy images from grayscale orthoimage were first used to generate a Digital Elevation Model (DEM). Afterwards, the DEM was used to separate ground and non-ground points, while the NDVI and entropy images were used to classify the structure lines extracted from grayscale images. Structural lines belonging to the building class were then used to extract the plane and edge of the roof. Finally, the roof planes and boundaries were extracted using the lines belonging to the building class. They further added texture information from the orthoimage and used an iterative region growing algorithm to extract the complete roof plane, which improved the completeness of building extraction [33]. Qin et al. [34] combined the high-resolution remotely sensed image and Digital Surface Model (DSM). The morphological index was first used to detect shadows and correct the NDVI. Then, the NDVI was incorporated through the reconstruction of the DEM using the top-hat algorithm to obtain the initial building mask. Finally, the building segments with high probability were consolidated by a graph cut optimization based on modified superpixel segmentation. The experimental results showed that this algorithm could extract buildings efficiently with 94% completeness, and the 87% correctness indicating its potential for many practical applications. Gilani et al. [35] developed a methodology using features from point cloud and orthoimage to extract and regularize the buildings. Vegetation elimination, building detection and extraction of their partially occluded parts were achieved by synthesizing the entropy, NDVI and elevation difference. Results indicated that the per-area completeness achieved was between 83% to 93%, while the per-area correctness was above 95%. Siddiqui et al. [36] transformed the LiDAR height information into intensity and then analyzed the gradient information in the image. In addition, a local color matching approach was introduced as a post-processing step to eliminate trees. Lai et al. [37] proposed a building extraction method by fusing the point cloud and texture features. The texture features were acquired using an elevation map. Chen et al. [38] also integrated high spatial resolution images and LiDAR point cloud data. In their method, an adaptive iterative segmentation method was adopted to effectively avoid over-segmentation or under-segmentation.

1.2. Motivation

Although more accurate building extraction results can be achieved by fusing multi-source remote sensing data, pre-registration is generally required [39]. Registration errors often exist during the fusion process, and how to provide registration accuracy remains an unsolved problem [40]. Therefore, building extraction with only LiDAR points is still a mainstream focus. Today, although the machine learning methods have been successfully applied for building extraction, large numbers of sample labeling are generally a prerequisite for the success of this kind of methods. Obviously, the sample labeling is always cumbersome. Thus, how to realize building extraction by classic methods is still in the focus of researchers. However, the classic building extraction methods based solely on airborne LiDAR points still involve the following difficulties and challenges:

i. The point-based building extraction methods generally involve high computation costs. Thus, it is difficult to process a large amount of LiDAR points.
ii. When encountering with different building environments, the performance of the methods generally varies greatly. The robustness of the building extraction methods is not good.
iii. The vegetation points adjacent to buildings are easily misclassified as building points, which results in low correctness of building extraction.

To solve these enumerated problems, a building extraction method from airborne LiDAR points based on multi-constraints graph segmentation was proposed in this paper. In the proposed method, the graph structure was first constructed based on the three-dimensional spatial features of points. Then, the graph segmentation was achieved by setting constraint conditions. In doing so, the point-based building extraction was transformed into the object-based building extraction to reduce the computation cost and improve the efficiency of the method. Subsequently, the initial building points were extracted by calculating the geometric morphological features of each segmented object. To improve the completeness of building extraction, this paper also proposed a multi-scale progressive growing optimization method to recover the omitted building points. Three publicly available datasets were adopted for testing the performance of the proposed method. The experimental results showed that the proposed method can achieve very good building extraction results.

2. Methodology

The flowchart of the proposed method is shown in Figure 1. In this paper, an improved morphological filtering method based on kriging interpolation proposed by Hui et al. [41] was first adopted to remove ground points. This filtering method can be seen as a hybrid model, which combined the morphology-based and the interpolation-based filtering methods. By removing objects and generating rough ground surface progressively, the terrain details can be protected successfully. Subsequently, the multi-constraints graph segmentation method was proposed to segment the non-ground points to obtain the object primitives. Hereafter, the point-based building extraction was transformed into the object-based building extraction, which reduced the computational cost and improved the implementation efficiency of the proposed method. Here, the initial building points are acquired by calculating the spatial geometric features of each object primitive. To improve the completeness of building extraction, this paper proposed a multi-scale progressive growth optimization method. This method was used to extract complete buildings by continuously recovering the omitted points that meet the setting rules. Detailed description of the proposed method is described in the subsequent sections as follows: Section 2.1, multi-constraints graph segmentation; Section 2.2, initial building points extraction based on spatial geometric features; and Section 2.3, building points optimization based on multi-scale progressive growing.

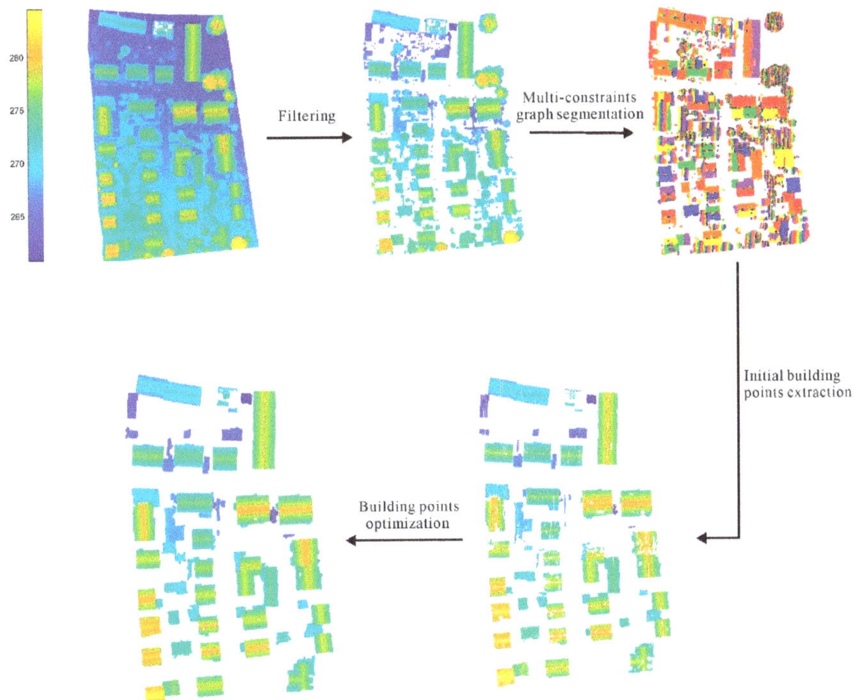

Figure 1. Flowchart of the proposed method. Filtering is first applied for removing the ground points. Then, the proposed multi-constrains graph segmentation is adopted to achieve the segmentation results. According to the geometric features, the initial building points can be obtained. Finally, an optimization step is applied to obtain the final building extraction results.

2.1. Multi-Constraints Graph Segmentation

To reduce the computation costs and improve the efficiency of the proposed method, this paper first transforms the point-based building extraction method into an object-based building extraction method. In this paper, the transformation is realized based on multi-constraints graph segmentation. Generally, the graph structure can be defined as Equation (1) [42]:

$$G = (V, E) \tag{1}$$

where G represents the graph, V is the set of nodes (v_i) and E denotes the corresponding edges ($e_{i,j}$). In this paper, v_i is made up of all points (p_i, $i = 1, 2, \ldots, N$) in the point cloud, while the edge $e_{i,j}$ connects the pair of neighboring points (p_i,p_j).

Today, with the fast development of the LiDAR system, the density of the LiDAR points can reach hundreds of points per square meter. For instance, the average point density of the acquired aerial LiDAR datasets of Dublin city center is 348.43 points/m^2 [17].

As a result, a huge amount of LiDAR data generally needs to be processed. If the graph is constructed using edges between every point, the built graph will be very complex and cost lots of computer memory. Moreover, it will also not be conducive to subsequent graph segmentation. To simplify the graph, this paper proposed several constraints for building the graph. The first constraint is that the edge $e_{i,j}$ only exists among neighboring points. This can be defined as Equation (2):

$$e_{i,j} = \begin{cases} 1 \ , & if \quad p_j \in Set_{p_i} \\ 0 \ , & otherwise \end{cases} \tag{2}$$

where Set_{p_i} is k nearest neighbors of point p_i. In Equation (2), it can be deduced that only if p_j is an adjacent point to p_i, there will be an edge connection between two points. k is a constant which represents the number of neighboring points. If the number of points is massive or the calculating capability of computer hardware is limited, k should not be set too large. Otherwise, the computation costs will increase. In this paper, k is set to 10.

Aiming at achieving the accurate building object primitives, two other constraints were set after the graph construction. On the one hand, the points within an identical object primitive should own similar normal vectors. That is, the edge between p_i and p_j can be reserved only if the angle between the normal vectors of the two points is less than a threshold. This can be defined as Equation (3):

$$e_{i,j} = \begin{cases} 1 \ , & if \quad \theta(p_i, p_j) \le \varsigma \\ 0 \ , & otherwise \end{cases} \tag{3}$$

where $\theta(p_i, p_j)$ is the angle between the normal vectors of p_i and p_j. The normal vector of a point can be estimated from the covariance matrix of its neighboring points using Principal Component Analysis (PCA). The eigenvector corresponding to the minimum eigenvalue of the covariance matrix is treated as the normal vector at the point. ς is the angle threshold. In this paper, ς is set to 5° to maintain the similarity of points within the same object primitives.

When the normal vector constraint is applied, the points located on the same plane will be divided into the same object primitives, such as the building roofs, as shown in Figure 2a. Conversely, some adjacent points that are not in the same plane will be divided into multiple object primitives, such as vegetation. However, if only the normal vector constraint is adopted, some vegetation points that are adjacent to buildings will be misclassified as building points, as shown in Figure 2b,c. It is because the normal vectors of these vegetation points may be similar to the ones of the building points, which may lead to misjudgment. To solve this problem, the third constraint of graph segmentation that is the longest edge constraint is defined as Equation (4):

$$e_{i,j} = \begin{cases} 1 \ , & if \quad Dist(p_i, p_j) \le mean(Dist(Set_{p_i})) + std(Dist(Set_{p_i})) \\ 0 \ , & otherwise \end{cases} \tag{4}$$

where $Dist(p_i, p_j)$ is the Euclidean distance between p_i and p_j. $mean(Dist(Set_{p_i}))$ represents the mean distance from point p_i to all its neighbors. $std(Dist(Set_{p_i}))$ is the standard deviation of the distance from point p_i to all its neighbors. This constraint will limit the range of the longest edge and avoids the problem that some adjacent vegetation points can be misclassified as building points.

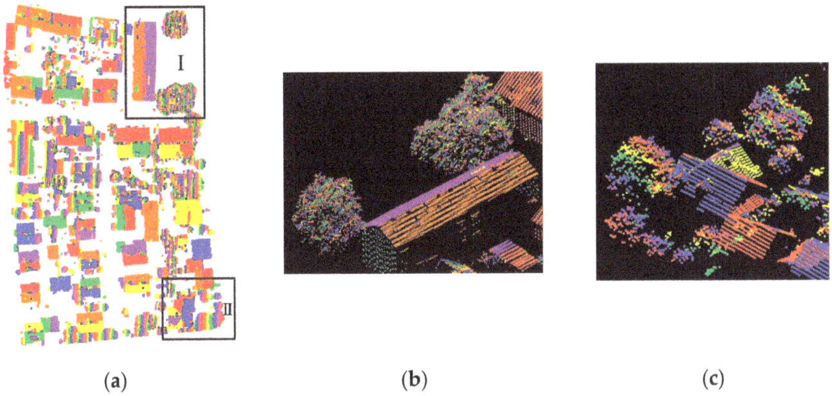

Figure 2. Graph segmentation based on multi-constraints. (**a**) The result of graph segmentation based on multi-constraints; (**b**) enlarged version of the area I in (**a**); (**c**) enlarged version of the area II in (**a**). The colors are assigned randomly according to different segmentation objects. Different colors represent different segmentation objects.

2.2. Initial Building Points Extraction Based on Spatial Geometric Features

As shown in Figure 2a, the point cloud is divided into multiple object primitives after the multi-constraints graph segmentation. Since the angle of normal vectors and the longest edge constraints were adopted in the multi-constraints graph segmentation, the building roofs can be correctly divided into independent object primitives, as shown in Figure 2b,c. However, some other objects without similar normal vectors, such as bushes, vegetation and fences, are generally divided into multiple small object primitives, which will lead to "over-segmentation". According to this characteristic, the initial building points can be extracted based on the spatial geometric features of the extracted object primitives. In this paper, the roughness and size of object primitives were selected to extract the initial building point cloud.

In this paper, the object primitive roughness ($roughness_{obj_i}$) is defined as the mean distance residual between the point and the best-fit plane of the points within the same object primitive. This has been defined in Equation (5) as:

$$\begin{cases} roughness_{obj_i} = \frac{\sum_{i=1}^{n} roughness_{p_i}}{n} \\ roughness_{p_i} = \frac{|Ax_{p_i}+By_{p_i}+Cz_{p_i}+D|}{\sqrt{A^2+B^2+C^2}} \end{cases} \quad (5)$$

where $roughness_{p_i}$ is the roughness of p_i in an object primitive obj_i. $roughness_{p_i}$ is defined as the distance residual between the point to the fitting plane ($Ax + By + Cz + D = 0$).

Generally speaking, for a relatively flat roof, the object primitive roughness of a building tends to be smaller than that of the pseudo plane generated by dense vegetation points. Compared with the over-segmented object primitives, building object primitives often contain more points. Meanwhile, the non-building object primitives are generally divided into several small object primitives since the direction of their normal vectors are not consistent. Therefore, the non-building object primitives can be removed by limiting the size of object primitives to further improve the accuracy of the initial building points extraction.

2.3. Building Points Optimization Based on Multi-Scale Progressive Growing

Although most of the building points can be correctly extracted based on the spatial geometric features of the object primitives, there are still some omitted building points, which will cause larger omission error. As shown in Figure 3, the omitted building points are mainly located on the ridge and edge area of the roof. These points are easily divided into different object primitives after multi-constraints graph segmentation since their spatial geometric features are generally quite different from their adjacent points. As a result, these ridge or edge points are wrongly eliminated, which can result in low completeness of building extraction.

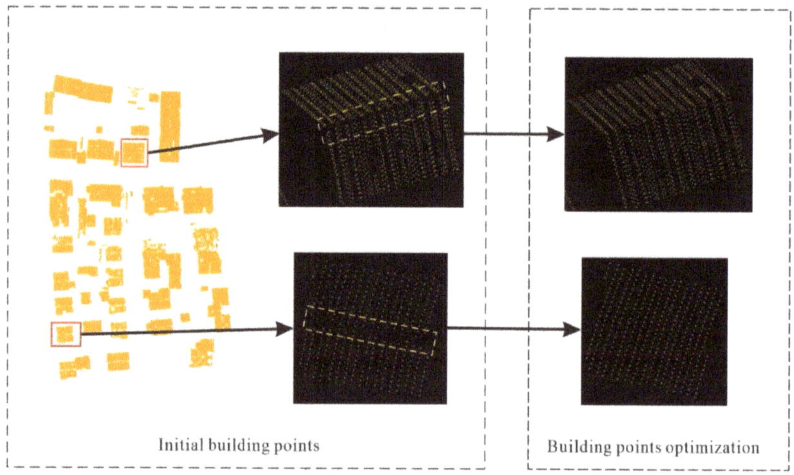

Figure 3. Building points optimization.

To improve the completeness of the building points extraction, this paper proposed a building points optimization method based on multi-scale progressive growing. The pseudocode of the proposed method is presented in Algorithm 1.

Algorithm 1. Building points optimization based on multi-scale progressive growing.

Input: Initial building points: $Point\_set = \{p_i | p_i \in U \parallel p_i \in \complement_U\}, i = 1, 2, \cdots, N$
p_i is a random point, U is the building point set, $\complement_U$ is the complementary set, which represents a non-building point set.
Scale sets: $s = \{s_1, s_2, \cdots, s_K\}, s_1 > s_2 > \cdots > s_K$

for *iter* = 1 to K
  $s = s_{iter}$
  for i = 1 to N
   if $p_i \in U$
    Find the neighboring point set of p_i under the scale of s:
$Set_{p_i} = \left\{p_j \middle| Dist\|p_j, p_i\| \le s, j = 1, 2, \cdots, M\right\}$
    for j = 1 to M
     if $p_j \in Set_{p_i} \&\& p_j \in \complement_U$
      Calculate the distance ($dist_{p_j}$) between p_j and the fitting plane of the object primitives where p_i is
      Calculate the angle ($\theta\left(p_j, p_i\right)$) between the normal vector of p_j and p_i
       if $dist_{p_j} \le th1 \parallel \theta\left(p_j, p_i\right) \le \xi$
        $p_j \in U$
   Update building point set U and non-building point set $\complement_U$
end

Output: Building points set U

From Algorithm 1, it can be found that this paper mainly adopted a multi-scale progressive strategy to gradually optimize the building points extraction results. Through experimental analysis, three fixed scales are enough to acquire complete building points. In doing so, not only the problem of overgrowth can be solved but also the implementation efficiency can be improved. In this paper, the scale sets are defined as s_1 (2 m), s_2 (1.5 m) and s_3 (0.5 m), respectively. In each step, the neighboring points set of each point is obtained within the current scale. If p_j is a non-building point in the neighboring points set (Set_{p_i}) of p_i, the distance residual ($dist_{p_j}$) between p_j and the fitting plane of the object primitive should be calculated. If p_j is an omitted building point, the distance residual will be less than the threshold $th1$, which is set to 0.3 m in this paper. In addition, building point clouds often have consistent normal vectors. Thus, if p_j is an omitted building point, the angle between the normal vectors of p_j and p_i should be less than the angle threshold ξ which is set to 10° in this paper.

3. Experimental Results and Analysis

3.1. Experimental Datasets

To evaluate the performance of the proposed method, three publicly available datasets provided by ISPRS were tested [43]. The datasets were obtained by Leica ALS50 with a 45° field of view and a 500 m mean flight height above the ground. The obtained points' accuracy is approximately 0.1 m in horizontal and vertical directions. The average strip overlap is 30%, and the average point density is 4–7 points/m^2. The three datasets are located in Vaihingen and contain three areas (Area1, Area2, Area3). As shown in Figure 4, the main objects in the three areas are powerline, low vegetation, impervious surfaces, car, building, shrub, tree and fence and so on. The buildings in Area1 are with complex shapes and different sizes as shown in Figure 4a. Thus, Area1 can be used to test the extraction accuracy of the proposed method on complex buildings. Area2 is characterized by buildings surrounding with dense vegetation (Figure 4b). Adjacent vegetation often causes great interference on the building extraction. Therefore, Area2 can verify whether the proposed method effectively eliminates the interference of adjacent vegetation or not. In Area3, as shown in Figure 4c, low vegetation is the main challenge to the building extraction. How to eliminate the pseudo planes formed by the low vegetation is still an unresolved problem. Thus, it can be concluded that the three datasets are a good representation for testing the effectiveness and robustness of the proposed method for building extraction in different environments.

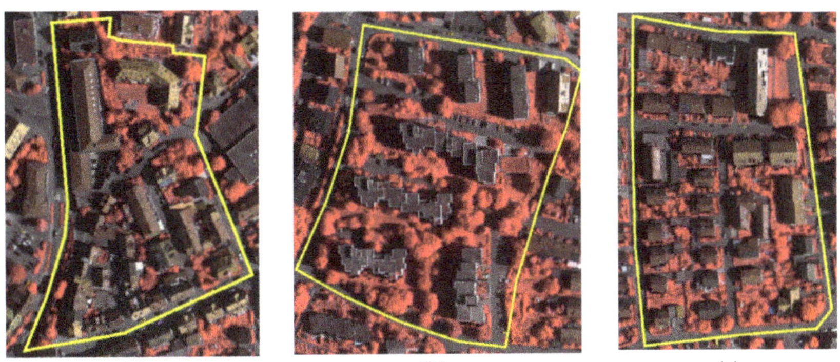

(a) (b) (c)

Figure 4. The tested three datasets provided by the ISPRS. (**a**) Area1; (**b**) Area2; (**c**) Area3.

3.2. Experimental Results and Analysis

Figure 5 shows the building extraction results using the proposed method. From Figure 5, most of the buildings can be extracted correctly so as to achieve good building extraction accuracy in the three experimental areas. As shown in Figure 5a,d, although there are many buildings with complex shapes and different sizes in Area1, the proposed method can extract them correctly. Thus, it can be concluded that the proposed method is robust towards building with different shapes and sizes. However, there are still some omitted building points in Area1 (the blue points in Figure 5a,d), which are mainly located at the edge of the building. It is because these buildings are generally with low elevation. Due to the height difference, the low buildings cannot form a complete object primitive along with the adjacent buildings but form an independent object primitive. Since the size constraint of object primitives is adopted in the initial building extraction step, these small object primitives will be wrongly removed. Although there is dense adjacent vegetation in Area2, the proposed method effectively eliminates its interference on building extraction. However, there are still some points that are wrongly classified as buildings, as shown by the red points in Figure 5b,e. It is because the buildings in Area2 are stacked and the facades of some buildings are connected with the roofs. As a result, they are easily misclassified as roof points. The building extraction accuracy of the proposed method in Area3 is relatively high due to the simple features of the buildings in it. Some omitted points are located near the chimneys because of the spatial differences among the chimneys with the roof (Figure 5c,f).

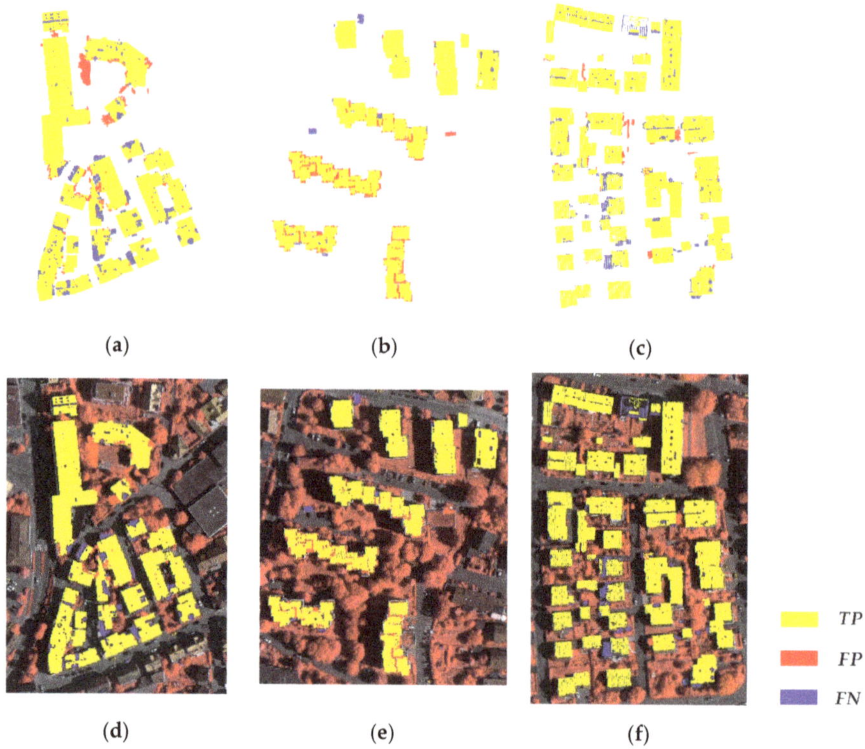

Figure 5. Building extraction results of the proposed method. (**a**) Extraction results for Area1; (**b**) extraction results for Area2; (**c**) extraction results for Area3; (**d**) the extraction results of Area1 integrating with the corresponding orthophoto image; (**e**) the extraction results of Area2 integrating with the corresponding orthophoto image; (**f**) the extraction results of Area3 integrating with the corresponding orthophoto image. Yellow represents correctly extracted buildings (TP), red represents wrongly extracted buildings (FP), and blue represents omitted buildings (FN).

For quantitative evaluation, four indicators proposed by Rutzinger et al. [44] were adopted to evaluate the precision of the building detection results at both pixel-based and object-based levels. They are completeness ($Comp$), correctness ($Corr$), quality ($Quality$) and F1 score (F_1).

Completeness represents the percentage of buildings in the reference that were detected. Correctness indicates how well the detected buildings match the reference. Completeness tends to evaluate the ability of building detection, while correctness evaluates the correct detection ability of the proposed method. Quality and F1 score are another two indicators, which can integrate completeness and correctness to reflect the effectiveness of the method. The definitions of the above four indicators are defined in Equations (6)–(9):

$$Comp = \frac{TP}{TP + FN} \tag{6}$$

$$Corr = \frac{TP}{TP + FP} \tag{7}$$

$$Quality = \frac{Comp \times Corr}{Comp + Corr - Comp \times Corr} \tag{8}$$

$$F_1 = \frac{2 \times Comp \times Corr}{Comp + Corr} \tag{9}$$

This paper evaluated the performance of the proposed method at pixel-based and object-based levels, respectively. In the pixel-based evaluation, TP represents the number of correctly extracted building points, FN is the number of omitted building points, FP represents the number of wrongly detected building points. In the object-based evaluation, the object can be accepted as TP if it has a 50% minimum overlap with the reference data. FN represents the number of omitted building objects, FP is the number of objects whose overlap ratio is less than 50% with the reference data [45].

To evaluate the performance of the proposed method objectively, this paper selected other ten methods that have also used these public datasets provided by ISPRS for comparative analysis. Among the ten methods, the first three methods belong to the kind of machine learning method, while the other seven methods belong to the kind of classic methods. Maltezos et al. [11] first created a multi-dimensional feature vector using the raw LiDAR data and the seven additional features. Then, a CNN was used to nonlinearly transform the input data into abstract forms of representations. After that, a training set was used to learn the parameters of the CNN model to perform the building extraction. Doulamis et al. [46] developed a radial base function kernel Support Vector Machine (SVM) classifier. The classifier adopted an efficient recursive weight estimation algorithm to make the network response adaptive. Protopapadakis et al. [47] proposed a nonlinear scheme of a typical feed-forward artificial neural network with a hidden layer. When the appropriate features were fed to the detection model, the optimal parameters of the detector structure were selected by an island genetic algorithm. Awrangjeb and Fraser [24] formed a "building mask" using the ground points, and then extracted building and vegetation objects with the co-planarity between adjacent points. After that, vegetation plane primitives were removed by using the information of area and neighborhood characteristics to complete the building extraction. Nguyen et al. [45] presented an unsupervised classification method called super-resolution-based snake model to extract buildings by combining the spectral features of images and LiDAR point cloud. Niemeyer et al. [48] established the classification model with a conditional random field approach. This method utilized a nonlinear decision surface to separate the object clusters in feature space reliably, and then extracted buildings. Wei et al. [49] presented an integrated method to comprehensively evaluate the feature relevance of point cloud and image data. Firstly, point cloud and image data were co-registered. After that, all data points were grid-fitted to facilitate acquiring spatial context information per pixel/point. Then, spatial-statistical and radiometric features could be extracted using a cylindrical volume neighborhood of point. Finally, the AdaBoost classifier combined with contribution ratio was used to label the points. Moussa and EI-Sheimy [50] fused the aerial images data with single return LiDAR data to extract buildings for an urban area. Then, they segmented the entire DSM data into objects based on height variation. After that, the area, average height, and vegetation index of each object were adopted to exactly classify the objects. Yang et al. [51] defined the Gibbs energy model of building objects within the framework of reversible-jump Markov Chain Monte Carlo to describe the building points. Then, they found an optimal energy configuration using simulated annealing. Finally, the detected building objects were refined to eliminate false detection. Gerke and Xiao [52] introduced a new segmentation approach by making use of geometric and spectral data. They quantified the point cloud into voxels according to the geometric and textural features of optical images, and then, buildings are extracted from voxels by using the supervised classification method based on random forest. Note that the methods proposed by Maltezos et al. [11], Doulamis et al. [46] and Protopapadakis et al. [47] only provide the completeness, correctness and quality at per-area level. The F1 score can be calculated according to Equation (9). Thus, in Tables 1–3, we only compared their building performance at per-area level. It is the same with Figures 6 and 7, only the average quality and F1 score of the three methods at per-area level were compared with other methods.

Table 1. Accuracy comparison of building extraction in Area1. The experimental results of the ten methods are obtained from the corresponding references. Bold font represents the highest value among the comparison results. "×" denotes that the results are not provided by the reference. Note that the methods proposed by Maltezos et al. (2019), Doulamis et al. (2003) and Protopapadakis et al. (2016) do not provide the completeness, correctness and quality at per-object level.

	Methods	Per-Area (%)				Per-Object (%)			
		Comp	*Corr*	*Quality*	F_1	*Comp*	*Corr*	*Quality*	F_1
Area1	Maltezos et al. (2019)	79.80	91.50	74.40	85.25	×	×	×	×
	Doulamis et al. (2003)	68.80	94.00	65.90	79.45	×	×	×	×
	Protopapadakis et al. (2016)	92.20	68.00	64.30	78.27	×	×	×	×
	Awrangjeb and Fraser (2014)	92.70	88.70	82.90	90.66	83.80	96.90	81.61	89.88
	Nguyen et al. (2020)	90.42	94.20	85.65	92.27	83.78	**100.00**	83.78	91.17
	Niemeyer et al. (2012)	87.00	90.10	79.40	88.52	83.80	75.60	65.96	79.49
	Wei et al. (2012)	89.80	92.20	83.46	90.98	89.20	97.10	86.89	92.98
	Moussa and El-Sheimy (2012)	89.10	**94.70**	84.87	91.81	83.80	**100.00**	83.80	91.19
	Yang et al. (2013)	87.90	91.20	81.03	89.52	81.10	96.80	78.98	88.26
	Gerke and Xiao (2014)	91.20	90.30	83.06	90.75	86.50	91.40	79.99	88.88
	The proposed method	**93.04**	91.61	**85.74**	**92.32**	**97.22**	90.34	**88.07**	**93.65**

Table 2. Accuracy comparison of building extraction in Area2. The experimental results of the ten methods are obtained from the corresponding references. Bold font represents the highest value among the comparison results. "×" denotes that the results are not provided by the reference. Note that the methods proposed by Maltezos et al. (2019), Doulamis et al. (2003) and Protopapadakis et al. (2016) do not provide the completeness, correctness and quality at per-object level.

	Methods	Per-Area (%)				Per-Object (%)			
		Comp	*Corr*	*Quality*	F_1	*Comp*	*Corr*	*Quality*	F_1
Area2	Maltezos et al. (2019)	87.70	**96.00**	84.60	91.66	×	×	×	×
	Doulamis et al. (2003)	83.10	92.30	77.60	87.46	×	×	×	×
	Protopapadakis et al. (2016)	90.80	90.50	82.90	90.65	×	×	×	×
	Awrangjeb and Fraser (2014)	91.50	91.00	83.90	91.25	85.70	84.60	74.20	85.15
	Nguyen et al. (2020)	93.47	94.75	88.87	94.11	78.57	**100.00**	78.57	88.00
	Niemeyer et al. (2012)	93.80	91.40	86.19	92.58	78.60	52.40	45.86	62.88
	Wei et al. (2012)	92.50	93.90	87.26	93.19	78.60	**100.00**	78.60	88.02
	Moussa and El-Sheimy (2012)	93.20	95.40	89.19	94.29	78.60	**100.00**	78.60	88.02
	Yang et al. (2013)	88.80	94.00	84.04	91.33	78.60	**100.00**	78.60	88.02
	Gerke and Xiao (2014)	94.00	89.00	84.22	91.43	78.60	42.30	37.93	55.00
	The proposed method	**96.86**	92.93	**90.21**	**94.85**	**93.33**	96.55	**90.32**	**94.91**

Table 3. Accuracy comparison of building extraction in Area3. The experimental results of the ten methods are obtained from the corresponding references. Bold font represents the highest value among the comparison results. "×" denotes that the results are not provided by the reference. Note that the methods proposed by Maltezos et al. (2019), Doulamis et al. (2003) and Protopapadakis et al. (2016) do not provide the completeness, correctness and quality at per-object level.

	Methods	Per-Area (%)				Per-Object (%)			
		Comp	*Corr*	*Quality*	F_1	*Comp*	*Corr*	*Quality*	F_1
Area3	Maltezos et al. (2019)	88.20	93.70	83.20	90.87	×	×	×	×
	Doulamis et al. (2003)	82.90	92.90	78.00	87.62	×	×	×	×
	Protopapadakis et al. (2016)	**96.70**	84.50	82.20	90.19	×	×	×	×
	Awrangjeb and Fraser (2014)	93.90	86.30	81.70	89.94	78.60	97.80	77.23	87.16
	Nguyen et al. (2020)	91.00	93.02	85.18	92.00	83.93	97.92	82.46	90.39
	Niemeyer et al. (2012)	93.80	93.70	88.24	93.75	82.10	90.20	75.38	85.96
	Wei et al. (2012)	86.80	92.50	81.09	89.56	75.00	**100.00**	75.00	85.71
	Moussa and El-Sheimy (2012)	87.00	95.20	83.34	90.92	66.10	**100.00**	66.10	79.59
	Yang et al. (2013)	85.20	89.50	77.46	87.30	73.20	97.60	71.91	83.66
	Gerke and Xiao (2014)	89.10	92.50	83.10	90.77	75.00	78.20	62.30	76.57
	The proposed method	91.54	**97.59**	**89.52**	**94.46**	**92.16**	94.09	**87.12**	**93.12**

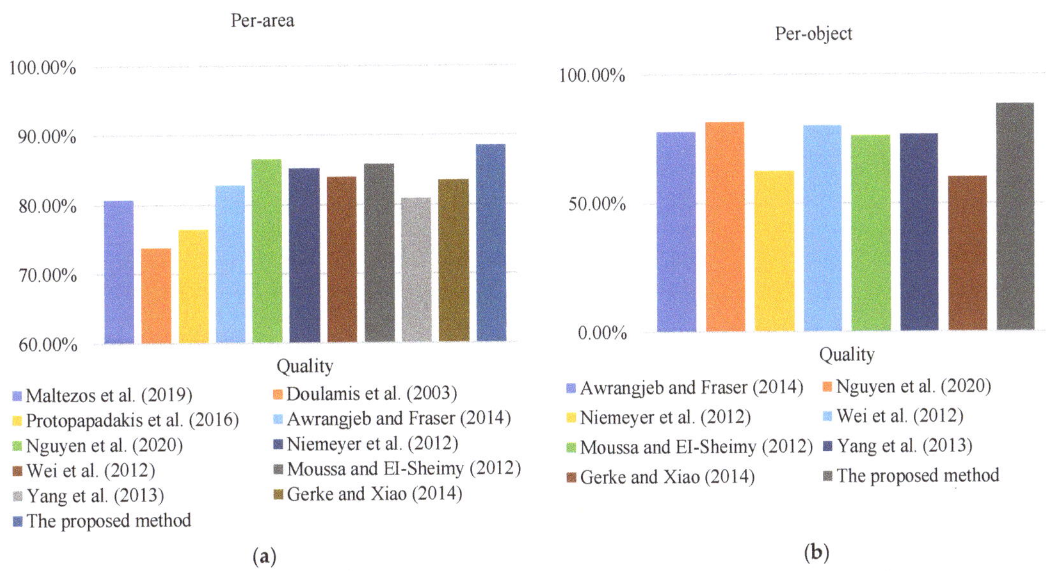

Figure 6. Average quality comparison between the proposed method and ten other methods. (**a**) Per-area level; (**b**) per-object level. The average qualities of the ten methods are obtained from the corresponding references. The methods proposed by Maltezos et al. (2019), Doulamis et al. (2003) and Protopapadakis et al. (2016) do not provide the quality at per-object level.

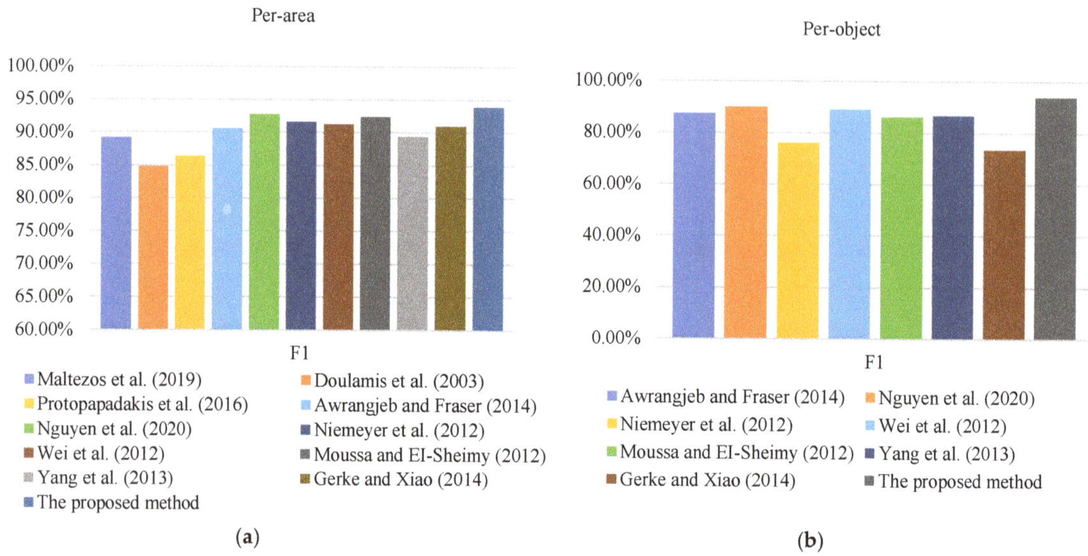

Figure 7. Average F1 score comparison between the proposed method and ten other methods. (**a**) Per-area level; (**b**) per-object level. The F1 score of the ten methods are obtained from the corresponding references. The methods proposed by Maltezos et al. (2019), Doulamis et al. (2003) and Protopapadakis et al. (2016) do not provide the quality at per-object level.

Tables 1–3 show the accuracy comparison of the proposed method with the ten methods described above in the three areas (Area1, Area2 and Area3) at per-area and per-object levels. From the comparison, it can be found that the satisfying results in the three areas are obtained by the proposed method. All the four indicators (completeness, correctness, quality and F1 score) of the proposed method of the three testing areas are higher than 85%, and most of them are higher than 90%. The results indicate that the proposed method can obtain highly desirable accuracy in different environments and can robustly extract buildings. In the three areas, the proposed method achieved the best on six out of eight indicators with four indicators at per-area and per-object levels, respectively. Therefore, compared with the other investigated methods, the proposed method has the best building extraction performance. In Area1, the proposed method achieved the highest completeness (Table 1) at both per-area and per-object levels, which demonstrates that the proposed method has a strong capability of building detection. Compared with the other two areas, the proposed method performed best in Area2, and all indicators were higher than 90%, as shown in Table 2. Especially in the per-object evaluation, the quality of the proposed method (90.32%) is significantly better than the other methods. This indicates that the proposed method can effectively overcome the interference caused by dense vegetation around buildings, and can achieve the correct extraction of buildings. In Area3, the per-area correctness of the proposed method can reach 97.59% (Table 3). It shows that the proposed method can detect buildings correctly. Overall, the completeness and correctness of the proposed method are relatively balanced, which indicates that the proposed method can extract as many buildings as possible while ensuring the correctness of the extraction results.

Figures 6 and 7 show the comparison of average quality and average F1 score of the proposed method with the other ten methods in the three study areas. In terms of the average quality, the proposed method achieved the best results in both per-area and per-object evaluations. Especially, at per-object level, the average quality of the proposed method is obviously better than that of the other seven methods. In addition, the proposed method also obtained the highest average F1 score. Both at per-area and per-object levels, the average F1 score of the proposed method is more than 90%. Consequently, the proposed method performed well in the three kinds of building environments and thus can be considered as relatively robust.

4. Discussion

In the multi-constraints graph segmentation, two parameters are involved, namely the number of neighboring points k and the angle threshold ς. The number of neighboring points determines the size of neighboring set (Equation (1)), while the angle threshold (Equation (2)) directly affects segmentation results. These two parameters determine the selection of edges directly and have a distinct influence on the results of graph construction. Concretely, the larger the k value, the more edges will be accepted, thereby complicating the graph, increasing the computational costs, and reducing the efficiency of the method. In the last two decades, some techniques have been proposed to determine the k neighbors, such as the lowest entropy or highest similarity [53]. In terms of the lowest entropy, the Shannon entropy has to be calculated for every point. The optimal radius can be determined by calculating the lowest entropy. In terms of the highest similarity, a similarity index is defined as the ratio of neighbors whose dimensionality labelling is same with that of the center point. The optimal radius can be obtained by finding the highest similarity index for each point. Although these optimal radius selection techniques can help to determine the k neighbors, it will involve too much calculation. Obviously, the computational burden will increase greatly and the optimal radius selection process will be time consuming. To facilitate the implementation of the proposed method, this paper set a fixed value for the neighbors. Through experimental analysis, the graph complexity and the efficiency can be balanced when the value of k is set to 10.

The angle threshold ς was set to divide the points on the same plane into the same object primitives, while the points on different planes were split into multiple object primitives. Figure 8a–c shows the results of multi-constraints graph segmentation results with ς taken as $1°$, $10°$ and $15°$, respectively. Figure 8d shows the reference segmentation results. Although vegetation points were divided into multiple object primitives, the same building roof was also separated into several small object primitives when ς was set to $1°$, as the black rectangle shown in Figure 8a. This over-segmentation phenomenon is not conducive to the subsequent extraction of initial building points which are based on the size of object primitives. When ς was set to $10°$, some points on different roof planes of the building failed to be separated into different object primitives, as shown in Figure 8b as blue rectangle. When the value of ς continues to increase ($\varsigma = 15°$), a more under-segmented roof planes were observed, as shown in Figure 8c. If the points from different roof planes are segmented into same object primitives, there will be a large fitting error in calculating the roughness of each object primitive. Consequently, it will be difficult to discriminate buildings from other object primitives (such as dense vegetation). The experimental results showed that the proper segmentation results can be achieved when the angle threshold ς is set to $5°$, as shown in Figure 2a. In this case, not only the vegetation points can be divided into multiple object primitives, but also the building points from the same plane can be separated into the same object primitives.

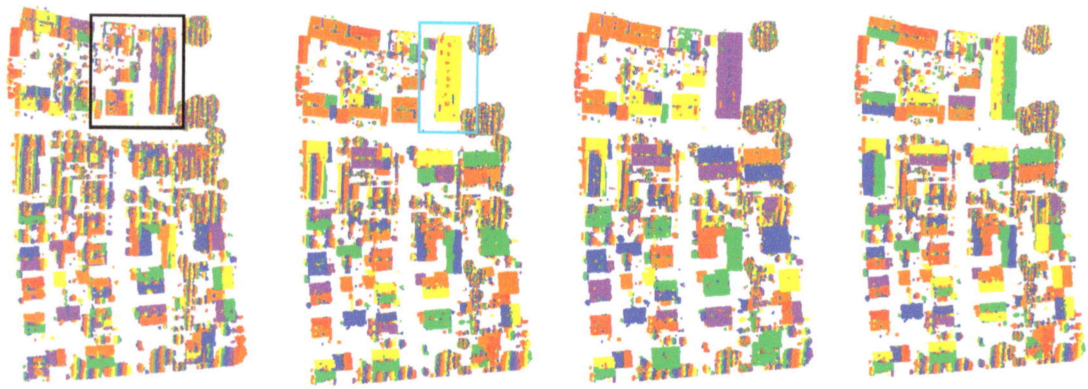

(a) (b) (c) (d)

Figure 8. Graph segmentation results with different ς. (**a**) $\varsigma = 1°$; (**b**) $\varsigma = 10°$; (**c**) $\varsigma = 15°$; (**d**) the reference segmentation results. Different object primitives are colored with different colors.

Another important parameter involved in this paper is the threshold for the object primitive roughness, which was set in the step of initial building extraction. Figure 9a–c are the initial building extraction results for Area3 with the roughness thresholds of 0.02, 0.04 and 0.06, respectively. In Figure 9a, it can be found that when the value of this threshold was too small, many independent buildings were not successfully detected. It is because that part of the points located at the boundary of buildings were omitted, resulting in the low completeness of the initial building points. Figure 9b shows the initial buildings extracted in this paper with the roughness threshold of 0.04. It illustrates that, with the proper value of the threshold, more buildings could be detected, and the misclassification of some non-building points can be avoided. When the threshold is set too large, it can be found that mainly dense vegetation points are misclassified as buildings, as shown in Figure 9c. Therefore, setting a too large roughness threshold will reduce the correctness of the result.

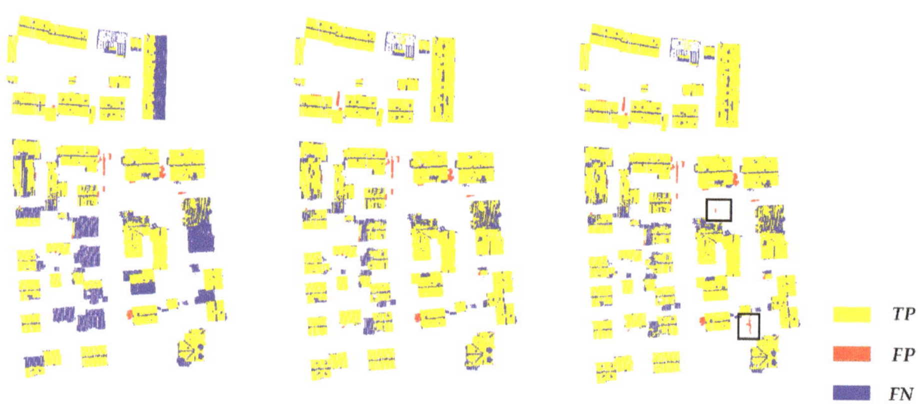

(a) (b) (c)

Figure 9. Initial building points extraction results with different roughness threshold. (**a**) The threshold is 0.02; (**b**) the threshold is 0.04; (**c**) the threshold is 0.06. Yellow represents correctly extracted buildings (TP), red represents wrongly extracted buildings (FP), and blue represents omitted buildings (FN).

To validate the applicability of the proposed method, two other public datasets provided by OpenTopography were selected for further testing [54]. The datasets are located in the north of Palmerston, New Zealand, with a point density of 22 points/m^2. The ground objects include buildings, impervious surface, low vegetation, trees and shrubs and so on, as shown in Figure 10. Figure 10a is the true-color image of the first dataset named as S1. It can be seen that the buildings in S1 are densely distributed with regular shapes, surrounded by dense vegetation. Figure 10b is the point cloud data of this area. Figure 10c is the true-color image of the second dataset named as S2, which contains many buildings with complex shapes and large sizes. Figure 10d shows the corresponding point cloud data of this area.

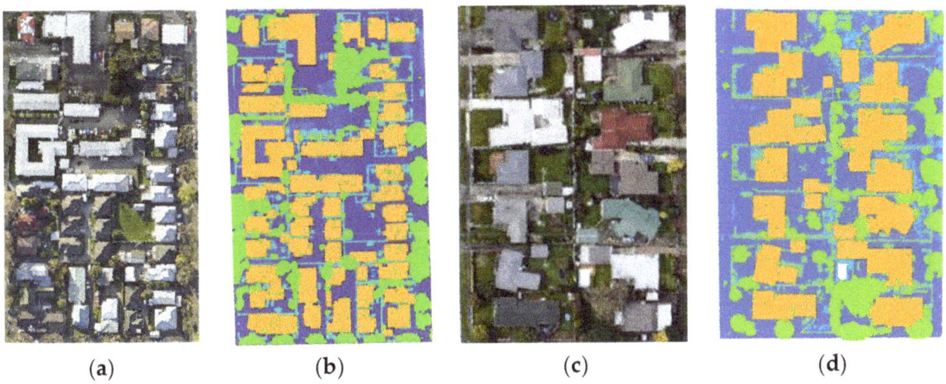

Figure 10. The study areas of OpenTopography datasets. (**a**) The true-color image of S1; (**b**) the point cloud data of S1, which is colored according to different labels; (**c**) the true-color image of S2; (**d**) the point cloud data of S2, which is colored according to different labels.

The building extraction results of the two areas mentioned above are shown in Figure 11. Figure 11b,d are the reference building extraction results from the orthophoto images of S1 and S2, respectively. It can be found that the proposed method can achieve satisfying results in both areas (S1 and S2) because most of the building points were accurately extracted. However, there still exist some points on the roof plane that were not effectively detected as shown as blue points in the black rectangle of Figure 11a. It is mainly because there are some small roof planes in S1, which makes the object primitives formed by these blue points smaller so that they cannot be effectively identified in the initial building extraction. As shown in Figure 11c, there are a few wrongly detected points in S2, such as the red points in the black rectangle. It is caused by the overlapping of some high vegetation and buildings. A small number of vegetation points around the edge of buildings are divided into the same object primitives with their adjacent buildings, thereby being misclassified. Table 4 shows the quantitative evaluation results of the two areas. All the indicators are higher than 90%. Especially, the completeness and F1 score of the proposed method are higher than 95%. Thus, it can be concluded that the proposed method can effectively detect buildings with different sizes and shapes.

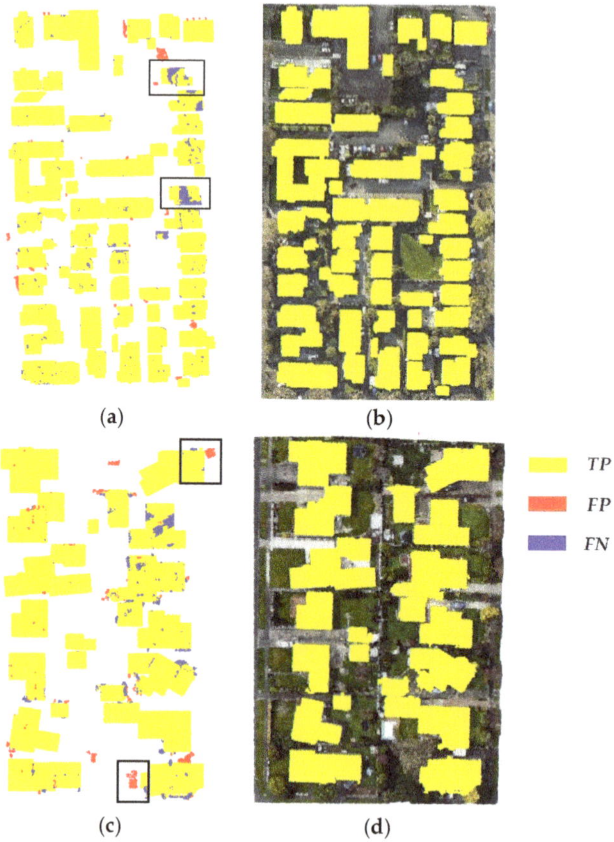

Figure 11. Building extraction results of OpenTopography datasets. (**a**) The building extraction result for S1; (**b**) the reference building extraction results of S1 from orthophoto image; (**c**) the building extraction result for S2; (**d**) the reference building extraction results of S2 from orthophoto image. Yellow represents correctly extracted buildings (TP), red represents wrongly extracted buildings (FP), and blue represents omitted buildings (FN).

Table 4. Accuracy of building extraction in OpenTopography datasets.

	Per-Area (%)				Per-Object (%)			
	Comp	*Corr*	*Quality*	F_1	*Comp*	*Corr*	*Quality*	F_1
S1	96.54	99.28	95.87	97.89	98.25	98.61	96.90	98.43
S2	96.09	98.34	94.56	97.20	100.00	91.43	91.43	95.52

5. Conclusions

Building extraction from airborne LiDAR point clouds is a significant step in the applications of point cloud post-processing, such as urban three-dimensional model construction and digital urban management. To solve the problems existing in the building extraction, such as huge computational cost, poor adaptability to different building environments and the interference of adjacent vegetation, this paper proposed a building extraction method from airborne LiDAR data based on multi-constraints graph segmentation. In this paper, the graph structure of the point cloud was first built. Afterwards, the object primitives were obtained based on the multi-constraints graph segmentation. In doing so, the point-based building extraction is transformed into the object-based building extraction to improve the efficiency of the method. After that, the initial building point cloud is extracted based on the different spatial geometric features of each object primitive. To improve the completeness of building extraction, this paper proposed a multi-scale progressive growth optimization method to recover some omitted building points located on the ridge and edge areas. Three public datasets with different building environments provided by ISPRS were adopted for the testing. The experimental results showed that the proposed method can achieve good building extraction performance in all the three testing areas. Compared with other ten famous building extraction methods, the proposed method also performed the best. Two other publicly available datasets provided by the OpenTopography were also used for further testing. The experimental results showed that the proposed method has strong robustness and promising performance in building extraction. In addition, the completeness and correctness achieved by this method were relatively high and balanced. It reveals that the proposed method can detect more buildings while ensuring the accuracy of the results. However, the implementation of the proposed method needs setting of some parameters, such as the number of neighboring points k and the angle threshold ς. To ease the implementation of the proposed method, this paper set fixed values for the parameters. However, to obtain better building extraction results, it is necessary to adjust the parameters according to the point clouds with different densities. In future work, we will try to improve the automation of the proposed method to make the parameters adjusting automatically according to the particularity of each point cloud.

Author Contributions: Z.H. conceived the original idea of the study and drafted the manuscript. Z.L. and P.C. performed the experiments and made the experimental analysis. Y.Y.Z. and J.F. contributed to the revision of the manuscript. All authors have read and agreed to the published version of the manuscript.

Funding: This work was supported by the China Post-Doctoral Science Foundation (2019M661858), the National Natural Science Foundation of China (NSF) (41801325, 42161060, 41861052), the Natural Science Foundation of Jiangxi Province (20192BAB217010), Education Department of Jiangxi Province (GJJ170449), Key Laboratory for Digital Land and Resources of Jiangxi Province, East China University of Technology (DLLJ201806), East China University of Technology Ph.D. Project (DHBK2017155) for their financial support.

Institutional Review Board Statement: Not applicable.

Informed Consent Statement: Not applicable.

Data Availability Statement: The first three publicly available datasets were provided by the ISPRS commissions. https://www2.isprs.org/commissions/comm2/wg4/benchmark/ (accessed on 22 July 2021). The last two datasets were provided by OpenTopography. https://portal.opentopography.org (accessed on 22 July 2021).

Acknowledgments: The authors would like to thank the ISPRS Working Group and OpenTopography for providing the experimental datasets. Moreover, the authors would also like to thank the anonymous reviewers for their constructive comments for improving the manuscript.

Conflicts of Interest: The authors declare no conflict of interest.

References

1. Du, S.; Zhang, Y.; Zou, Z.; Xu, S.; He, X.; Chen, S. Automatic building extraction from LiDAR data fusion of point and grid-based features. *ISPRS J. Photogramm. Remote Sens.* **2017**, *130*, 294–307. [CrossRef]
2. Huang, X.; Zhang, L. Morphological building/shadow index for building extraction from high-resolution imagery over urban areas. *IEEE J. -STARS* **2012**, *5*, 161–172. [CrossRef]
3. Toth, C.; Jozkow, G. Remote sensing platforms and sensors: A survey. *ISPRS J. Photogramm. Remote Sens.* **2016**, *115*, 22–36. [CrossRef]
4. Luo, L.; Wang, X.; Guo, H.; Lasaponara, R.; Zong, X.; Masini, N.; Wang, G.; Shi, P.; Khatteli, H.; Chen, F.; et al. Airborne and spaceborne remote sensing for archaeological and cultural heritage applications: A review of the century (1907–2017). *Remote Sens. Environ.* **2019**, *232*, 111280. [CrossRef]
5. Tarsha Kurdi, F.; Awrangjeb, M.; Munir, N. Automatic filtering and 2D modeling of airborne laser scanning building point cloud. *Trans. GIS* **2021**, *25*, 164–188. [CrossRef]

6. Wen, C.; Li, X.; Yao, X.; Peng, L.; Chi, T. Airborne LiDAR point cloud classification with global-local graph attention convolution neural network. *ISPRS J. Photogramm. Remote Sens.* **2021**, *173*, 181–194. [CrossRef]
7. Zhang, P.; Du, P.; Lin, C.; Wang, X.; Li, E.; Xue, Z.; Bai, X. A hybrid attention-aware fusion network (HAFNET) for building extraction from high-resolution imagery and LiDAR data. *Remote Sens.* **2020**, *12*, 3764. [CrossRef]
8. Liu, K.; Ma, H.; Ma, H.; Cai, Z.; Zhang, L. Building extraction from airborne LiDAR data based on min-cut and improved post-processing. *Remote Sens.* **2020**, *12*, 2849. [CrossRef]
9. He, Y.; Xu, G.; Kaufmann, H.; Wang, J.; Ma, H.; Liu, T. Integration of InSAR and LiDAR technologies for a detailed urban subsidence and hazard assessment in Shenzhen, China. *Remote Sens.* **2021**, *13*, 2366. [CrossRef]
10. Zhou, Z.; Gong, J. Automated residential building detection from airborne LiDAR data with deep neural networks. *Adv. Eng. Inform.* **2018**, *36*, 229–241. [CrossRef]
11. Maltezos, E.; Doulamis, A.; Doulamis, N.; Ioannidis, C. Building extraction from LiDAR data applying deep convolutional neural networks. *IEEE Geosci. Remote Sens.* **2019**, *16*, 155–159. [CrossRef]
12. Ni, H.; Lin, X.; Zhang, J. Classification of ALS point cloud with improved point cloud segmentation and random forests. *Remote Sens.* **2017**, *9*, 288. [CrossRef]
13. Nahhas, F.H.; Shafri, H.Z.M.; Sameen, M.I.; Pradhan, B.; Mansor, S. Deep learning approach for building detection using LiDAR–orthophoto fusion. *J. Sens.* **2018**, *2018*, 1–12. [CrossRef]
14. Huang, J.; Zhang, X.; Xin, Q.; Sun, Y.; Zhang, P. Automatic building extraction from high-resolution aerial images and LiDAR data using gated residual refinement network. *ISPRS J. Photogramm. Remote Sens.* **2019**, *151*, 91–105. [CrossRef]
15. Li, D.; Shen, X.; Yu, Y.; Guan, H.; Li, J.; Zhang, G.; Li, D. Building extraction from airborne multi-spectral LiDAR point clouds based on graph geometric moments convolutional neural networks. *Remote Sens.* **2020**, *12*, 3186. [CrossRef]
16. Yuan, Q.; Shafri, H.Z.M.; Alias, A.H.; Hashim, S.J.B. Multiscale semantic feature optimization and fusion network for building extraction using high-resolution aerial images and LiDAR data. *Remote Sens.* **2021**, *13*, 2473. [CrossRef]
17. Zolanvari, S.M.I.; Ruano, S.; Rana, A.; Cummins, A.; Da Silva, R.E.; Rahbar, M.; Smolic, A. DublinCity: Annotated LiDAR point cloud and its applications. In Proceedings of the BMVC 30th British Machine Vision Conference, Cardiff, UK, 9 September 2019.
18. Costantino, D.; Angelini, M.G. Features and ground automatic extraction from airborne LiDAR data. *ISPRS-Int. Arch. Photogramm. Remote Sens. Spat. Inf. Sci.* **2011**, *38*, 19–24. [CrossRef]
19. Crosilla, F.; Macorig, D.; Scaioni, M.; Sebastianutti, I.; Visintini, D. LiDAR data filtering and classification by skewness and kurtosis iterative analysis of multiple point cloud data categories. *Appl. Geogr.* **2013**, *5*, 225–240. [CrossRef]
20. Ywata, M.S.Y.; Dal Poz, A.P.; Shimabukuro, M.H.; de Oliveira, H.C. Snake-based model for automatic roof boundary extraction in the object space integrating a high-resolution aerial images stereo pair and 3D roof models. *Remote Sens.* **2021**, *13*, 1429. [CrossRef]
21. Dorninger, P.; Pfeifer, N. A comprehensive automated 3D approach for building extraction, reconstruction, and regularization from airborne laser scanning point clouds. *Sensors* **2008**, *8*, 7323–7343. [CrossRef] [PubMed]
22. Poullis, C.; You, S. Photorealistic large-scale urban city model reconstruction. *IEEE Trans. Vis. Comput. Graph.* **2009**, *15*, 654–669. [CrossRef] [PubMed]
23. Sun, S.; Salvaggio, C. Aerial 3D building detection and modeling from airborne LiDAR point clouds. *IEEE J. -STARS* **2013**, *6*, 1440–1449. [CrossRef]
24. Awrangjeb, M.; Fraser, C. Automatic segmentation of raw LIDAR data for extraction of building roofs. *Remote Sens.* **2014**, *6*, 3716–3751. [CrossRef]
25. Fan, H.; Yao, W.; Fu, Q. Segmentation of sloped roofs from airborne LiDAR point clouds using ridge-based hierarchical decomposition. *Remote Sens.* **2014**, *6*, 3284–3301. [CrossRef]
26. Ural, S.; Shan, J. A min-cut based filter for airborne LiDAR data. *ISPRS-Int. Arch. Photogramm. Remote Sens. Spat. Inf. Sci.* **2016**, *49*, 395–401. [CrossRef]
27. Zou, X.; Feng, Y.; Li, H.; Zhu, J. An adaptive strips method for extraction buildings from light detection and ranging data. *IEEE Trans. Geosci. Remote Sens.* **2017**, *14*, 1651–1655. [CrossRef]
28. Cai, Z.; Ma, H.; Zhang, L. A building detection method based on semi-suppressed fuzzy c-means and restricted region growing using airborne LiDAR. *Remote Sens.* **2019**, *11*, 848. [CrossRef]
29. Wang, Y.; Jiang, T.; Yu, M.; Tao, S.; Sun, J.; Liu, S. Semantic-based building extraction from LiDAR point clouds using contexts and optimization in complex environment. *Sensors* **2020**, *20*, 3386. [CrossRef]
30. Vosselman, G.; Maas, H. *Airborne and Terrestrial Laser Scanning*; Whittles Publishing: Dunbeath, UK, 2014.
31. Zhou, G.; Zhou, X. Seamless fusion of LiDAR and aerial imagery for building extraction. *IEEE Trans. Geosci. Remote Sens.* **2014**, *52*, 7393–7407. [CrossRef]
32. Awrangjeb, M.; Zhang, C.; Fraser, C.S. Automatic reconstruction of building roofs through effective integration of LiDAR and multispectral imagery. *ISPRS Ann. Photogramm. Remote Sens. Spat. Inf. Sci.* **2012**, *1*, 203–208. [CrossRef]
33. Awrangjeb, M.; Zhang, C.; Fraser, C.S. Automatic extraction of building roofs using LiDAR data and multispectral imagery. *ISPRS J. Photogramm. Remote Sens.* **2013**, *83*, 1–18. [CrossRef]
34. Qin, R.; Fang, W. A hierarchical building detection method for very high resolution remotely sensed images combined with DSM using graph cut optimization. *Photogramm. Eng. Remote Sens.* **2014**, *80*, 873–883. [CrossRef]
35. Gilani, S.; Awrangjeb, M.; Lu, G. An automatic building extraction and regularisation technique using LiDAR point cloud data and orthoimage. *Remote Sens.* **2016**, *8*, 258. [CrossRef]

36. Siddiqui, F.; Teng, S.; Awrangjeb, M.; Lu, G. A robust gradient based method for building extraction from LiDAR and photogrammetric imagery. *Sensors* **2016**, *16*, 1110. [CrossRef]
37. Lai, X.; Yang, J.; Li, Y.; Wang, M. A building extraction approach based on the fusion of LiDAR point cloud and elevation map texture features. *Remote Sens.* **2019**, *11*, 1636. [CrossRef]
38. Chen, S.; Shi, W.; Zhou, M.; Min, Z.; Chen, P. Automatic building extraction via adaptive iterative segmentation with LiDAR data and high spatial resolution imagery fusion. *IEEE J. -STARS* **2020**, *13*, 2081–2095. [CrossRef]
39. Chen, J.; Qiu, X.; Ding, C.; Wu, Y. CVCMFF net: Complex-valued convolutional and multifeature fusion network for building semantic segmentation of InSAR images. *IEEE Trans. Geosci. Remote Sens.* **2021**, 1–14. [CrossRef]
40. Pan, S.; Guan, H.; Yu, Y.; Li, J.; Peng, D. A comparative land-cover classification feature study of learning algorithms: DBM, PCA, and RF using multispectral LiDAR data. *IEEE J. -STARS* **2019**, *12*, 1314–1326. [CrossRef]
41. Hui, Z.; Hu, Y.; Yevenyo, Y.Z.; Yu, X. An improved morphological algorithm for filtering airborne LiDAR point cloud based on multi-level kriging interpolation. *Remote Sens.* **2016**, *8*, 35. [CrossRef]
42. Wang, D.; Takoudjou, S.M.; Casella, E. LeWoS: A universal leaf-wood classification method to facilitate the 3D modelling of large tropical trees using terrestrial LiDAR. *Methods Ecol. Evol.* **2020**, *11*, 376–389. [CrossRef]
43. ISPRS Test Project on Urban Classification, 3D Building Reconstruction and Semantic Labeling. Available online: https://www2.isprs.org/commissions/comm2/wg4/benchmark/ (accessed on 22 July 2021).
44. Rutzinger, M.; Rottensteiner, F.; Pfeifer, N. A comparison of evaluation techniques for building extraction from airborne laser scanning. *IEEE J. -STARS* **2009**, *2*, 11–20. [CrossRef]
45. Nguyen, T.H.; Daniel, S.; Guériot, D.; Sintès, C.; Le Caillec, J. Super-resolution-based snake model—an unsupervised method for large-scale building extraction using airborne LiDAR data and optical image. *Remote Sens.* **2020**, *12*, 1702. [CrossRef]
46. Doulamis, A.D.; Doulamis, N.D.; Kollias, S.D. An adaptable neural-network model for recursive nonlinear traffic prediction and modeling of MPEG video sources. *IEEE Trans. Neural Netw.* **2003**, *14*, 150–166. [CrossRef]
47. Protopapadakis, E.; Schauer, M.; Pierri, E.; Doulamis, A.D.; Stavroulakis, G.E.; Böhrnsen, J.U.; Langer, S. A genetically optimized neural classifier applied to numerical pile integrity tests considering concrete piles. *Comput. Struct.* **2016**, *162*, 68–79. [CrossRef]
48. Niemeyer, J.; Rottensteiner, F.; Soergel, U. Conditional random fields for LiDAR point cloud classification in complex urban areas. *ISPRS Ann. Photogramm. Remote Sens. Spat. Inf. Sci.* **2012**, *1*, 263–268. [CrossRef]
49. Wei, Y.; Yao, W.; Wu, J.; Schmitt, M.; Stilla, U. Adaboost-based feature relevance assessment in fusing LiDAR and image data for classification of trees and vehicles in urban scenes. *ISPRS Ann. Photogramm. Remote Sens. Spat. Inf. Sci.* **2012**, *1*, 323–328. [CrossRef]
50. Moussa, A.; El-Sheimy, N. A new object based method for automated extraction of urban objects from airborne sensors data. *ISPRS-Int. Arch. Photogramm. Remote Sens. Spat. Inf. Sci.* **2012**, *39*, 309–314. [CrossRef]
51. Yang, B.; Xu, W.; Dong, Z. Automated extraction of building outlines from airborne laser scanning point clouds. *IEEE Geosci. Remote Sens.* **2013**, *10*, 1399–1403. [CrossRef]
52. Gerke, M.; Xiao, J. Fusion of airborne laserscanning point clouds and images for supervised and unsupervised scene classification. *Isprs J. Photogramm. Remote Sens.* **2014**, *87*, 78–92. [CrossRef]
53. Demantké, J.; Mallet, C.; David, N.; Vallet, B. Dimensionality based scale selection in 3D LiDAR point clouds. *ISPRS-Int. Arch. Photogramm. Remote Sens. Spat. Inf. Sci.* **2012**, *3812*, 97–102. [CrossRef]
54. OpenTopography. Available online: https://portal.opentopography.org (accessed on 22 July 2021).

remote sensing

MDPI

Article

Attention Enhanced U-Net for Building Extraction from Farmland Based on Google and WorldView-2 Remote Sensing Images

Chuangnong Li [1], Lin Fu [1,*], Qing Zhu [1], Jun Zhu [1], Zheng Fang [2], Yakun Xie [1], Yukun Guo [1] and Yuhang Gong [2]

[1] Faculty of Geosciences and Environmental Engineering, Southwest Jiaotong University, Chengdu 611756, China; sclicn@sina.com (C.L.); zhuq66@263.net (Q.Z.); zhujun@swjtu.edu.cn (J.Z.); yakunxie@my.swjtu.edu.cn (Y.X.); gyk@my.swjtu.edu.cn (Y.G.)

[2] Sichuan Center of Satellite Application Technology, Sichuan Institute of Land Science and Technology, Chengdu 610041, China; fangzheng5288@163.com (Z.F.); gongyuhang_geo@163.com (Y.G.)

* Correspondence: vge_fulin@my.swjtu.edu.cn

Citation: Li, C.; Fu, L.; Zhu, Q.; Zhu, J.; Fang, Z.; Xie, Y.; Guo, Y.; Gong, Y. Attention Enhanced U-Net for Building Extraction from Farmland Based on Google and WorldView-2 Remote Sensing Images. *Remote Sens.* **2021**, *13*, 4411. https://doi.org/10.3390/rs13214411

Academic Editors: Jiaojiao Tian, Qin Yan, Mohammad Awrangjeb, Beril Sirmacek and Nusret Demir

Received: 21 August 2021
Accepted: 29 October 2021
Published: 2 November 2021

Abstract: High-resolution remote sensing images contain abundant building information and provide an important data source for extracting buildings, which is of great significance to farmland preservation. However, the types of ground features in farmland are complex, the buildings are scattered and may be obscured by clouds or vegetation, leading to problems such as a low extraction accuracy in the existing methods. In response to the above problems, this paper proposes a method of attention-enhanced U-Net for building extraction from farmland, based on Google and WorldView-2 remote sensing images. First, a Resnet unit is adopted as the infrastructure of the U-Net network encoding part, then the spatial and channel attention mechanism module is introduced between the Resnet unit and the maximum pool and the multi-scale fusion module is added to improve the U-Net network. Second, the buildings found on WorldView-2 and Google images are extracted through farmland boundary constraints. Third, boundary optimization and fusion processing are carried out for the building extraction results on the WorldView-2 and Google images. Fourth, a case experiment is performed. The method in this paper is compared with semantic segmentation models, such as FCN8, U-Net, Attention_UNet, and DeepLabv3+. The experimental results indicate that this method attains a higher accuracy and better effect in terms of building extraction within farmland; the accuracy is 97.47%, the F1 score is 85.61%, the recall rate (Recall) is 93.02%, and the intersection of union (IoU) value is 74.85%. Hence, buildings within farming areas can be effectively extracted, which is conducive to the preservation of farmland.

Keywords: building extraction; farmland range; attention enhancement; U-Net network improvement; multi-source remote sensing image

1. Introduction

Farmland constitutes an important resource for human survival and development. Countries worldwide have each issued corresponding farmland protection policies or measures [1–6]. Farmland protection represents the basic guarantee to maintain the red line of 1.8 billion mu of farmland in China, but the phenomenon of illegal farmland occupation to build houses in rural areas is serious [7,8]. According to statistics, since 1 January 2013, more than 7 million mu of farmland has been occupied for house construction in China, which more than 600,000 mu has been occupied in the southwest region. Therefore, building extraction in farmland is of great significance for farmland protection [9–12].

Traditional field surveys provide high accuracy and reliable results but require considerable manpower, materials and financial resources. With the continuous development of remote sensing technology, the obtained remote sensing images with higher spatial resolution contain abundant building information. Extracting buildings from high-resolution remote sensing images has become a research hotspot [9,13–16]. Traditional building extraction methods, based on optical remote sensing images, mainly consider low-level semantic features such as color, texture, and shape to extract buildings. Methods of this kind include edge detection, region segmentation, corner detection, threshold segmentation, clustering, etc. [17–21]. However, these methods are affected by lighting conditions, sensor types, and building structures. Even if the high-resolution remote sensing images are rich in details, the complex types of features, pixel mixing, shadows and other problems within the farmland are serious, making the phenomenon of "same subject with different spectra" or "different subject with same spectra" more common. These methods are limited when solving the problem of building extraction under specific data conditions [14,22].

In recent years, deep-learning methods have continued to be developed, and various neural network models have been widely used for intelligent building extraction [23–26]. With complex structural features, single-pixel segmentation may destroy the integrity and structural features of the building. In response to this problem, scholars at home and abroad have further improved various semantic segmentation models and created networks for the extraction of buildings. Good results have been achieved [27–29]. For example, Li et al. proposed a rural building segmentation method based on Mask R-CNN and a histogram threshold, which achieved the high-performance semantic segmentation of buildings via a small number of samples [30]. Zhang et al. combined a neural network and edge detection to conduct building extraction experiments. After pixel classification through the neural network, edge detection was used to complete an accurate segmentation of building boundaries [31]. Wu et al. established a multi-constrained full convolutional neural network architecture based on FCN, and used multiple constraints to optimize the parameters of the middle layer in order to obtain more multi-scale features [32]; Lin et al. combined residual block and expanded convolution to construct a deep network architecture for building extraction, and improved computational efficiency through a certain accuracy loss [33]; Xu et al. combined Resnet and U-net networks to extract buildings from high-resolution remote sensing images, and integrated them using guided filters to eliminate noise, improve accuracy and optimize the output result [34]. Bai et al. proposed an improved faster R-CNN building extraction method, using dense residual network and region of interest alignment methods to solve the problem of regional mismatch, and further improve the effect of building detection [35]. Deng et al. used a new feature extraction method to combine an object suggestion network (MS-OPN) and object detection network (AODN) to construct multi-scale features for building extraction [36]. The above methods mainly focus on urban areas where buildings are dense. For these places, the features on the image are mainly buildings, and the occlusion of buildings is not considered. However, within the range of farmland, there are few buildings and most of the areas comprise non-construction land use. These methods are not suitable for effectively extracting small targets within the farmland, and attention needs to be paid to small target buildings.

The attention mechanism imitates human brain–eye vision, which can more accurately focus on and process the most important details, rather than establishing the whole visual content. Therefore, it is widely used in deep learning to improve the accuracy of target extraction [37,38]. Yang et al. combined a lightweight DenseNet and a spatial attention fusion module to construct a dense attention network for building extraction from remote sensing images [39]. Pan et al. combined spatial and channel attention mechanisms to detect buildings using a U-Net network [40]. Jiang et al. input a global co-attention mechanism, building an attention-guided Siamese network based on a pyramid feature to detect urban building changes and achieved good results [41]. Guo et al. proposed a building extraction structure based on attention and multiple losses, which further improved the sensitivity of the model and the feature extraction ability [42]. However, the building distribution within farmland is sparse, shielding effects such as clouds, rain, fog, and vegetation may occur, and the boundary may be blurred. The above existing methods encounter difficulties regarding the accurate extraction of building information in farmland. With the requirement of overcoming the problem of building occlusion, it is necessary to fuse multi-source remote sensing images to extract farmland buildings and fuse the extraction results.

Therefore, in response to the above problems, this paper proposes a method using an attention-enhanced U-Net for building extraction from farmland. First, the Resnet unit is introduced as the basic structure in the coding layer of the U-Net network, and the spatial and channel attention modules are added to the convolutional layer of the U-Net network to enhance the convolution process's attention, given in dispersed small targets. The buildings within farmland are better extracted and the U-Net network is improved. Then, we integrated WorldView-2 and Google remote sensing images (WorldView-2 images are provided by the Sichuan Provincial Bureau of Surveying and Mapping, and Google images are freely downloaded from the Internet). Taking the farmland boundaries given by the third national census dataset as the spatial semantic constraint, the buildings within farmland can be extracted and the influence of interference factors can be reduced. Finally, through morphological operations such as opening and closing operations, the extracted building boundaries are optimized and merged to enhance the accuracy of building extraction in farmland. The main chapters of this paper are arranged as follows: the second part introduces the main technical methods of the paper, including U-Net network improvement, intelligent building extraction from WorldView-2 and Google remote sensing images under boundary constraints, and boundary optimization processing operations. The third part mainly describes the source of the dataset, parameter settings, experimental results and discussion. The fourth part introduces the research conclusions and prospects for future work.

2. Methodology

2.1. Overall Framework

This paper proposes a method of attention enhanced U-Net for building extraction from farmland. The overall research idea is shown in Figure 1. The main content can be divided into two parts: (1) spatial-channel attention-enhanced building extraction: using the Resnet unit as the basic structure, adding a spatial-channel attention mechanism module and a multi-scale fusion module, the U-Net network is improved and the network's attention to small-building targets is enhanced; (2) building boundary optimization and fusion processing: with remote sensing images from WorldView-2 and Google as the input, the building extraction range is narrowed under farmland boundary constraints, an improved U-Net network is employed to extract buildings for farmland, morphological filtering methods are implemented, such as opening/closing operations to optimize the extraction results, and the optimized building extraction results are fused to achieve accurate building extraction results from the farmland.

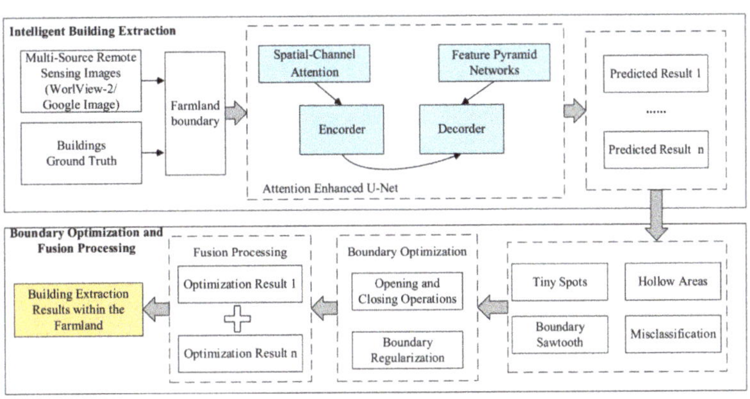

Figure 1. Overall procedural framework.

2.2. Improvement of the Attention-Enhanced U-Net Network

The U-Net network relies on skip connections to force the aggregation of same-scale feature maps of the encoder and decoder sub-networks, and the network performance is good. However, the types of features in the farmland are variable. These comprise not only buildings but also roads, woodlands, etc. Moreover, the building distribution is sparse, and various problems such as clouds, rain, fog, and vegetation shielding may ensue, resulting in an inability to extract small targets from complex backgrounds, incomplete building area extraction, and inaccurate boundaries. To overcome the limitations of existing methods in the extraction of small building targets within the farmland range from a complex background, this paper improves the U-Net network, as shown in Figure 2. The upper part represents the encoding structure, and the lower part represents the decoding structure. In the U-Net network coding stage, Resnet and attention models are added to enhance the network's attention to small target buildings. Since continuous up-sampling will cause the loss of detail in the building information, this paper adds a multi-scale fusion module in the boundary stage to ensure the local detail characteristics of the buildings are retained.

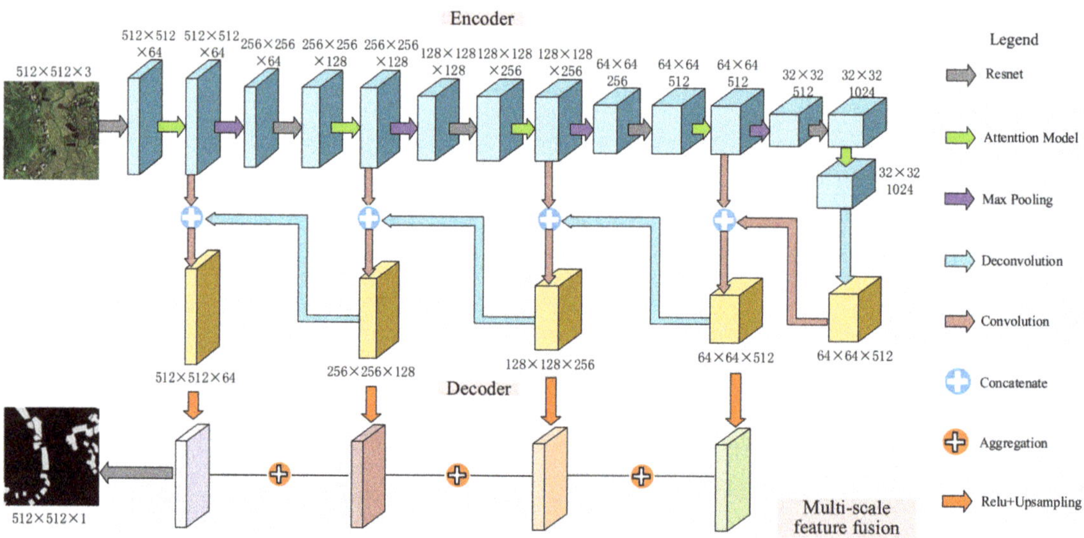

Figure 2. Improvement of the attention-enhanced U-Net network.

The encoding part adopts the Resnet unit as the basic structure, as shown in Figure 3. Compared to a traditional convolution layer, the Resnet residual network achieves convergence more easily and avoids the performance degradation issues caused by an increase in network depth. To prevent overfitting, batch normalization (BN) and rectified linear unit (ReLU) activation function layers are added, based on the residual network, to establish a refined residual network. The nonlinear expression ability of the network model is thus improved, and the features extracted from remote sensing images become richer.

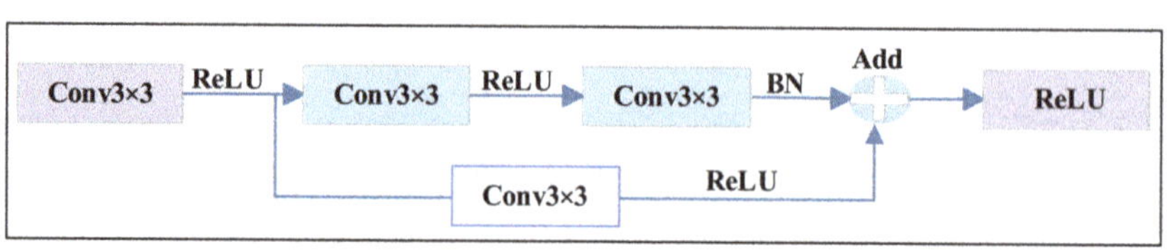

Figure 3. Schematic diagram of the Resnet unit.

Buildings are scattered throughout farmland. To enable the model to focus on scattered buildings, in the coding stage, the small target buildings within the farmland are focused upon, and the details of important targets are paid more attention to, according to the patterns of human vision. Meanwhile, the attention-mechanism module is adopted between each Resnet unit and the maximum pooling layer, which mainly includes two parts: channel and spatial attention. Channel attention is based on the relationships among the extracted features, and the channel weight is modeled to determine what the target object is. Spatial attention uses the spatial relationship of features to remodel the weight of spatial location pixels and determine which location represents the information. The local and global information can be aggregated by the attention-mechanism module, the relationship between buildings and the background can be captured, the feature weight of each channel and spatial location can be adaptively adjusted, and the network feature-extraction capacity with regard to small building targets and the ability to better grasp complex scenes can be enhanced.

As shown in Figure 4, the attention-mechanism model designed in this paper can be used to locate the area where small building targets are scattered in a given remote sensing image and suppress useless information. First, channel attention is modeled for input feature mapping, the global average pool and maximum pool layers are processed to obtain an input feature map, and a multi-layer perceptron is constructed. The weight of each channel is automatically obtained via self-learning and is then multiplied by the input feature map to obtain a channel attention feature map. Second, spatial attention modeling is conducted, and the above channel attention feature map is applied as an input to perform the spatial convolution operation and obtain the corresponding attention weights at different spatial positions on the feature map. Then, the feature map processed by the spatial and channel mixed-attention mechanism is multiplied by the input feature map, for adaptive feature refinement. In the process of training and prediction, the model can better focus on the most important feature channels and spatial positions in remote sensing images, which thus improves the model's detection performance.

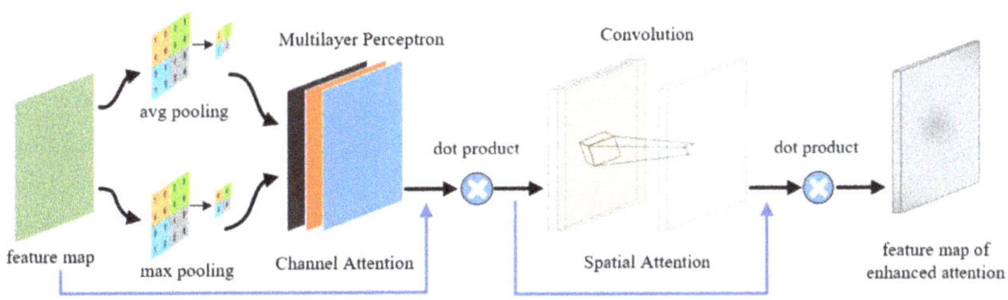

Figure 4. Attention mechanism model.

In the decoding structure, the feature map is reconstructed through the deconvolution layer, the feature map after the deconvolution layer is skip-connected with the attention-enhanced feature map included at the encoding stage, the depth of the feature map is reduced relying on a conventional convolution layer, and the size of the feature map is gradually expanded. Finally, to further improve the network performance, which entails the perception ability of multi-scale buildings in the farmland, especially small buildings and building edge information, and considering that shallow features have high resolution, but rich details and deep features have low resolution but offer rich semantic information, multi-scale feature fusion is carried out, to acquire both deep and shallow features after up-sampling and nonlinear processing. This strategy can ensure that the network will not ignore texture, edge, and other image details, while extracting global semantic information to obtain building texture, shape and spatial context features and provide more precise building segmentation results.

2.3. Building Extraction from Multi-Source Remote Sensing Images under Boundary Constraints

Buildings within the farmland may be obscured, the building extraction range may require narrowing, and the extraction accuracy may necessitate improvement. To resolve the above problems, this paper fully employs the advantages of satellite images and proposes an intelligent building extraction method from multi-source remote sensing images under boundary constraints, considering different satellite remote sensing images of the same area, as shown in Figure 5. WorldView-2, Google, and other image series can be used to assess whether buildings are located within the boundary range, through the boundary constraints of farmland. If buildings are not located within the range, they are directly discarded. With regard to buildings within the boundary range, an improved U-Net neural network model with an enhanced attention mechanism is adopted to intelligently extract building contours and the relative position information.

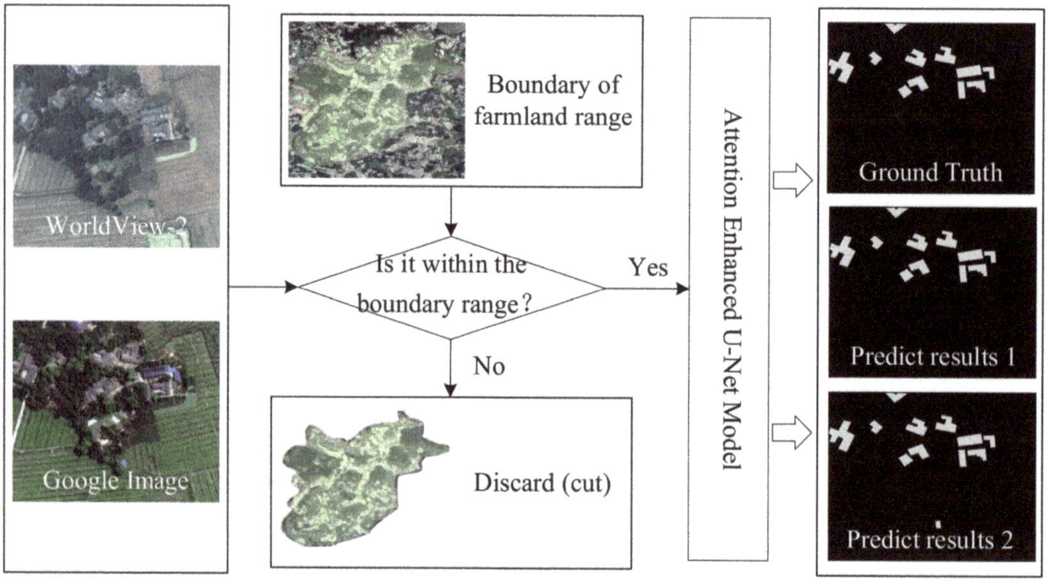

Figure 5. Building extraction based on the boundary constraints of farmland range.

2.4. Building Boundary Optimization and Fusion Processing

There are certain problems in the extraction results obtained with neural network-based models, such as very small spots, hollow areas, boundary sawtooth features and misclassification issues. Moreover, the building extraction results based on multi-source remote sensing images are varied. Therefore, this paper proposes a building boundary optimization and fusion processing method that is based on morphological filtering, as shown in Figure 6, to improve the accuracy of building extraction.

First, in terms of the neural network extraction results containing fine patches, a filter based on geometric operations is applied to execute the opening operation. As expressed in Equation (1), isolated points and burrs in tiny spots are removed in this manner. Regarding the extraction results including hollow areas, the closing operation of a morphological filtering operation is implemented, as expressed in Equation (2), to fill any cracks or hollow areas in the extraction results:

$$I \circ S = (I \ominus S) \oplus S \tag{1}$$

$$I \bullet S = (I \oplus S) \ominus S \tag{2}$$

where I denotes the original image extracted by the network, and S denotes the structural element of the filter.

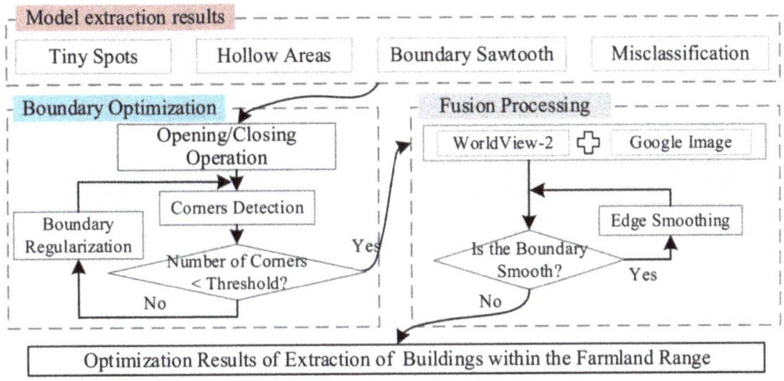

Figure 6. Building boundary optimization and fusion processing.

Then, based on the results of the morphological filtering operation, refer to the method proposed by Xie et al. to judge whether the number of building contour corners is smaller than the set threshold. If the set conditions are met, the multi-source remote sensing image extraction results are fused. Conversely, the building boundary should be optimized until the corner number conditions are satisfied [43]. Finally, the extraction results of the WorldView-2 and Google images after boundary optimization are compared pixel by pixel, and the pixels representing the building at the current location are merged as the prediction results, so as to realize the fusion processing of the extraction results of WorldView-2 and Google remote sensing images and solve the problem of misclassification in the extraction results. Gridlines are created, with the considered building outlines as boundaries. The results are further assessed to establish whether the building outline boundary requires further smoothing. If the conditions are met, the building extraction results for farmland are output. Otherwise, the boundary should be smoothed until the smoothing conditions are satisfied. The sawtooth effect in the extraction results is mitigated so that the extraction results are more precise and accurate.

3. Case Experiment Analysis

3.1. Case Area and Dataset

To verify the proposed method, the considered experimental data included five WorldView-2 satellite remote sensing image datasets, covering Qionglai city, Meishan city, Dayi County and Pujiang County of Sichuan Province, and Google images of the same areas, as shown in Figure 7. The image spatial resolution reached 0.5×0.5 m, with three bands, i.e., red, green and blue bands. The building pixel value was 1, and the pixel values of other features were set to 0. Due to the limited memory size of the adopted graphics card, each image was cropped with a sliding window exhibiting a size of 512×512 pixels.

Considering that the buildings in the images are relatively small, meaningless background slices in the training set were eliminated. The 5 satellite remote sensing images were randomly divided into a training set, validation set, and test set at a ratio of 6:2:2, and the data augmentation method was applied to expand the training dataset and improve the generalization ability of the model. Through comparative experiments, it was found that the addition of Gaussian noise and color perturbation during image data processing did not encourage model accuracy improvement. Therefore, the image processing procedure only involved rotation and flip operations, so that more morphological building features could be recognized, as shown in Figure 8. Finally, 6238 training sets, 2301 verification sets, and 1637 test sets were obtained, and no overlap occurred between the training data and test data.

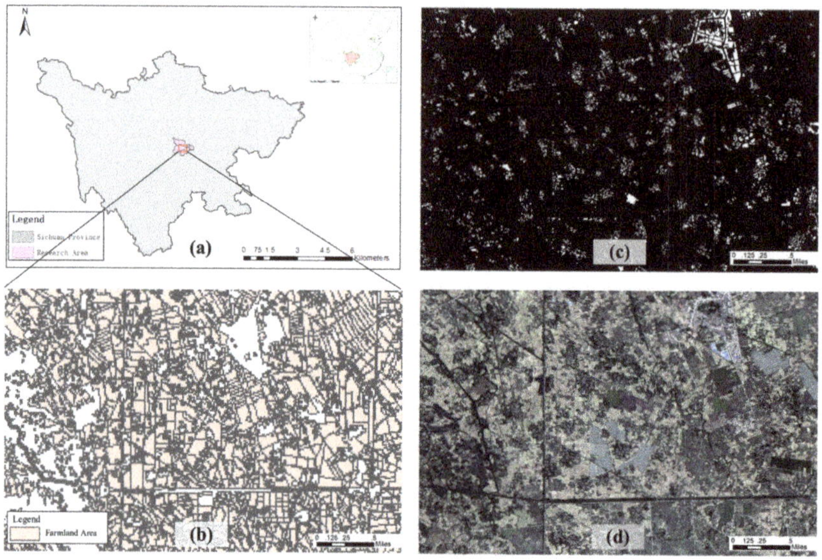

Figure 7. Case study experiment area. (**a**): Case area; (**b**): farmland area; (**c**): ground truth; (**d**): original images.

Figure 8. Image and label data.

3.2. Experimental Environment and Parameter Setting

All training and testing operations in this paper were performed on a Windows 10 system with an Intel (R) Core (TM) i9-10920x CPU @ 3.50 GHz processor, 64 GB of memory and an NVIDIA GeForce RTX 3090 GPU graphics card. The initial learning rate during model training was set at 0.001. By monitoring the loss value, the learning rate was reduced to the original value of 0.5 when the performance did not improve after 10 epochs. The experiment was trained using the built-in TensorFlow framework in Python version 3.6, the training batch size was 4 and a total of 60 epochs were iterated with a duration of 12 h. To verify the accuracy of the extraction results obtained using the method proposed in this paper, this paper chosed the accuracy, F1, recall, and an intersection of union (IoU) to evaluate the extraction results, where the accuracy is given by the proportion of correctly predicted pixels among the total pixels, as expressed in Equation (3). F1 is a comprehensive index to measure the model, as defined in Equation (4). Recall represents the proportion of correctly predicted buildings among the total buildings, as expressed in Equation (5), and the IoU represents the ratio of the intersection zone (the intersection between the predicted and true values) to the union zone (the union between the predicted and true values), as defined in Equation (6).

$$\text{Accuracy} = \frac{\text{TP} + \text{TN}}{\text{TP} + \text{TN} + \text{FP} + \text{FN}} \tag{3}$$

$$\mathrm{F1} = \frac{2\mathrm{TP}}{2\mathrm{TP} + \mathrm{FP} + \mathrm{FN}} \tag{4}$$

$$\mathrm{Recall} = \frac{\mathrm{TP}}{\mathrm{TP} + \mathrm{FN}} \tag{5}$$

$$\mathrm{IoU} = \frac{\mathrm{TP}}{\mathrm{TP} + \mathrm{FP} + \mathrm{FN}} \tag{6}$$

where the true positive (TP), false positive (FP), true negative (TN) and false negative (FN) rates evaluate the pixel classification results through comparison of the extracted building pixels to the available ground-truth points. TP denotes the number of correctly extracted building pixels, FP denotes the number of erroneously detected building pixels, TN denotes the number of correctly extracted non-building pixels, and FN denotes the number of missed building pixels.

3.3. Experimental Results and Analysis

As shown in Figure 9, the results for the buildings extracted using the attention-enhanced U-Net model proposed in this paper showed that our model can effectively extract buildings within the scope of farmland. However, the buildings in the dataset are scattered and staggered in the farmland, which leads to the problems of unclear edges and holes at certain map locations, as shown in Figure 9c. With the application of boundary optimization and data fusion methods, the building extraction results are optimized, as shown in Figure 9d. Comparing the red boxes in Figure 9c,d, we can observe that the building pattern in Figure 9d contains no small holes and that the boundary is smooth. In particular, the optimization process can eliminate the very small spots previously found in the extraction results and smooth the boundary, so as to obtain a more complete extracted shape and clearer boundary, respectively.

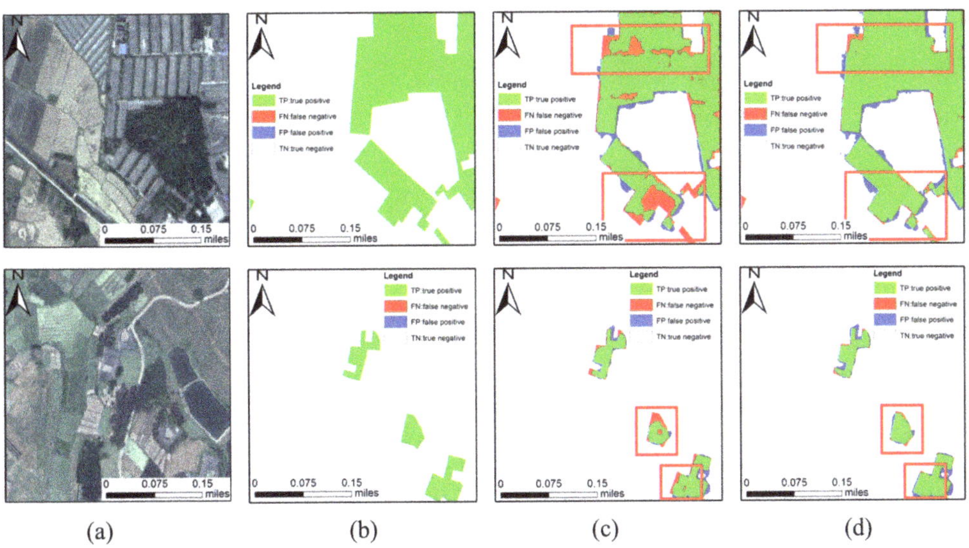

Figure 9. Experimental results. (**a**): Original images; (**b**): ground truth; (**c**): ours model extraction results; (**d**): post processing results.

To achieve the accurate extraction of building targets, boundary optimization and fusion processing of the building patterns extracted with the network model are executed. The experimental results are listed in Table 1. The IoU value after optimization reaches 74.85%, and the F1 score reaches 85.61%, which are 6.13% and 4.14% higher, respectively, than the values without optimization, thus demonstrating that the post-processing optimization method designed in this paper can effectively improve building extraction accuracy.

Table 1. Quantitative comparative analysis of case experiment results.

Method	Accuracy	F1	Recall	IoU
Our model	96.96%	81.47%	82.72%	68.72%
Post-processing	97.47%	85.61%	93.02%	74.85%

3.4. Discussion

To verify the accuracy and applicability of the proposed method, several groups of comparative experiments were established to compare this method with FCN8, U-Net and Attention_UNet, DeepLabv3 + [44] and other models qualitatively and quantitatively, where Attention_UNet includes a convolutional block attention module (CBAM) module based on the U-Net model [38].

3.4.1. Comparative Experiments of Building Extraction

Figure 10 shows the results obtained by the proposed method, FCN8, U-Net and Attention_UNet, DeepLabv3+ and the other considered models based on WorldView-2 images, where green indicates the positive extraction area, red indicates the missing extraction area and blue indicates the false extraction area. As shown in Figure 10b,c, the building boundaries extracted with the U-Net and FCN8 models are relatively fuzzy, and the building omission phenomenon is obvious with the FCN8 model, while incorrect extraction obviously occurs with the U-Net model. As shown in Figure 10d, the extraction results obtained with the U-Net model containing the attention mechanism are notably better than those obtained with the original U-Net and FCN8 models, but the extraction results remain insufficiently fine. Furthermore, Figure 10e shows that there are many holes, missing extraction and incorrect extraction results among the results obtained with the DeepLabv3+ model. The black box in Figure 10 reveals that the building pixels extracted with the network model designed in this paper are closer to the real image, and compared to the other four network models, the positive extraction area accounts for the majority of the image, which demonstrates that the proposed model can finely extract scattered small buildings from farmland images.

To quantitatively analyze the building extraction results, we considered the building labels drawn by manual visual interpretation as a reference, and the four evaluation indicators of Accuracy, Recall, IoU, and F1 were adopted to evaluate the building extraction results for farmland protection, as indicated in Table 2. Compared with U-Net, FCN8, Attention_UNet, and DeepLabv3+, the model proposed in this paper achieves the highest accuracy in building extraction from remote sensing images, with the IoU value reaching 68.72% and the F1 score reaching 81.47%. The IoU value obtained with our model is 33.73%, 25.84%, 14.71% and 5.33% higher, respectively, than that obtained with the other four models. The F1 score is 29.63%, 13.29%, 11.33% and 3.87% higher, respectively. The results indicate that the shape of the building block extracted with the model developed in this paper is closer to the actual building block shape.

Table 2. Comparison of the building extraction results obtained with the different methods.

Method	Accuracy	F1	Recall	IoU
U-Net	88.99%	51.84%	73.31%	34.99%
FCN8	95.70%	68.18%	57.02%	42.88%
Attention_UNet	94.42%	70.14%	81.07%	54.01%
DeepLabv3+	96.60%	77.60%	72.78%	63.39%
Our model	96.96%	81.47%	82.72%	68.72%

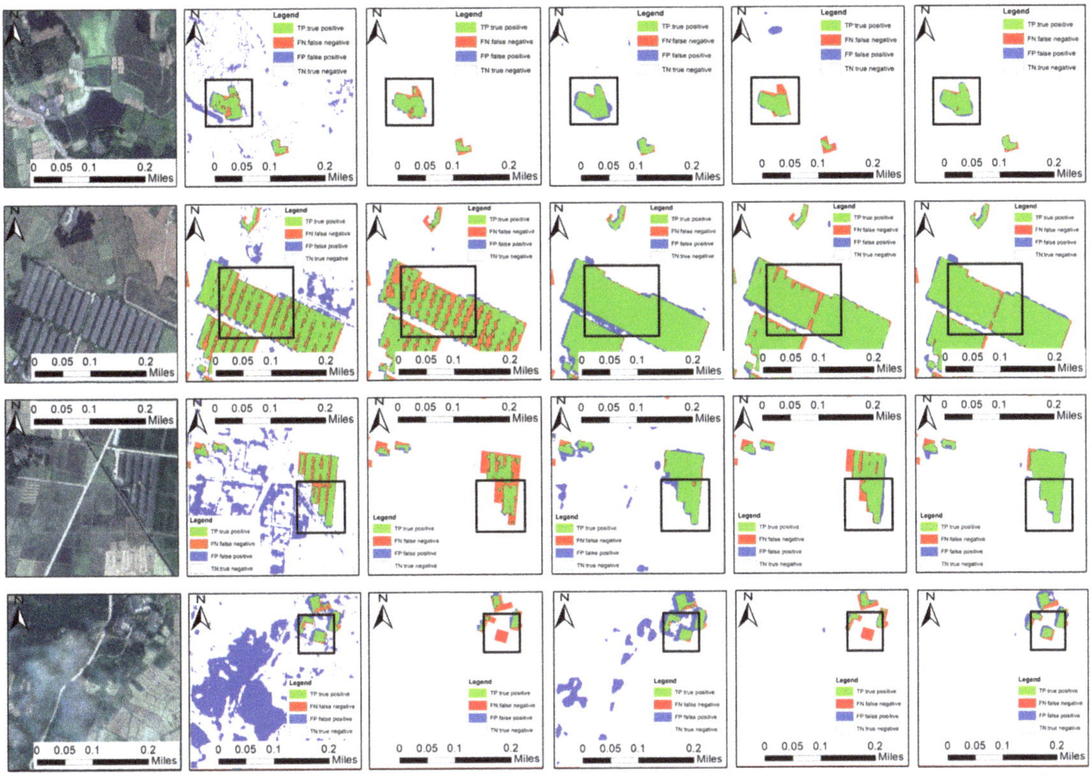

(a) Original images (b) U-Net (c) FCN8 (d) Attention_UNet (e) DeepLabv3+ (f) Our Model

Figure 10. Model comparison results. (**a**): Original images; columns (**b–e**): represent extracted results by U-Net, FCN8, Attention_UNet, DeepLabv3+; (**f**): represent our model extracted results.

3.4.2. Comparative Experiments of Boundary Optimization and Fusion

The results of building extraction using the method proposed in this paper, FCN8, U-Net, Attention_UNet and DeepLabv3+ are shown in Figure 11. According to Figure 11a, there are a large number of small spots showing wrong detection in the U-Net fusion result, and the missed detection is serious. According to Figure 11b–d, the FCN8 fusion presents obvious areas of missed detection. Meanwhile, the fusion results of Attention_UNet and DeepLabv3+ demonstrate obvious false detection areas around the boundary of the building's block. According to Figure 11e, compared with the other four network models, the fusion processing results of building patches extracted by this method account for the most positive inspection areas, and the building patches are more complete; that is, the fusion processing can eliminate the fine patches in the results and smooth the edges of the patches.

As shown in Table 3, compared with the other four network models, the model presented in this paper offers the highest accuracy after fusion processing; the IoU score is 74.85% and the F1 score is 85.61%, 36.69% and 32.06% higher than U-Net, respectively. Compared with FCN8, it is 5.33% and 7.55% higher, respectively. Compared with Attention_UNet, it is 17.71% and 13.55% higher, respectively. Compared with DeepLabv3+, it is 7.06% and 5.04% higher, respectively. These results show that the fusion processing of building blocks extracted by this model can greatly improve the accuracy of building extraction.

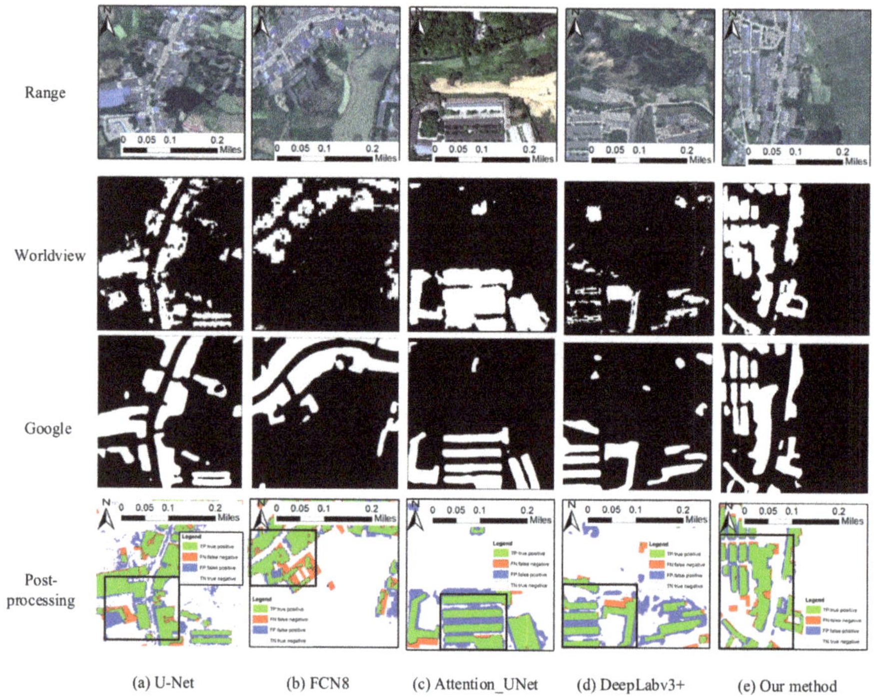

Figure 11. Boundary optimization and fusion comparison results. Columns (**a–d**) represent extracted results by U-Net, FCN8, Attention_UNet, DeepLabv3+; (**e**) represent extracted results by our method.

Table 3. Quantitative comparative analysis of fusion results.

Method	Accuracy	F1	Recall	IoU
U-Net	89.56%	53.55%	87.22%	38.16%
FCN8	96.88%	80.28%	84.27%	67.30%
Attention_UNet	94.40%	72.06%	89.78%	57.14%
DeepLabv3+	96.88%	80.57%	86.94%	67.79%
Our model	97.47%	85.61%	93.02%	74.85%

4. Conclusions and Future Work

Considering the problems of low extraction accuracy and unclear building boundaries when using existing methods for farmland, a method of attention enhanced U-Net for building extraction from farmland based on Google and WorldView-2 remote sensing images is proposed. The selected farmland range under test covers Qionglai city, Meishan city, Dayi County and Pujiang County of Sichuan Province, and case experiments were performed. The experimental results reveal the following: the accuracy is 97.47%, the F1 score is 85.61%, the recall rate is 93.02%, and the IoU value is 74.85%. All accuracy evaluation indicators are better than those obtained with U-Net, FCN8, Attention_UNet, DeepLabv3+ and other models, which verifies that the method proposed in this paper can effectively extract buildings on farmland. The main contributions of this paper are as follows: first, Resnet is adopted as the U-Net infrastructure, and a spatial and channel attention mechanism module, as well as a multi-scale fusion module, are added to improve the U-Net network and enhance the focus of attention on small building targets in the farmland. Secondly, the method developed uses WorldView-2 and Google remote sensing images to limit farmland boundaries, narrowing the extraction range of buildings and improving extraction accuracy. Finally, a building boundary optimization method based on morphological filtering is proposed, the extraction results are judged pixel by pixel, and the extraction results of the two images are merged. The method in this paper can effectively solve the problems offered by low accuracy of building extraction results, blurred boundaries, and so on, which can be attributed to the complex types of features, the sparse distribution of buildings, and building occlusion within farmland. Meanwhile, the method provides scientific and technical support for the investigation of buildings within the subject of farmland preservation, which is of great significance to maintaining farmland.

Despite the above achievements, the method presented in this paper also has certain limitations. For example, the buildings in mountainous areas are mostly low-rise bungalows and are relatively old, the boundary between the building and the surrounding ground objects is more blurred, and the method is adversely affected by clouds, rain, fog, and vegetation all the year round. Thus, it is difficult to accurately extract data regarding the buildings in these places. Meanwhile, if several buildings are adjacent and the boundaries are fuzzy, the method outlined in this paper finds it difficult to accurately determine the adjacent relationship of the buildings, and the method is likely to identify them as a single building. The boundary between adjacent buildings is not fully considered, resulting in several adjacent buildings being regarded as one complete building. The further development of remote sensing and interferometric synthetic aperture radar (InSAR) technology could overcome the influence of cloudy and rainy weather conditions and yield higher-resolution remote sensing images. Therefore, in future research, we will continue to integrate additional and higher-resolution remote sensing images, further develop generalization and complex-building abstraction methods, and study the processing methods for neighboring relationships between buildings, to improve building extraction accuracy in farmland.

Author Contributions: Conceptualization, C.L. and L.F.; methodology, C.L.; software, L.F.; validation, C.L., Y.X. and Y.G. (Yukun Guo); formal analysis, C.L. and L.F.; investigation, Y.G. (Yuhang Gong); resources, Z.F.; data curation, Y.X.; writing—original draft preparation, C.L. and L.F.; writing—review and editing, Q.Z.; visualization, C.L. and L.F.; supervision, Q.Z. and J.Z.; project administration, L.F.; funding acquisition, J.Z. All authors have read and agreed to the published version of the manuscript.

Funding: This research was funded by Project of Department of Natural Resources of Sichuan Province (grant number KJ-2020-4), Sichuan Science and Technology Program (grant 2020JDTD0003, 2020YFG0083).

Institutional Review Board Statement: Not applicable.

Informed Consent Statement: Not applicable.

Data Availability Statement: Not applicable.

Conflicts of Interest: The authors declare no conflict of interest.

References

1. Chen, Y.M.; Yao, M.R.; Zhao, Q.Q.; Chen, Z.J.; Jiang, P.H.; Li, M.C.; Chen, D. Delineation of a basic farmland protection zone based on spatial connectivity and comprehensive quality evaluation: A case study of Changsha City, China. *Land Use Policy* **2021**, *101*, 105145. [CrossRef]
2. Connell, D.J. The Quality of Farmland Protection in Canada: An Evaluation of the Strength of Provincial Legislative Frameworks. *Can. Plan. Policy Aménage. Polit. Can.* **2021**, *1*, 109–130.
3. Perrin, C.; Clément, C.; Melot, R.; Nougarèdes, B. Preserving farmland on the urban fringe: A literature review on land policies in developed countries. *Land* **2020**, *9*, 223. [CrossRef]
4. Perrin, C.; Nougarèdes, B.; Sini, L.; Branduini, P.; Salvati, L. Governance changes in peri-urban farmland protection following decentralisation: A comparison between Montpellier (France) and Rome (Italy). *Land Use Policy* **2018**, *70*, 535–546. [CrossRef]
5. Epp, S.; Caldwell, W.; Bryant, C. Farmland preservation and rural development in Canada. In *Agrourbanism*; Gottero, E., Ed.; GeoJournal Library; Springer: Cham, Switzerland, 2019; Volume 124, pp. 11–25.
6. Wu, Y.Z.; Shan, L.P.; Guo, Z.; Peng, L. Cultivated land protection policies in China facing 2030: Dynamic balance system versus basic farmland zoning. *Habitat Int.* **2017**, *69*, 126–138. [CrossRef]

7. Shao, Z.F.; Li, C.M.; Li, D.R.; Altan, O.; Zhang, L.; Ding, L. An accurate matching method for projecting vector data into surveillance video to monitor and protect cultivated land. *ISPRS Int. J. Geo-Inf.* **2020**, *9*, 448. [CrossRef]
8. Li, C.X.; Gao, X.; Xi, Z.L. Characteristics, hazards, and control of illegal villa (houses): Evidence from the Northern Piedmont of Qinling Mountains, Shaanxi Province, China. *Environ. Sci. Pollut. Res.* **2019**, *26*, 21059–21064. [CrossRef] [PubMed]
9. Shao, Z.F.; Tang, P.H.; Wang, Z.Y.; Saleem, N.; Yam, S. BRRNet: A fully convolutional neural network for automatic building extraction from high-resolution remote sensing images. *Remote Sens.* **2020**, *12*, 1050. [CrossRef]
10. Xie, J.L. *Research on Key Technologies of Rural Building Information Extraction Based on High Resolution Remote Sensing Images*; Southwest Jiaotong University: Chengdu, China, 2019.
11. Ji, S.P.; Wei, S.Q.; Lu, M. Fully convolutional networks for multisource building extraction from an open aerial and satellite imagery data set. *IEEE Trans. Geosci. Remote Sens.* **2018**, *57*, 574–586. [CrossRef]
12. You, Y.F.; Wang, S.Y.; Ma, Y.X.; Chen, G.S.; Wang, B. Building detection from VHR remote sensing imagery based on the morphological building index. *Remote Sens.* **2018**, *10*, 1287. [CrossRef]
13. Guo, H.N.; Shi, Q.; Du, B.; Zhang, L.P.; Wang, D.Z.; Ding, H.X. Scene-Driven Multitask Parallel Attention Network for Building Extraction in High-Resolution Remote Sensing Images. *IEEE Trans. Geosci. Remote Sens.* **2021**, *59*, 4287–4306. [CrossRef]
14. Liao, C.; Hu, H.; Li, H.F.; Ge, X.M.; Chen, M.; Li, C.N. Joint Learning of Contour and Structure for Boundary-Preserved Building Extraction. *Remote Sens.* **2021**, *13*, 1049. [CrossRef]
15. Yang, L.; Wang, H.; Yan, K.; Yu, X.Z. Building extraction of multi-source data based on deep learning. In Proceedings of the 2019 IEEE 4th International Conference on Image, Vision and Computing (ICIVC), Xiamen, China, 5–7 July 2019; pp. 296–300.
16. Sun, G.Y.; Huang, H.; Zhang, A.Z.; Li, F.; Zhao, H.M. Fusion of multiscale convolutional neural networks for building extraction in very high-resolution images. *Remote Sens.* **2019**, *11*, 227. [CrossRef]
17. Cheng, G.; Han, J.W. A survey on object detection in optical remote sensing images. *ISPRS J. Photogramm. Remote Sens.* **2016**, *117*, 11–28. [CrossRef]
18. Liasis, G.; Stavrou, S. Building extraction in satellite images using active contours and colour features. *Int. J. Remote Sens.* **2016**, *37*, 1127–1153. [CrossRef]
19. Ghaffarian, S.; Ghaffarian, S. Automatic building detection based on Purposive FastICA (PFICA) algorithm using monocular high resolution Google Earth images. *ISPRS J. Photogramm. Remote Sens.* **2014**, *97*, 152–159. [CrossRef]
20. Liu, Z.J.; Wang, J.; Liu, W.P. Building extraction from high resolution imagery based on multi-scale object oriented classification and probabilistic Hough transform. In Proceedings of the 2005 International Geoscience and Remote Sensing Symposium (IGARSS'05), Seoul, Korea, 29 July 2005; pp. 2250–2253.
21. Lin, C.G.; Nevatia, R. Building detection and description from a single intensity image. *Comput. Vis. Image Underst.* **1998**, *72*, 101–121. [CrossRef]
22. Zhang, H.; Zhao, H.; Zhang, X. High-resolution Image Building Extraction Using U-net Neural Network. *Remote Sens. Inf.* **2020**, *35*, 3547. [CrossRef]
23. Badrinarayanan, V.; Kendall, A.; Cipolla, R. SegNet: A deep convolutional encoder-decoder architecture for image segmentation. *IEEE Trans. Pattern Anal. Mach. Intell.* **2017**, *39*, 2481–2495. [CrossRef] [PubMed]
24. Noh, H.; Hong, S.; Han, B. Learning deconvolution network for semantic Segmentation. In Proceedings of the 2015 IEEE International Conference on Computer Vision (ICCV), Santiago, Chile, 7–13 December 2015; pp. 1520–1528.
25. Ronneberger, O.; Fischer, P.; Brox, T. U-Net: Convolutional networks for biomedical image segmentation. In *Proceedings of the International Conference on Medical Image Computing and Computer-Assisted Intervention*; Munich, Germany, 5–9 October 2015, Springer: Cham, Switzerland, 2015; pp. 234–241.
26. Long, J.; Shelhamer, E.; Darrell, T. Fully convolutional networks for semantic segmentation. In Proceedings of the IEEE Conference on Computer Vision and Pattern Recognition, Boston, MA, USA, 7–12 June 2015; pp. 3431–3440.
27. Yi, Y.N.; Zhang, Z.J.; Zhang, W.C.; Zhang, C.R.; Li, W.D. Semantic segmentation of urban buildings from VHR remote sensing imagery using a deep convolutional neural network. *Remote Sens.* **2019**, *11*, 1774. [CrossRef]
28. Ren, S.Q.; He, K.M.; Girshick, R.; Sun, J. Faster R-CNN: Towards real-time object detection with region proposal networks. *IEEE Trans. Pattern Anal. Mach. Intell.* **2017**, *39*, 1137–1149. [CrossRef] [PubMed]
29. He, K.M.; Gkioxari, G.; Dollár, P.; Girshick, R. Mask R-CNN. In Proceedings of the IEEE International Conference on Computer Vision, Venice, Italy, 22–29 October 2017; pp. 2961–2969.
30. Li, Y.; Xu, W.P.; Chen, H.H.; Jiang, J.H.; Li, X. A Novel Framework Based on Mask R-CNN and Histogram Thresholding for Scalable Segmentation of New and Old Rural Buildings. *Remote Sens.* **2021**, *13*, 1070. [CrossRef]
31. Zhang, L.L.; Wu, J.S.; Fan, Y.; Gao, H.M.; Shao, Y.H. An efficient building extraction method from high spatial resolution remote sensing images based on improved mask R-CNN. *Sensors* **2020**, *20*, 1465. [CrossRef] [PubMed]
32. Wu, G.; Shao, X.; Guo, Z.; Chen, Q.; Yuan, W.; Shi, X.; Xu, Y.W.; Shibasaki, R. Automatic building segmentation of aerial imagery using multi-constraint fully convolutional networks. *Remote Sens.* **2018**, *10*, 407. [CrossRef]
33. Lin, J.; Jing, W.; Song, H.; Chen, G. ESFNet: Efficient Network for Building Extraction from High-Resolution Aerial Images. *IEEE Access* **2019**, *7*, 54285–54294. [CrossRef]
34. Xu, Y.; Wu, L.; Xie, Z.; Chen, Z. Building extraction in very high resolution remote sensing imagery using deep learning and guided filters. *Remote Sens.* **2018**, *10*, 144. [CrossRef]

35. Bai, T.; Pang, Y.; Wang, J.C.; Han, K.N.; Luo, J.S.; Wang, H.Q.; Lin, J.Z.; Wu, J.; Zhang, H. An Optimized faster R-CNN method based on DRNet and RoI align for building detection in remote sensing images. *Remote Sens.* **2020**, *12*, 762. [CrossRef]
36. Deng, Z.; Sun, H.; Zhou, S.; Zhao, J.; Lei, L.; Zou, H. Multi-scale object detection in remote sensing imagery with convolutional neural networks. *ISPRS J. Photogramm. Remote Sens.* **2018**, *145*, 3–22. [CrossRef]
37. Ghaffarian, S.; Valente, J.; Voort, M.V.D.; Tekinerdogan, B. Effect of Attention Mechanism in Deep Learning-Based Remote Sensing Image Processing: A Systematic Literature Review. *Remote Sens.* **2021**, *13*, 2965. [CrossRef]
38. Woo, S.; Park, J.; Lee, J.Y.; Kweon, I.S. CBAM: Convolutional Block Attention Module. In *Proceedings of the European Conference on Computer Vision*; Munich, Germany, 8–14 September 2018, Springer: Cham, Switzerland, 2018; pp. 3–19.
39. Yang, H.; Wu, P.; Yao, X.; Wu, Y.; Wang, B.; Xu, Y. Building Extraction in Very High Resolution Imagery by Dense-Attention Networks. *Remote Sens.* **2018**, *10*, 1768. [CrossRef]
40. Pan, X.; Yang, F.; Gao, L.; Chen, Z.; Zhang, B.; Fan, H.; Ren, J. Building extraction from high-resolution aerial imagery using a generative adversarial network with spatial and channel attention mechanisms. *Remote Sens.* **2019**, *11*, 917. [CrossRef]
41. Jiang, H.W.; Hu, X.Y.; Li, K.; Zhang, J.M.; Gong, J.Q.; Zhang, M. PGA-SiamNet: Pyramid Feature-Based Attention-Guided Siamese Network for Remote Sensing Orthoimagery Building Change Detection. *Remote Sens.* **2020**, *12*, 484. [CrossRef]
42. Guo, M.Q.; Liu, H.; Xu, Y.Y.; Huang, Y. Building extraction based on U-Net with an attention block and multiple losses. *Remote Sens.* **2020**, *12*, 1400. [CrossRef]
43. Xie, Y.K.; Zhu, J.; Cao, Y.G.; Feng, D.J.; Hu, M.J.; Li, W.L.; Zhang, Y.H.; Fu, L. Refined Extraction of Building Outlines From High-Resolution Remote Sensing Imagery Based on a Multifeature Convolutional Neural Network and Morphological Filtering. *IEEE J. Sel. Top. Appl. Earth Obs. Remote Sens.* **2020**, *13*, 1852–1855. [CrossRef]
44. Chen, L.C.; Zhu, Y.K.; Papandreou, G.; Schroff, F.; Adam, H. Encoder-decoder with atrous separable convolution for semantic image segmentation. In *Proceedings of the European Conference on Computer Vision*; Munich, Germany, 8–14 September 2018, Springer: Cham, Switzerland, 2018; pp. 833–851.

remote sensing

MDPI

Article

Parameter-Free Half-Spaces Based 3D Building Reconstruction Using Ground and Segmented Building Points from Airborne LiDAR Data with 2D Outlines

Marko Bizjak *, Borut Žalik and Niko Lukač

Faculty of Electrical Engineering and Computer Science, University of Maribor, Koroška cesta 46, SI-2000 Maribor, Slovenia; borut.zalik@um.si (B.Ž.); niko.lukac@um.si (N.L.)
* Correspondence: m.bizjak@um.si

Abstract: This paper aims to automatically reconstruct 3D building models on a large scale using a new approach on the basis of half-spaces, while making no assumptions about the building layout and keeping the number of input parameters to a minimum. The proposed algorithm is performed in two stages. First, the airborne LiDAR data and buildings' outlines are preprocessed to generate buildings' base models and the corresponding half-spaces. In the second stage, the half-spaces are analysed and used for shaping the final 3D building model using 3D Boolean operations. In experiments, the proposed algorithm was applied on a large scale, and its' performance was inspected on a city level and on a single building level. Accurate reconstruction of buildings with various layouts were demonstrated and limitations were identified for large-scale applications. Finally, the proposed algorithm was validated on an ISPRS benchmark dataset, where a RMSE of 1.31 m and completeness of 98.9% were obtained.

Keywords: building model; reconstruction; half-space; LiDAR data; urban scale

Citation: Bizjak, M.; Žalik, B.; Lukač, N. Parameter-Free Half-Spaces Based 3D Building Reconstruction Using Ground and Segmented Building Points from Airborne LiDAR Data with 2D Outlines. *Remote Sens.* **2021**, *13*, 4430. https://doi.org/10.3390rs13214430

Academic Editors: Jiaojiao Tian, Qin Yan, Mohammad Awrangjeb, Beril Sirmacek and Nusret Demir

Received: 29 September 2021
Accepted: 2 November 2021
Published: 3 November 2021

Publisher's Note: MDPI stays neutral with regard to jurisdictional claims in published maps and institutional affiliations.

1. Introduction

As a part of the digitalisation of cities, 3D building models are essential and provide more accurate spatial and environmental analysis for various applications [1–3]. The availability of building models on a large scale is usually very limited, as for most buildings the corresponding 3D models do not exist due to their age, or are not in the public domain. Manual modelling of existing buildings is not practical on a large scale as it is very time consuming and costly. Hence, research is oriented towards solutions that reconstruct buildings automatically [4]. The common approach to automatic building reconstruction is by using point clouds [5], which can be obtained by various remote sensing technologies (e.g., stereo imaging, laser scanning). One of the more widely used technologies for this purpose is LiDAR (Light Detection and Ranging), which is the technology of active remote sensing, where laser light is used to acquire the geometry of the observed surface. It is usually mounted on an aircraft to scan large geographic areas. The quality of the point cloud obtained during such acquisition affects the accuracy of the reconstructed building models and other spatial analyses directly [6]. With the fast development of remote sensing technologies and, consequently, larger availability of the resulting data, the reconstruction of 3D building models from point clouds gained popularity in recent years [5,7,8]. In general, the building reconstruction algorithms from point clouds can be data-driven, model-driven, or hybrid-driven [5]. An additional type of algorithms that employ the machine learning approach to building reconstruction [9–11] is becoming popular, which is accelerated by the availability of training datasets for this purpose, such as the dataset presented by Wichmann et al. [12].

Data-driven algorithms are bottom-up based, where basic geometric shapes (i.e., planes, cylinders, cones ...) are detected in point clouds first. The final building's shape is then obtained with establishing a topological structure by analysing the dependence of adjacent basic geometric shapes. Data-driven algorithms can be further divided by the type of faces that are considered. Most algorithms consider flat faces only [13–18], while some maintain curved faces as well [19,20]. There are various approaches to the analysis of basic geometric shapes, such as by the adjacency matrix [16], region growing [17], adjacency graph [21], or by direct processing of planar faces and their neighbours for estimating the common edges [14]. Data-driven algorithms are sensitive to under- or over-segmentation, which can cause problems for the analysis of basic geometric shapes. When the extracted basic geometric shapes are incomplete or noisy, which is a common occurrence for low-quality input data in complex scenes, the reconstructed building model can be of bad quality. Vosselman and Dijkman [22] used building outlines and LiDAR data for building reconstruction, where building outlines are partitioned into segments, while considering intersection and height jump lines. Segments are later merged back together until each face corresponds to one segment to obtain the 3D model. Li et al. [23] presented a new framework based on TIN (Triangulated Irregular Network) and label maps to automatically create building models from LiDAR data. TIN-based roof primitives detection supports varying point density and label maps are processed by a graph-cut to provide a good representation of roof faces. Wang et al. [24] developed a new methodology for building reconstruction based on structural and closed constraints. A surface optimisation scheme is adopted to enforce consistency between polygonal surfaces of the building and geometric structures. Zhang et al. [25] perform reconstruction of building models using unclassified LiDAR data, where the points are classified in the first two steps. In the final third step the building models are generated on the basis of the 2D topology of roof facets and estimated dominant directions. Shan et al. [26] introduced a framework for building model reconstruction where point cloud segmentation and building reconstruction are described as a minimisation problem of the corresponding energy functions. The initial segmentation is optimised by a global energy function that takes distances of LiDAR points to planes, spatial smoothness and the number of planes into account. After segmentation the reconstruction is performed by partitioning the building into volumetric cells, which is followed by determination of building surfaces and their topology. Building models are obtained with a global energy function, which is minimised using the min-cut theorem. Tarsha Kurdi et al. [27] developed a methodology that is performed in two steps. In the first step 2D building outlines are generated automatically on the basis of a neighbourhood matrix, while detecting inner roof plane boundaries as well. In the second step the 3D building models are generated, where, after fitting and refining of roof planes, the roof plane boundaries are transformed to 3D by the analysis of relationships between neighbouring planes. Later, they [28] improved the building outline modelling by filtering the point cloud with the bias of a Z-coordinate histogram.

A reverse, top-down based approach is typical for model-driven algorithms, where the building shapes are estimated by parametric fitting of shape candidates to the input point cloud. The result of fitting is a building model with a roof that is defined by parameters for the roof shape of the candidate that is best-fitted to the point cloud. This type of algorithms are usually faster than data-based, are easier to implement and can be used for datasets with a low point cloud density (1.2 pts/m^2 [29]). However, the main restriction of such approach is the limited collection of possible candidates that do not cover all possible roof shapes. In case a building's roof shape is in part or entirely different from every existing candidate in the library, the reconstruction will either fail, or generate a model with large errors. Model-driven algorithms differ mainly regarding the manner in which candidates are fitted to the point cloud. Poullis et al. [30] fitted geometric shapes while considering constraints found in architecture, Huang et al. [31] used statistical analysis, and Henn et al. [29] employed a modified RANSAC algorithm for this purpose.

The components of both approaches are combined in hybrid-driven algorithms in order to reduce the weaknesses of both. There are two main types of such reconstruction [5]. The first is by dividing buildings into smaller parts, based on the edges of the buildings' outlines, jump edges and roof ridges. Smaller parts can be fitted to the candidates of roof shapes with a higher success rate [5]. Building parts are then combined to obtain a whole building by 3D Boolean operations as a part of CSG (Constructive Solid Geometry) [32–34]. Kada and Wichmann [34] estimate building shapes with half-spaces, where the model is divided into smaller convex parts, which are then combined. The concave shape can be obtained through a correct division, which is only considered for a limited number of building layouts. The second type of reconstruction is by the RTG (Roof Topology Graph), which is established over basic geometric shapes and can contain additional information about edges. Verma et al. [35] establish RTG over segmented planar roof faces. They determine building's geometry by searching for subgraphs within RTG, that exist in the roof candidate database. More advanced RTG approaches were introduced by Xiaong et al. [36,37], where it was demonstrated, that it is possible to present the topology of any roof using a graph with minimal cycles in addition to nodes and edges [36]. Later, they improved the algorithm with a graph edit dictionary, which was used to reduce typical errors in RTG [37].

A new half-spaces based algorithm for building reconstruction from point clouds is introduced in this paper. In contrast to the related algorithms, which divide buildings' 2D outlines into smaller parts and then process them while taking only convex shapes into account, the proposed algorithm performs reconstruction without division, while also considering concave parts of the building's roof. Additionally, no assumptions about the building layout are made, which allows processing of buildings on a large-scale. This is achieved in two stages, where the input data is processed first to obtain the definition of each building's base model and the corresponding half-spaces. The second stage generates a building shape by performing 3D Boolean operations over the analysed half-spaces.

The remainder of the paper is divided into three Sections. The next Section describes both stages of the proposed algorithm in detail. The third Section presents the results over a large geographic area with the complementary discussion, and the final Section concludes this work.

2. Methodology

The proposed algorithm for building reconstruction is performed in two stages as shown in Figure 1. In the first stage, the input data is preprocessed, where the buildings' 2D outlines and the classified airborne LiDAR point cloud are used to obtain base building models and the corresponding definitions of half-spaces through segmentation. These serve as the input to the second stage, where the half-spaces are classified and processed by 3D Boolean operations to obtain the final building shape, where convex and concave parts of the final shape are considered. The model of each building is bounded by floor, exterior wall and roof faces. Only roofs without height jumps are considered. A height jump can be described as an edge of a roof's face, which height is different from a neighbouring roof's face. The following subsections describe both stages in detail.

2.1. Data Preprocessing

The input data to the proposed algorithm represent 2D building outlines and the airborne LiDAR point cloud. Both types of data are considered to be georeferenced using the same coordinate system and, therefore, aligned. The points of the LiDAR data that are classified as ground or building are considered in this work. In case the input LiDAR point cloud classification is missing or is of bad quality, there are various methods for classification available (for an overview, see [38]). As there are many possible classes of LiDAR points, a method that focuses on ground and building points [39] should be used for this purpose. Only building points that are located within the 2D building outline are considered for reconstructing the roof of the corresponding building model.

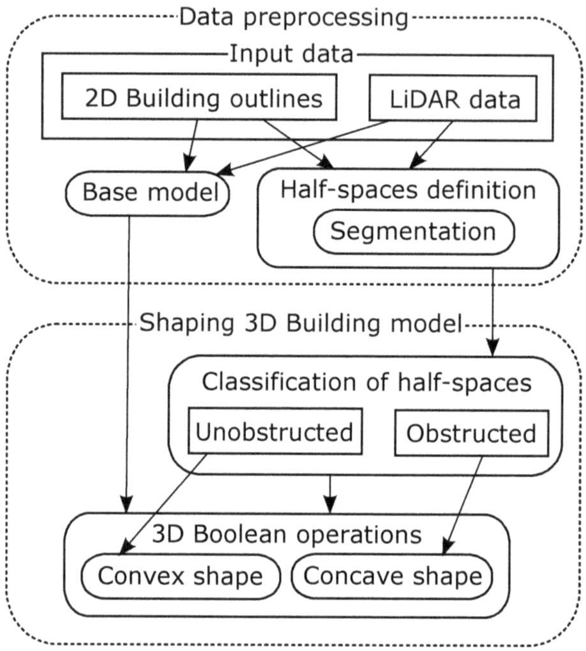

Figure 1. Workflow of the proposed algorithm.

2.1.1. Generating Base Models

The base model of each building is used in the second stage for estimating its final roof shape. It is bounded by the floor face, top bounding face and exterior wall faces that are obtained from the 2D building outline. Floor face g is determined by placing the polygon of the corresponding 2D building outline to the height of the lowest LiDAR ground point in the direct proximity of the building. The top bounding face g' is determined in the same way as g, only it is placed above the highest LiDAR building point, so it does not limit the final building model. Each exterior wall face s_i is defined by the points that lie on sides g and g' as shown in Figure 2. The base model M is given as a watertight model, bounded by a set of faces: $\{g, g', s_1, s_2, ...s_n\}$, where n is the number of sides of the building outline. An example of a generated base model from a 2D building outline is illustrated in Figure 2.

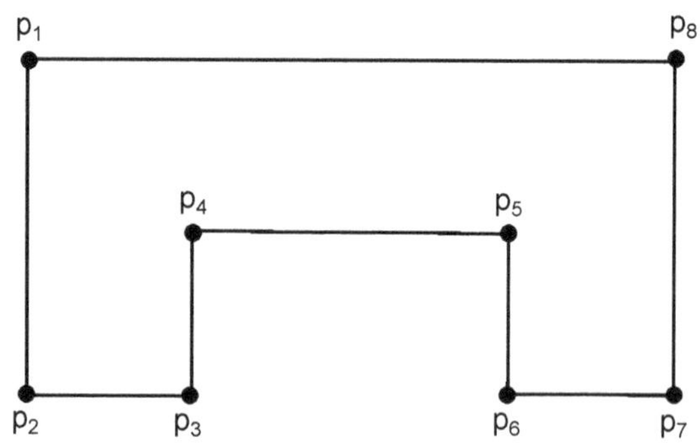

(a)

Figure 2. *Cont.*

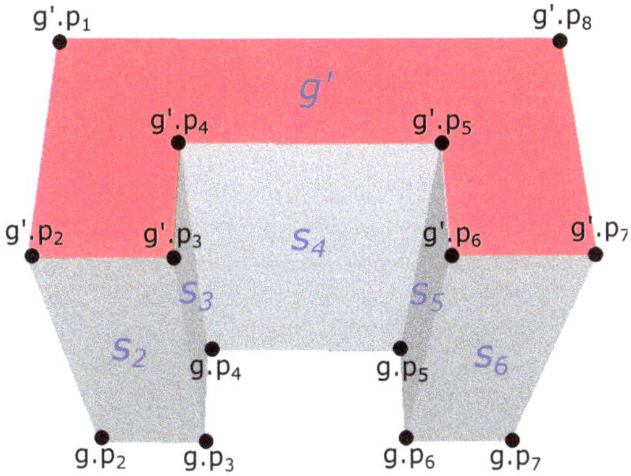

(**b**)

Figure 2. For a building outline (**a**) a base model is generated (**b**), where visible faces and points are marked.

2.1.2. Half-Spaces' Definition

A half-space is a core element for shaping the roof in this work. It is obtained from the LiDAR point cloud, which is first segmented into sets of LiDAR points that describe each individual planar roof face. For the segmentation of planar faces various segmentation methods are applicable. Considering the type of input LiDAR point cloud, segmentation methods that take the LiDAR point cloud density into account [40–43] are more suitable for this purpose. The reason for this is variable density of the input point cloud, which occurs due to different laser scanner technology and height of the aerial vehicle that performs point cloud acquisition. Segmentation that takes variable point cloud density into account should, therefore, be used for the best results. The overview of segmentation methods is out of the scope of this paper, for more information see [44]. An example segmentation of a LiDAR point cloud that belongs to the same building as the base model from Figure 2 is shown in Figure 3.

The roof of each building will be shaped by a set of half-spaces $H = \{H_i\}$. Half-space H_i is defined by the corresponding set of LiDAR points S_i and a plane P_i that is calculated from S_i. H_i is given as $xP_i.a + yP_i.b + zP_i.c + d > 0$. The definition of each half-space is obtained from the segmented LiDAR point cloud, where sets of segmented LiDAR points S_i that describe individual roof faces are taken. The plane P_i that is best-fitted to the corresponding set of LiDAR points is determined as $P_i = [a, b, c, d] = \text{LSqFit}(S_i)$, where $P_i.c > 0$ and LSqFit is a function that fits a plane to a set of LiDAR points by least-square fitting [45].

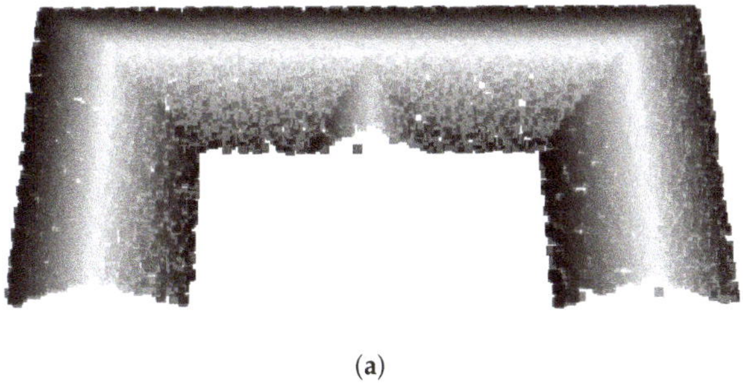

(**a**)

Figure 3. *Cont.*

(**b**)

Figure 3. A segmentation of a LiDAR point cloud, where the LiDAR points on top (**a**) are coloured by their height and the LiDAR points on the bottom (**b**) in regard to which roof face they correspond to. The point cloud belongs to the same building as the base model from Figure 2.

2.2. Shaping 3D Building Models by 3D Boolean Operations

During this stage, base models are shaped by 3D Boolean operations based on half-spaces to obtain the final 3D building models. The half-spaces of the previous stage are classified as obstructed or unobstructed for further analysis and shaping by slicing the base models.

Classification of Half-Spaces

Each half-space H_i, of which the corresponding set of LiDAR points S_i is partly contained within the base model M, is classified as unobstructed or obstructed in this step. A half-space is unobstructed, if it does not contain any LiDAR points of other half-spaces:

$$H_i \in \begin{cases} H^O, & \exists p \in S_j : p \in H_i, j \neq i \\ H^U, & \text{else} \end{cases}, \tag{1}$$

where $H^O \subset H$ is a set of obstructed half-spaces and $H^U \subset H$ a set of unobstructed half-spaces. As only planar roof faces are considered, small variations in faces' shapes can be neglected, and in Equation (1) the number of LiDAR points are reduced from S_j to $C_j \subset S_j$. The set C_j only contains points that are located on the convex hull of perpendicularly projected LiDAR points from S_j to the plane P_j. In case H_i contains a point from another half-space, it is classified as obstructed, which means that H_i is obstructed by a different half-space on a concave part of the roof. As a result, it cannot be used directly for shaping the building roof and needs to be analysed further.

2.3. Performing 3D Boolean Operations with Half-Spaces

Unobstructed half-spaces describe the convex shape of the building's roof as they are not obstructed by any other half-spaces. Therefore, they can be used directly for shaping the building model. This is performed by subtracting all corresponding open half-spaces from the building's base model M:

$$\forall H_i \in H^U : M = M \setminus H_i \tag{2}$$

Figure 4 shows an example of shaping the model with unobstructed half-spaces.

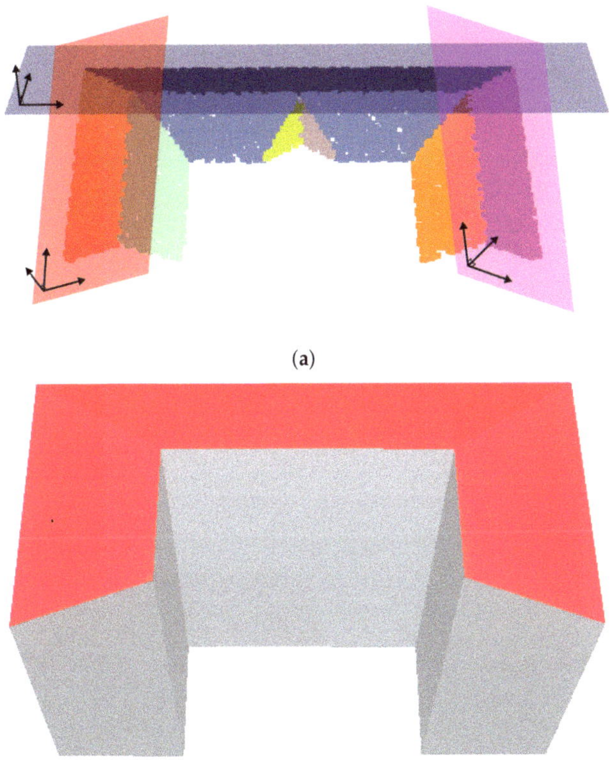

(**a**)

(**b**)

Figure 4. Illustration of shaping the building model with unobstructed half-spaces, where on the top (**a**) unobstructed half-spaces are shown, which are then subtracted from the base model (**b**) from Figure 2b.

Each obstructed half-space may not be directly useful for shaping the model. As an example, a pair of half-spaces' corresponding planes might be parallel at different heights or oriented in such a way that they intersect outside of the building model. Therefore, obstructed half-spaces are processed by determining the smallest slice s_k, if it exists, for each half-space $H_k \in H^O$. A slice s_k is determined by the intersection between H_k and all half-spaces $H_l \in H^O, k \neq l$, that are visible from H_k. Half-spaces H_k and H_l are visible, if a line between a point from C_k and a point from C_l exists that does not intersect with any other roof face, and if $\exists p \in C_k \land \exists p' \in C_l$, such that:

$$(\vec{p'} - \vec{p}) \cdot \widehat{P_k.n} > 0 \text{ and } (\vec{p} - \vec{p'}) \cdot \widehat{P_l.n} > 0, \tag{3}$$

where $\widehat{P_k.n}$ and $\widehat{P_l.n}$ are normalised normal vectors of the corresponding planes. A slice is valid if half-spaces that determine the slice intersect within the model. In case a slice is not valid, it is processed as empty. The final model M is obtained by subtracting all slices:

$$\forall H_k \in H^O : M = M \setminus s_k \tag{4}$$

Figure 5 illustrates shaping the building with obstructed half-spaces for the model from Figure 4b. The final building model is shown in Figure 5e.

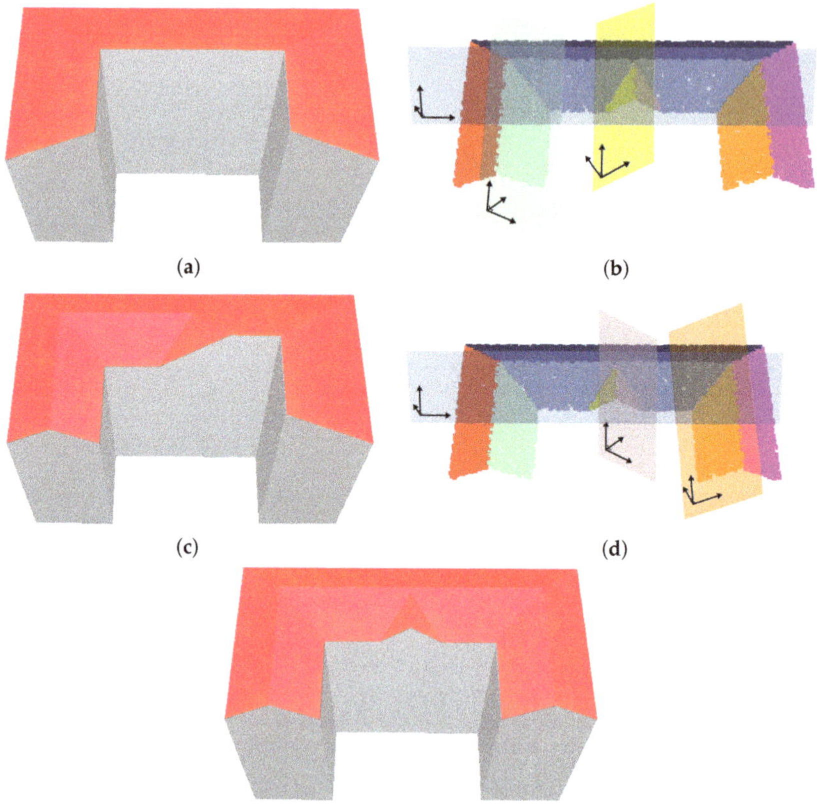

(**a**) (**b**) (**c**) (**d**) (**e**)

Figure 5. Illustration of shaping the model with obstructed half-spaces, where (**a**) shows the input model from the previous step, and (**b**) illustrates a slice that will be subtracted from the model. The slice is determined by three obstructed half-spaces that determine the smallest possible slice for one of the half-spaces. The result of the subtraction of the first slice is shown in (**c**). This is followed by the next slice (**d**), after which the final building model is obtained (**e**).

3. Results

To analyse the proposed algorithm's large-scale performance, it was applied on the basis of airborne LiDAR data and 2D buildings' outlines. The input data cover a 6.153 km^2 large geographic area of the city of Maribor, Slovenia (bounding box: 46°33′8.54301″N 15°37′29.23113″E, 46°34′13.15573″N 15°39′52.97941″E). The provided LiDAR data's average density is 11.3 pts/m^2 and the considered 14,227,123 building points and 31,420,386 ground points are shown in Figure 6. Buildings' outlines were obtained from a public spatial database maintained by The Surveying and Mapping Authority of the Republic of Slovenia. There were 5254 buildings' outlines contained entirely within the bounding box of the area.

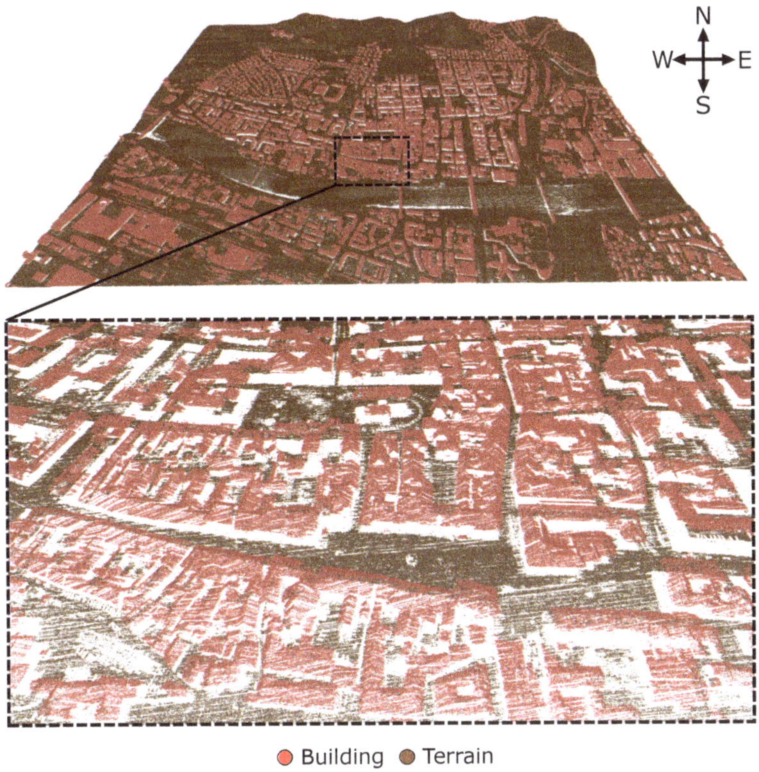

Figure 6. Input LiDAR point cloud.

First, the building points for each corresponding building's outline were segmented. For this, a graph-based segmentation [41] that takes point cloud density and local curvature of faces into account was selected, as roofs of buildings are not always completely flat. The segmentation establishes the initial topology over the point cloud as an undirected graph by the k-nearest neighbour approach, and is controlled by the local curvature t_θ and distance t_d thresholds. The building points from the input LiDAR point cloud for each corresponding building's outline were segmented using the settings given in Table 1. A minimum of 50 points were required for each segment ($t_{CC} = 50$), which was set to avoid too small roof faces. The resulting segmentation of building points from Figure 6 is shown in Figure 7.

Table 1. Parameters used for segmentation of the point cloud.

Parameter	Value
k	20
t_θ	2.0
t_{CC}	50
t_d [m]	2

The building models were generated next. A building model is deemed geometrically valid, if the building's floor face remains intact which, as such, does not necessarily imply an accurately reconstructed model. In some cases, especially with height jumps, the obtained building models were invalid. To achieve a higher validity rate the number of obstructed half-spaces was limited to 10 for cases with height jumps. In such cases, only unobstructed half-spaces were considered for shaping the model. An example of handling these cases is shown in Figure 8.

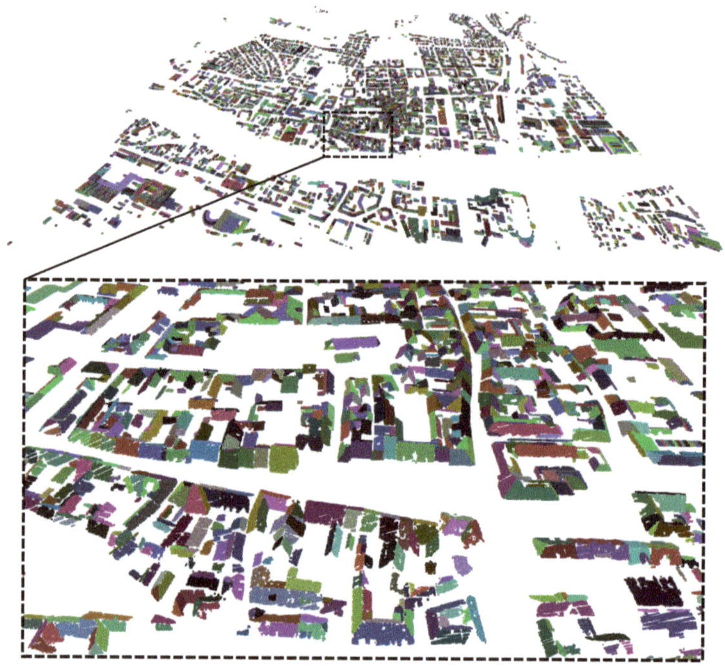

Figure 7. Segmented building points from Figure 6.

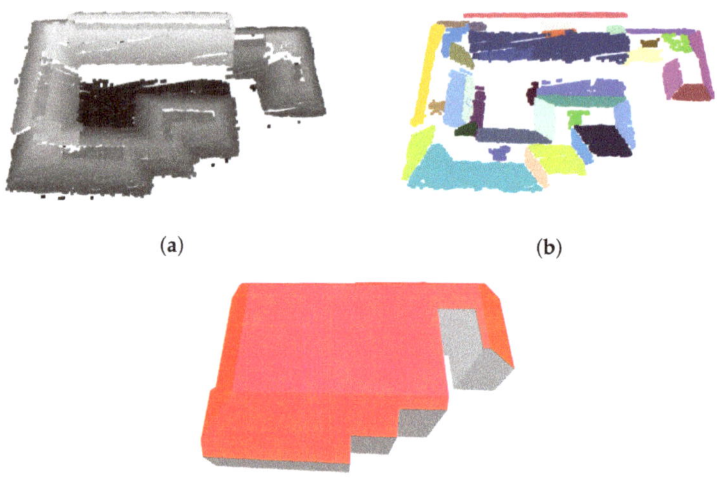

Figure 8. Illustration of shaping a large building with height jumps and too many obstructed half-spaces where from (**a**) the input point cloud segments of points (**b**) were obtained and only unobstructed half-spaces were considered for the reconstruction of the building (**c**).

The reconstruction produced a geometrically valid model for 4817 building outlines in 92% of cases. The resulting buildings' reconstruction is shown in Figure 9.

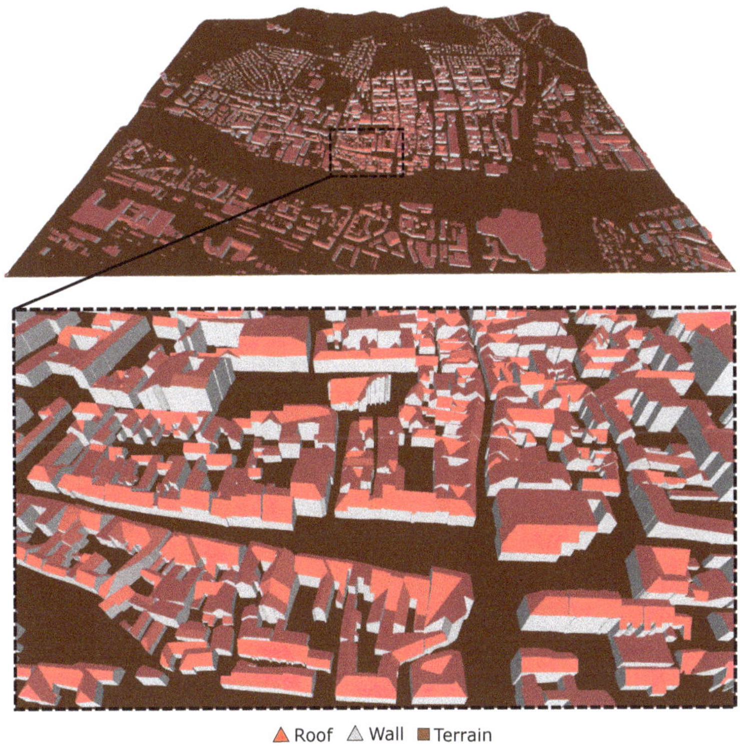

Figure 9. Reconstructed building models for the input point cloud from Figure 6.

In practice, many buildings' roofs contain height jumps and, as they are not considered by the proposed method, several geometrically invalid models are to be expected.

Next, we examined various cases of buildings from the reconstructed city model with roofs of different types without height jumps, which are shown in Figure 10.

Figure 10a shows a relatively simple case that is described by eight half-spaces of a single building, of which five are obstructed. The next case from Figure 10b shows an example of two terraced buildings, where the same roof faces are shared between buildings. In such cases, where there are no geometrical features available to determine the border between two buildings, it is crucial to have buildings' outlines already available. The building shown in Figure 10c has a roof with four ridges at different heights or orientations, and a horizontal part of the roof, which were incorporated into the model correctly from 17 half-spaces. The following case, shown in Figure 10d, has a flat face at the highest point of the building model. The building is described by 17 segments of points, where gaps can be observed in the shape of some of them. The gaps occur due to small objects located at the roof that are either too small or curved. The final case, Figure 10e, illustrates the reconstruction of a building with a complex roof that is described by 30 half-spaces, of which only one (top horizontal half-space) is unobstructed. There were at least 14 slices performed to obtain the correct shape, including slices determined by four half-spaces. The presented cases from Figure 10 demonstrate the ability of the proposed algorithm for the reconstruction of buildings of various layouts.

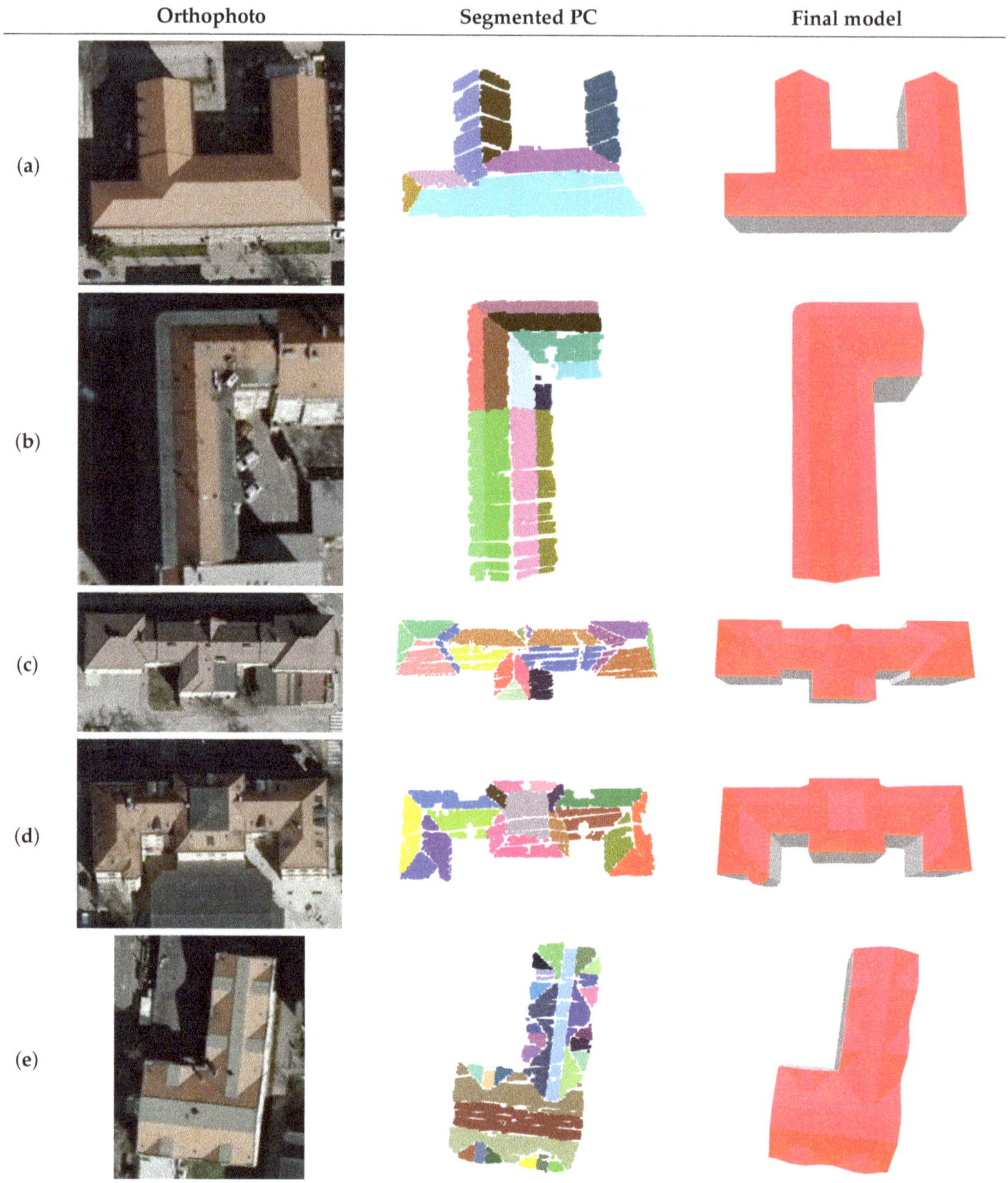

Figure 10. Illustrations of shaping several cases of buildings of various layouts with roofs without height jumps, where for the buildings (**left**) segments of points (**middle**) were obtained from the input point cloud for the reconstruction of buildings (**right**).

Validation

The proposed methodology was validated with the International Society for Photogrammetry and Remote Sensing (ISPRS) benchmark dataset, specifically the third area of the data captured over Vaihingen, Germany [46,47]. Building outlines and LiDAR data, where building points were classified manually and the density is 4 pts/m^2, were used from the benchmark as input data. The output building models of the proposed algorithm were compared by height difference with the ground truth data from the benchmark, as shown in Figure 11. Ground truth data is presented as building roof faces, given as 3D polylines. The height difference was estimated over a regular grid with 10 cm resolution. Statistics for the comparison is given in Table 2, where RMSE (Root Mean Square Error) was estimated over height differences, completeness specifies the proportion of the building area covered by geometrically valid building models, and $e_{0.5}$ designates the proportion of the building area, where the height difference is lower than 0.5 m.

Table 2. Validation of the obtained results from the comparison with ISPRS benchmark data.

Metric	Value
RMSE [m]	1.31
Completeness [%]	98.9
$e_{0.5}$ [%]	79.4

The majority of the difference manifested over areas, where height jumps are present. Some difference can be observed over parts of roofs with small elements (e.g., chimneys or small dormers) or parts of the dataset, where the point cloud is too sparse.

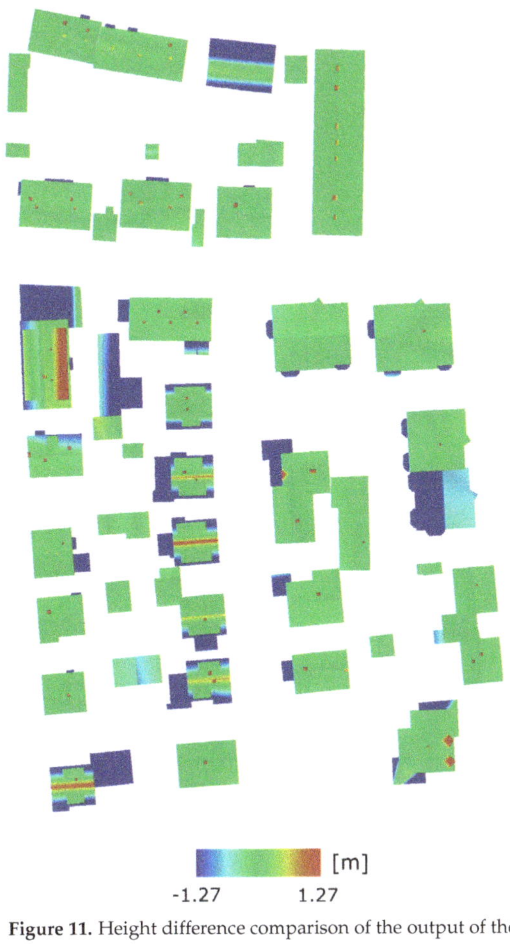

Figure 11. Height difference comparison of the output of the proposed algorithm with the ground truth data of the ISPRS benchmark.

4. Discussion

The main challenges of applying the proposed algorithm to large-scale datasets include the presence of building roofs with height jumps and the availability and quality of the input data. Building roofs with height jumps are a common occurrence on a large scale and, as the proposed method does not consider them, had to be taken into account to obtain as complete city model as possible. For cases with height jumps, shaping the building model with obstructed half-spaces caused many building models to become invalid, as the subtraction of some slices affected the floor face of the building model. This was particularly apparent for buildings with many small obstructed half-spaces, where half-spaces may be oriented towards a part of a roof, where only unobstructed half-spaces are present. In such case a slice can cut through a part of a building model that would otherwise remain intact. For this reason the number of obstructed half-spaces was limited to keep a validity rate over 90%.

The availability of the input data represents an important drawback that should be taken into account. Building outlines could be, to an extent, generated from the LiDAR data [28], however some drop in accuracy can not be avoided. On the other hand, LiDAR data is essential and may not be available at a selected location. In such cases, the point cloud data could be obtained using an aerial or UAV (Unmanned Aerial Vehicle) photogrammetric survey. Even though the proposed algorithm was developed with aerial LiDAR data in mind, it could be applied to any 3D point cloud. Another aspect of the input data to consider is quality. It includes density and accuracy of the point cloud data, which largely depend on the laser scanner quality. The publicly available LiDAR data is often of low density, which means that smaller roof faces are much harder to detect. The required density for the correct reconstruction depends on the minimum size of roof faces that are desired to be included in the generated building models. In theory, there need to be at least 3 segmented points on a roof face to determine the corresponding plane. However, due to variations in measurement accuracy that affect plane orientation and the fact that as the shape of the roof face is important for further analysis as well, a higher point count is beneficial. In our experience, the minimal point cloud density, using the selected segmentation algorithm for faces larger than 10 m^2, was 1.5 pts/m^2. Building outline datasets are often acquired manually, which means that human error can be present as well. Another option is to use cartographic outline datasets from public databases, as was done in this work. However, it should be noted that existing outline datasets can be misaligned, inconsistent, or out of date. Some buildings might have been demolished, changed or replaced, and outlines of new buildings could be missing. Any discrepancies in the input data are presented directly in building models.

The comparison with the Vaihingen dataset has shown that, apart from the acknowledged lack of height jump processing, the proposed algorithm demonstrated satisfactory performance. This is confirmed by a large proportion of the building area, where the height error was under 0.5 m. When comparing the results with related work [47] in terms of RMSE, the proposed method yields higher value, which is largely attributed to height jumps, where every error accumulates substantially over a large surface. In addition, as completeness was reported higher than in related work at nearly 100%, which comes as a result of a reliable building reconstruction and using building outlines as input, the RMSE for the proposed algorithm accumulates extensively over a sparse part of the dataset. Building outlines provided additional completeness over sparse part of the dataset. In such cases, related work, in contrast to the presented algorithm, did not generate the building model over the entire building outline due to lack of data.

Moreover, as shown in Figure 10, it can be observed how a LiDAR point cloud can contain relatively large empty spaces between scan lines in practice. It is important to keep this in mind when choosing the segmentation method and the appropriate parameters. Apart from the segmentation of point clouds, no additional parameters are required for controlling the reconstruction. This simplifies the use of the proposed algorithm significantly, especially as the segmented point cloud could also be provided as an input to the algorithm.

5. Conclusions

This work presents a novel algorithm, based on half-spaces, for 3D building reconstruction that processes airborne LiDAR data and buildings' outlines to generate building models. It performs in two stages, where the input data is preprocessed first to obtain buildings' base models and the corresponding half-spaces. In the final stage, 3D building models are finalised by shaping their roof using 3D Boolean operations over the analysed half-spaces.

In experiments, the presented method has shown promising reconstruction performance for the considered type of buildings. As in practice, on a large scale, there are many building roofs with height jumps, some constraints were required to obtain a more complete 3D city model. For a more accurate reconstruction, the height jumps should be considered and incorporated in future work, which could be explored by splitting the buildings' outlines. Another possible improvement could be the integration of curved faces support, where special attention should be given to classification and limiting the reach of a curved face when using 3D Boolean operations to shape the building model. Moreover, as the segmentation performance greatly affects the algorithm's output, the impact of various segmentation algorithms with different parameter settings could be investigated as well.

The proposed algorithm's large-scale applicability is highly beneficial for urban simulations, or as a component of various urban analytical processes, especially those that consider environmental impact, which is a growing global concern. Apart from the segmentation part, for which any appropriate segmentation algorithm can be used, the proposed algorithm is parameter-free, which simplifies its use strongly and enhances adoption potential.

Author Contributions: Conceptualization, M.B.; methodology, M.B. and N.L.; software, M.B. and N.L.; investigation, M.B.; formal analysis, M.B., N.L. and B.Ž; data curation, M.B.; writing—original draft preparation, M.B.; writing—review and editing, M.B., N.L. and B.Ž; visualization, M.B. and N.L.; supervision, N.L.; project administration, B.Ž.; funding acquisition, B.Ž. All authors have read and agreed to the published version of the manuscript.

Funding: The authors acknowledge the financial support from the Slovenian Research Agency (Research Funding No. P2-0041 and Research Project No. L7-2633).

Acknowledgments: Thanks to the Slovenian Environment Agency and the German Society for Photogrammetry, Remote Sensing and Geoinformation (DGPF) for providing LiDAR data. Moreover, the authors thank The Surveying and Mapping Authority of the Republic of Slovenia for buildings' data.

Conflicts of Interest: The authors declare no conflict of interest.

References

1. Biljecki, F.; Stoter, J.; Ledoux, H.; Zlatanova, S.; Çöltekin, A. Applications of 3D City Models: State of the Art Review. *ISPRS Int. J. Geo-Inf.* **2015**, *4*, 2842–2889. [CrossRef]
2. Bizjak, M.; Žalik, B.; Štumberger, G.; Lukač, N. Large-scale estimation of buildings' thermal load using LiDAR data. *Energy Build.* **2021**, *231*, 110626. [CrossRef]
3. Ali, U.; Shamsi, M.H.; Hoare, C.; Mangina, E.; O'Donnell, J. Review of urban building energy modeling (UBEM) approaches, methods and tools using qualitative and quantitative analysis. *Energy Build.* **2021**, *246*, 111073. [CrossRef]
4. Wang, R. 3D building modeling using images and LiDAR: A review. *Int. J. Image Data Fusion* **2013**, *4*, 273–292. [CrossRef]
5. Wang, R.; Peethambaran, J.; Dong, C. LiDAR Point Clouds to 3D Urban Models: A Review. *IEEE J. Sel. Top. Appl. Earth Obs. Remote Sens.* **2018**, *11*, 606–627. [CrossRef]
6. Biljecki, F.; Heuvelink, G.B.; Ledoux, H.; Stoter, J. The effect of acquisition error and level of detail on the accuracy of spatial analyses. *Cartogr. Geogr. Inf. Sci.* **2018**, *45*, 156–176. [CrossRef]
7. Rottensteiner, F.; Sohn, G.; Gerke, M.; Wegner, J.D.; Breitkopf, U.; Jung, J. Results of the ISPRS benchmark on urban object detection and 3D building reconstruction. *ISPRS J. Photogramm. Remote Sens.* **2014**, *93*, 256–271. [CrossRef]

8. Tomljenovic, I.; Höfle, B.; Tiede, D.; Blaschke, T. Building extraction from Airborne Laser Scanning data: An analysis of the state of the art. *Remote Sens.* **2015**, *7*, 3826–3862. [CrossRef]
9. Axelsson, M.; Soderman, U.; Berg, A.; Lithen, T. Roof Type Classification Using Deep Convolutional Neural Networks on Low Resolution Photogrammetric Point Clouds From Aerial Imagery. In Proceedings of the 2018 IEEE International Conference on Acoustics, Speech and Signal Processing (ICASSP), Calgary, AB, Canada, 15–20 April 2018; pp. 1293–1297.
10. Zhang, L.; Zhang, L. Deep Learning-Based Classification and Reconstruction of Residential Scenes From Large-Scale Point Clouds. *IEEE Trans. Geosci. Remote Sens.* **2018**, *56*, 1887–1897. [CrossRef]
11. Yu, D.; Ji, S.; Liu, J.; Wei, S. Automatic 3D building reconstruction from multi-view aerial images with deep learning. *ISPRS J. Photogramm. Remote Sens.* **2021**, *171*, 155–170. [CrossRef]
12. Wichmann, A.; Agoub, A.; Schmidt, V.; Kada, M. RoofN3D: A Database for 3D Building Reconstruction with Deep Learning. *Photogramm. Eng. Remote Sens.* **2019**, *85*, 435–443. [CrossRef]
13. Vosselman, G. Building Reconstruction Using Planar Faces In Very High Density Height Data. *Int. Arch. Photogramm. Remote Sens.* **1999**, *32*, 87–92.
14. Dorninger, P.; Pfeifer, N. A Comprehensive Automated 3D Approach for Building Extraction, Reconstruction, and Regularization from Airborne Laser Scanning Point Clouds. *Sensors* **2008**, *8*, 7323–7343. [CrossRef] [PubMed]
15. Elberink, S.O.; Vosselman, G. Building reconstruction by target based graph matching on incomplete laser data: Analysis and limitations. *Sensors* **2009**, *9*, 6101–6118. [CrossRef]
16. Sampath, A.; Shan, J. Segmentation and reconstruction of polyhedral building roofs from aerial LiDAR point clouds. *IEEE Trans. Geosci. Remote Sens.* **2010**, *48*, 1554–1567. [CrossRef]
17. Chen, Y.; Cheng, L.; Li, M.; Wang, J.; Tong, L.; Yang, K. Multiscale grid method for detection and reconstruction of building roofs from airborne LiDAR data. *IEEE J. Sel. Top. Appl. Earth Obs. Remote Sens.* **2014**, *7*, 4081–4094. [CrossRef]
18. Chen, D.; Wang, R.; Peethambaran, J. Topologically Aware Building Rooftop Reconstruction From Airborne Laser Scanning Point Clouds. *IEEE Trans. Geosci. Remote Sens.* **2017**, *55*, 7032–7052. [CrossRef]
19. Lafarge, F.; Mallet, C. Creating large-scale city models from 3D-point clouds: A robust approach with hybrid representation. *Int. J. Comput. Vis.* **2012**, *99*, 69–85. [CrossRef]
20. Zhou, Q.Y.; Neumann, U. 2.5D building modeling by discovering global regularities. In Proceedings of the IEEE Computer Society Conference on Computer Vision and Pattern Recognition, Providence, RI, USA, 16–21 June 2012; pp. 326–333. [CrossRef]
21. Chen, D.; Zhang, L.; Mathiopoulos, P.T.; Huang, X. A methodology for automated segmentation and reconstruction of urban 3-D buildings from ALS point clouds. *IEEE J. Sel. Top. Appl. Earth Obs. Remote Sens.* **2014**, *7*, 4199–4217. [CrossRef]
22. Vosselman, G.; Dijkman, S. 3D building model reconstruction from point clouds and ground plans. *Int. Arch. Photogramm. Remote Sens.* **2001**, *34*, 37–44.
23. Li, M.; Rottensteiner, F.; Heipke, C. Modelling of buildings from aerial LiDAR point clouds using TINs and label maps. *ISPRS J. Photogramm. Remote Sens.* **2019**, *154*, 127–138. [CrossRef]
24. Wang, S.; Cai, G.; Cheng, M.; Marcato, J., Jr.; Huang, S.; Wang, Z.; Su, S.; Li, J. Robust 3D reconstruction of building surfaces from point clouds based on structural and closed constraints. *ISPRS J. Photogramm. Remote Sens.* **2020**, *170*, 29–44. [CrossRef]
25. Zhang, K.; Yan, J.; Chen, S.C. A Framework for Automated Construction of Building Models from Airborne LiDAR Measurements. In *Topographic Laser Ranging and Scanning: Principles and Processing*, 2nd ed.; Shan, J., Toth, C., Eds.; CRC Press: Boca Raton, FL, USA, 2018; pp. 563–585.
26. Shan, J.; Yan, J.; Jiang, W. Global Solutions to Building Segmentation and Reconstruction. In *Topographic Laser Ranging and Scanning: Principles and Processing*, 2nd ed.; Shan, J., Toth, C., Eds.; CRC Press: Boca Raton, FL, USA, 2018; pp. 459–484.
27. Tarsha Kurdi, F.; Awrangjeb, M.; Liew, A.W.C. Automated Building Footprint and 3D Building Model Generation from Lidar Point Cloud Data. In Proceedings of the 2019 Digital Image Computing Techniques and Applications (DICTA), Perth, Australia, 2–4 December 2019; pp. 75–82.
28. Tarsha Kurdi, F.; Awrangjeb, M.; Munir, N. Automatic filtering and 2D modeling of airborne laser scanning building point cloud. *Trans. GIS* **2021**, *25*, 164–188. [CrossRef]
29. Henn, A.; Gröger, G.; Stroh, V.; Plümer, L. Model driven reconstruction of roofs from sparse LIDAR point clouds. *ISPRS J. Photogramm. Remote Sens.* **2013**, *76*, 17–29. [CrossRef]
30. Poullis, C.; You, S. Photorealistic large-scale Urban city model reconstruction. *IEEE Trans. Vis. Comput. Graph.* **2009**, *15*, 654–669. [CrossRef] [PubMed]
31. Huang, H.; Brenner, C.; Sester, M. A generative statistical approach to automatic 3D building roof reconstruction from laser scanning data. *ISPRS J. Photogramm. Remote Sens.* **2013**, *79*, 29–43. [CrossRef]
32. Haala, N.; Brenner, C. Virtual city models from laser altimeter and 2D map data. *Photogramm. Eng. Remote Sens.* **1999**, *65*, 787–795.
33. Kada, M.; McKinley, L. 3D Building Reconstruction from LIDAR based on a Cell Decomposition Approach. In Proceedings of the CMRT09: Object Extraction for 3D City Models, Road Databases and Traffic Monitoring—Concepts, Algorithms and Evaluation, Paris, France, 3–4 September 2009; Volume XXXVIII, pp. 47–52.
34. Kada, M.; Wichmann, A. Feature-Driven 3d Building Modeling Using Planar Halfspaces. *ISPRS Ann. Photogramm. Remote Sens. Spat. Inf. Sci.* **2013**, *II-3/W3*, 37–42. [CrossRef]

35. Verma, V.; Kumar, R.; Hsu, S. 3D Building Detection and Modeling from Aerial LiDAR Data. In Proceedings of the 2006 IEEE Computer Society Conference on Computer Vision and Pattern Recognition, New York, NY, USA, 17–22 June 2006; Volume 2, pp. 2213–2220. [CrossRef]
36. Xiong, B.; Oude Elberink, S.; Vosselman, G. A graph edit dictionary for correcting errors in roof topology graphs reconstructed from point clouds. *ISPRS J. Photogramm. Remote Sens.* **2014**, *93*, 227–242. [CrossRef]
37. Xiong, B.; Jancosek, M.; Oude Elberink, S.; Vosselman, G. Flexible building primitives for 3D building modeling. *ISPRS J. Photogramm. Remote Sens.* **2015**, *101*, 275–290. [CrossRef]
38. Yan, W.Y.; Shaker, A.; El-Ashmawy, N. Urban land cover classification using airborne LiDAR data: A review. *Remote Sens. Environ.* **2015**, *158*, 295–310. [CrossRef]
39. Mongus, D.; Lukač, N.; Žalik, B. Ground and building extraction from LiDAR data based on differential morphological profiles and locally fitted surfaces. *ISPRS J. Photogramm. Remote Sens.* **2014**, *93*, 145–156. [CrossRef]
40. Rabbani, T.; den Heuvel, F.; Vosselmann, G. Segmentation of point clouds using smoothness constraint. *Int. Arch. Photogramm. Remote Sens. Spat. Inf. Sci.* **2006**, *36*, 248–253.
41. Bizjak, M. The segmentation of a point cloud using locally fitted surfaces. In Proceedings of the 18th Mediterranean Electrotechnical Conference: Intelligent and Efficient Technologis and Services for the Citizen, MELECON, Limassol, Cyprus, 18–20 April 2016; pp. 1–6. [CrossRef]
42. Czerniawski, T.; Sankaran, B.; Nahangi, M.; Haas, C.; Leite, F. 6D DBSCAN-based segmentation of building point clouds for planar object classification. *Autom. Constr.* **2018**, *88*, 44–58. [CrossRef]
43. Li, L.; Yao, J.; Tu, J.; Liu, X.; Li, Y.; Guo, L. Roof plane segmentation from airborne LiDAR data using hierarchical clustering and boundary relabeling. *Remote Sens.* **2020**, *12*, 1363. [CrossRef]
44. Nguyen, A.; Le, B. 3D point cloud segmentation: A survey. In Proceedings of the 6th IEEE Conference on Robotics, Automation and Mechatronics (RAM), Manila, Philippines, 12–15 November 2013; pp. 225–230.
45. Bevington, P.R.; Robinson, D.K. *Data Reduction and Error Analysis for the Physical Sciences*, 3rd ed.; McGraw–Hill: New York, NY, USA, 2002.
46. Cramer, M. The DGPF-Test on Digital Airborne Camera Evaluation Overview and Test Design. *Photogramm.—Fernerkund.—Geoinf.* **2010**, *2010*, 73–82. [CrossRef]
47. Rottensteiner, F.; Sohn, G.; Jung, J.; Gerke, M.; Baillard, C.; Benitez, S.; Breitkopf, U. The ISPRS benchmark on urban object classification and 3D building reconstruction. *ISPRS Ann. Photogramm. Remote Sens. Spat. Inf. Sci.* **2012**, *I-3*, 293–298. [CrossRef]

remote sensing MDPI

Article

Progress Guidance Representation for Robust Interactive Extraction of Buildings from Remotely Sensed Images

Zhen Shu [1], Xiangyun Hu [1,2,*] and Hengming Dai [1]

1 School of Remote Sensing and Information Engineering, Wuhan University, Wuhan 430079, China; zhenshu1994@whu.edu.cn (Z.S.); hengmingdai@whu.edu.cn (H.D.)
2 Institute of Artificial Intelligence in Geomatics, Wuhan University, Wuhan 430079, China
* Correspondence: huxy@whu.edu.cn; Tel.: +86-27-6877-1528; Fax: +86-27-6877-8086

Abstract: Accurate building extraction from remotely sensed images is essential for topographic mapping, cadastral surveying and many other applications. Fully automatic segmentation methods still remain a great challenge due to the poor generalization ability and the inaccurate segmentation results. In this work, we are committed to robust click-based interactive building extraction in remote sensing imagery. We argue that stability is vital to an interactive segmentation system, and we observe that the distance of the newly added click to the boundaries of the previous segmentation mask contains progress guidance information of the interactive segmentation process. To promote the robustness of the interactive segmentation, we exploit this information with the previous segmentation mask, positive and negative clicks to form a progress guidance map, and feed it to a convolutional neural network (CNN) with the original RGB image, we name the network as PGR-Net. In addition, an adaptive zoom-in strategy and an iterative training scheme are proposed to further promote the stability of PGR-Net. Compared with the latest methods FCA and f-BRS, the proposed PGR-Net basically requires 1–2 fewer clicks to achieve the same segmentation results. Comprehensive experiments have demonstrated that the PGR-Net outperforms related state-of-the-art methods on five natural image datasets and three building datasets of remote sensing images.

Keywords: building extraction; interactive segmentation network; deep learning; iterative training; remote sensing images

Citation: Shu, Z.; Hu, X.; Dai, H. Progress Guidance Representation for Robust Interactive Extraction of Buildings from Remotely Sensed Images. *Remote Sens.* **2021**, *13*, 5111. https://doi.org/10.3390/rs13245111

Academic Editor: Devrim Akca

Received: 16 November 2021
Accepted: 13 December 2021
Published: 16 December 2021

1. Introduction

The extraction of buildings from remotely sensed images is essential for topographic mapping and urban planning. Although automatic building extraction methods have been investigated for decades, they are still difficult to achieve sufficient performance to meet the requirements for fully automated use. Conventional methods mainly exploit empirically designed features to recognize buildings, such as color, texture, and shadow, etc. Due to the limitation of hand-crafted features, these methods usually produce frustrating results in complex scenes. In recent years, with the development of deep learning techniques, building segmentation performance has been lifted a lot by various deep convolutional neural networks (DCNNs), such as U-Net [1], SegNet [2], DeepLabV3+ [3]. These networks take RGB images as input and directly output the probability map of buildings in the image. By learning from massive amounts of training samples, they can achieve performance far beyond conventional methods. However, these CNN-based automatic building extraction algorithms are suffering from poor generalization ability, which means a well-trained network can only make good predictions on images with a similar distribution of the training data. Furthermore, the acquirement of pixel-wised annotated data itself is time-consuming and expensive, and the accuracy of the segmentation results is also far from the requirement of actual use.

Under these circumstances, the study of interactive building extraction is of great value and importance. Fully-automatic extraction methods are characterized by the prior constraints, such as shape and appearance of the buildings, and the output results are automatically generated and presented in front of the users before further processing. The main difference between fully-automatic and semi-automatic building extraction methods is that the latter can accept human supervision as additional input to ensure the quality of the output results. A good interactive segmentation method is always aimed at reducing the user effort. Actually, there was a significant amount of research before the advent of deep learning techniques. An earlier well-known method is intelligent scissors [4], which focuses on the boundary property for object extraction. Afterward, a graph model-based interactive image segmentation algorithm was studied a significant amount. Boykov and Jolly [5] utilize scribbles to estimate the probability of the foreground/background of the target object. The task is formulated as a graph partition problem and solved by a min-cut/max-flow algorithm [6]. Veksler [7] integrates a star-convexity shape into a graph-cut segmentation, and Gulshan et al. [8] further improve the results with multiple stars and geodesics distances. Rother et al. [9] take the bounding box as input and utilize a Gaussian mixture model for foreground and background prediction. Yu et al. [10] use a Markov Random Field (MRF) to segment objects with loosely bounded boxes. In addition, Grady [11] uses the label of the seed firstly reached by a random walker to mark unlabeled pixels. Limited by the capacity of these hand-crafted features, the amount of user inputs are still required in complex scenarios, such as low contrast and poor illumination.

Recently, deep learning techniques have also been applied to interactive object extraction and achieved great success in the field of computer vision. Xu et al. [12] first propose a CNN-based model for interactive segmentation and devise a click simulation strategy for the training of the network. They transform the user-provided clicks into Euclidean distance maps and concatenate them with RGB images as the input to the network. This work is then extended and improved by many other works in different aspects. For example, Mahadevan et al. [13] introduce an iterative way to improve the heuristic sampling strategy during the training stage. Li et al. [14] select the optimal result among multiple diverse solutions to reduce the user efforts. Majumder and Yao [15] combine superpixels and class-independent object proposals with user-provided clicks to generate more informative guidance maps for the performance promotion of interactive segmentation systems. Jang and Kim [16] propose a backpropagating refinement scheme (BRS) to guarantee the correct prediction in user-annotated positions. Sofiiuk et al. [17] further propose a feature backpropagating refinement scheme (f-BRS) to alleviate the computational burden of the forward and backward pass. Lin et al. [18] consider that the first click contains location guidance of the main body of the target object and put forward a first click attention module to make better use of the first click.

Generally speaking, user inputs are typically given as clicks [12–19], scribbles [5] and bounding boxes [9,10]. Compared with the other two modes of interactions, the click-based way is relatively simple and can reduce the burden of the annotators. In this work, we focus on the study of click-based interactive extraction of buildings. In this setting, users sequentially provide positive points on the foreground or negative points on the background to interact with the model until the segmented results are satisfied. To feed the interaction information to the CNN-based interactive segmentation model, an important issue is how to encode the user-provided clicks. Most of the existing methods follow [12] to simulate positive and negative clicks and transform them into a two-channel guidance map by Euclidean distance [12,14,16,17,19] or gaussian masks [18], and we utilize a satellite image in CrowdAI [20] dataset to demonstrate this in Figure 1. These two encoding methods are simple and straightforward representations of user-provided click points, which are convenient for the simulation of the training samples and the batch training of the network. In addition, they are also flexible in dealing with objects of multi-parts or weird shapes in natural scenes. However, such two-channel representations lack enough information to make an interactive segmentation network maintain good stability. In Figure 2, we present an image segmentation example of a baseline network in a natural scene that only utilizes guidance maps transformed from the user-provided clicks by Euclidean distance. We can see that the building is almost perfectly segmented after the first click. However, after the second click is added for further mask refinement, the segmentation result is severely degraded. It is because the predictions are independent at each step both in the training and inference stages. The guidance map treats all clicks independently, the network basically predicts the mask of the target object by the distribution of the positive and negative click points. Furthermore, the Gaussian masks of clicks or Euclidean distance maps are a kind of "weak" guidance, which is not conducive to the stability of mask refinement.

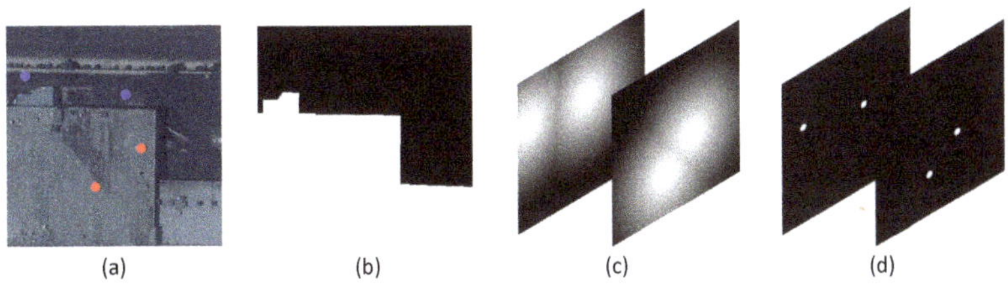

Figure 1. (**a**) Original image and corresponding positive and negative clicks. (**b**) The gt mask, (**c**) guidance maps transformed by Euclidean distance transform. (**d**) guidance maps transformed by Gaussians.

(**a**) 1 click (**b**) 2 clicks (**c**) gt mask

Figure 2. A failure case of existing methods that utilize guidance maps transformed from the user-provided clicks by Euclidean distance.

In our opinion, an interactive segmentation process is a coarse-to-fine process, which is carried out with two objectives, the fast estimation of the scale of the target object and the continuous refinement of the predicted masks. These two objectives conflict in the extent of the change to the previous segmentation mask. The former focuses on flexibility, and the latter emphasizes stability. A robust interactive segmentation system should progressively improve the segmentation mask with as little oscillation as possible because a false prediction will require additional clicks to revise. We argue that existing guidance maps are flexible representations for dealing with complex objects in natural scenes. However, compared with objects (multi-parts, elongated) in natural scenarios, buildings in overhead remote sensing images tend to have relatively regular shapes. For these "easy" buildings, the stability of the interactive segmentation process is critical to the improvement of performance. Motivated by the above circumstances, in this work, we focus on developing a robust interactive building extraction method based on CNN. To promote the stability of the interactive segmentation network, we firstly combine the previous segmentation map, which is considered as a kind of "strong" guidance, with existing distance-based guidance maps. In addition, we observe that annotators often tend to click around the center of the largest misclassified region. Thus, in most cases, the distance of the newly added click to the boundary of the previous segmentation mask can provide instructive progress information of the interactive segmentation process. This distance can be easily obtained during the inference stage, and we call this distance the indication distance. We make use of this distance and transform it into another guidance map to increase the stability of the interactive segmentation model. Moreover, we propose an adaptive zoom-in strategy and an iterative training strategy for further performance promotion of the algorithm. Comprehensive experiments show that our method is effective in both natural scenes and remote sensing images. Especially, compared with the latest state-of-the-art methods, FCA [18] and f-BRS [17], our approach basically requires 1–2 fewer clicks to achieve the same segmentation results on three building datasets of remote sensing images, which significantly reduces the workload of users. Furthermore, we propose an additional metric for the further evaluation of the robustness of the proposed interactive segmentation network, and the experimental results demonstrate that our approach yields better stability over other methods.

Our contributions can be summarized as follows:

- We analyze the benefits of a segmentation mask to improve the stability of network prediction, and we combine it with existing distance-based guidance maps to promote the performance of the interactive segmentation system.
- We also propose an adaptive zoom-in scheme during the inference phase, and we propose an iterative training strategy for the training of an interactive segmentation network.
- We achieve state-of-the-art performance on five widely used natural image datasets and three building datasets. In particular, our approach significantly reduces the user interactions in the interactive extraction of buildings. Comprehensive experiments demonstrate the good robustness of our algorithm.

The remainder of this article is arranged as follows: Section 2 describes details of the proposed method; the corresponding experimental assessment and discussion of the obtained results are shown in Sections 3 and 4, respectively; Section 5 presents our concluding remarks.

2. Materials and Methods

In this section, we provide the details of the proposed algorithm for the interactive extraction of buildings in remote sensing images. Firstly, we introduce the datasets utilized in our study in Section 2.1. Then, we describe the detail of the proposed PGR-Net in Section 2.2. Finally, the implementation detail is presented in Section 2.3.

2.1. Datasets

A CNN-based interactive segmentation network is characterized by class-agnostic object extraction, which requires a dataset with diversity for the training to ensure the performance and generalization ability. In this study, in order to train the proposed PGR-Net, we follow [14,16,17] to adopt Semantic Boundaries Dataset (SBD [21]) as the training data and evaluate on five natural image datasets. Moreover, to verify the effectiveness of the proposed PGR-Net on buildings, we select three building datasets for detailed evaluation. The details of these datasets are described as follows.

2.1.1. Natural Image Dataset

We use Semantic Boundaries Dataset (SBD) to train our model, and test on five datasets to evaluate the performance of our algorithm, the details of utilized datasets are described as follows:

- **SBD** [21]: The dataset contains 8498 training images and 2820 test images. Following [16,17], we use the training set of this dataset to train our network, and the test set, which contains 6671 instances, is utilized for the evaluation of our algorithm.
- **GrabCut** [9]: The dataset consists of 50 images with a single object mask provided for each image. It is used as a common benchmark for most interactive segmentation algorithms.

- **Berkeley** [22]: The dataset contains 200 training images and 100 test images. It contains some challenging segmentation scenarios such as low contrast between the foreground and the background. We use 100 object masks on 96 test images as in [16] for the evaluation.
- **DAVIS** [23]: This is a benchmark dataset for video segmentation. It contains 50 videos with high-quality annotated masks. We sample the same 10% frames as [17,18] for the evaluation.
- **MSCOCO** [24]: This dataset is a large instance segmentation dataset, which contains 80 object categories. We sample 10 images per category to compose a test set of 800 images as in [17].

2.1.2. Building Dataset

We utilize three building segmentation datasets to validate the effectiveness of our algorithm; two of them are publicly available (AIRS [25], CrowdAI [20]) and the other is annotated by our team, named EVLabBuilding. The details of these datasets are as follows:

- **AIRS** [25]: This is a large-scale aerial imagery dataset for roof segmentation. It provides high-quality roof annotations for 7.5 cm resolution images. We crop the provided training images into 480 × 480 image patches and select 3398 images for testing.
- **CrowdAI** [20]: This is a large dataset for the extraction of building footprints in satellite images. The dataset annotates buildings at the instance level, and each individual building is annotated in a polygon format according to MS COCO standards. All the images are 300 × 300 pixels with a resolution of 0.3 m. We use the provided small subset of the validation set, which contains 1820 images, for testing.
- **EVLabBuilding**: It is a mixture dataset of aerial images of Guangzhou and Zhengzhou, China. It contains 40 images with resolutions ranging from 0.15 to 0.3 m. The buildings are annotated at the instance level by Earth Vision Lab, Wuhan University. We crop the images into 512 × 512 pixels and finally produce 3669 patches for testing.

In Figure 3, we present some example images of each dataset. As it can be seen, images in the AIRS dataset are of high quality. The CrowdAI dataset contains many small buildings and the images are also blurry. The scenes in the EVLabBuilding dataset are usually very messy, which is very challenging for the extraction of buildings. It is noted that for each image, we randomly choose one building instance from it for evaluation. For a fair comparison, we determined these instances in advance and used them for the evaluation of all algorithms.

(**a**) AIRS (**b**) CrowdAI (**c**) EVLabBuilding

Figure 3. Example images in AIRS, CrowdAI and EVLabBuilding datasets.

Furthermore, achieving the same IoU score is often more difficult for small objects. In order to facilitate a more detailed analysis of the algorithm, we further divide the test set into three subcategories according to the size of the buildings. Specifically, we classify the buildings into Small Buildings ($\alpha \leq 20^2$), Medium Buildings ($20^2 \leq \alpha \leq 60^2$) and Large Buildings ($\alpha > 60^2$); here α denotes the area of the building (number of pixels). The details of the divided subcategories of each dataset are listed in Table 1.

Table 1. Details of divided subcategories of AIRS, CrowdAI and EVLabBuilding datasets.

Dataset	Small	Medium	Large	All
AIRS	222	672	2504	3398
CrowdAI	388	966	466	1820
EVLabBuilding	348	1064	2257	3669

2.2. Methods

In this section, we first introduce the preparatory concept *indication distance* of the proposed PGR-Net in Section 2.2.1. Afterward, we present the input and the structure of the PGR-Net in Section 2.2.2. In Section 2.2.3, we show how to simulate the training samples for the training of the PGR-Net. Finally, the adaptive zoom-in technique and the iterative training strategy are described in Sections 2.2.4 and 2.2.5, respectively.

2.2.1. Indication Distance

Our approach is based on the assumption that annotators are always accustomed to clicking around the center of the main misclassified regions for the segmentation mask refinement. Under this circumstance, we notice that the minimal distance of the newly added click to the boundary of the previous mask has a good indication of the segmentation progress, and here we refer to this distance as the indication distance. In Figure 4, we show an example to illustrate our point. We can easily infer that the indication distance is large when the previous segmentation mask is far from the ground truth, otherwise this distance is relatively small.

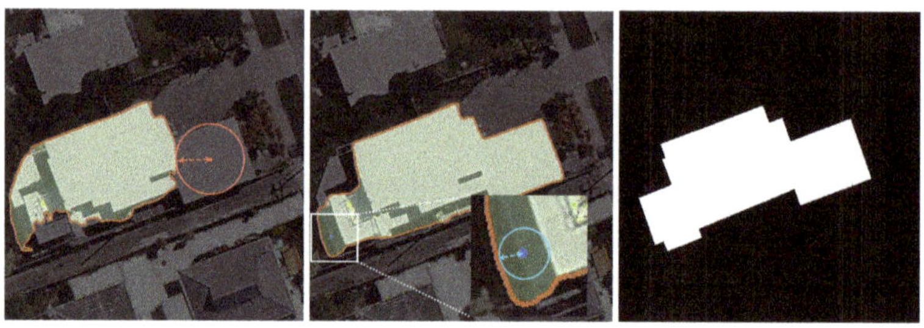

Figure 4. Illustration of indication distance. For poorly segmented results, the indication distance of the newly added click is large; otherwise, it is relatively small.

In some special cases, as shown in Figure 5, for objects with holes, multi-parts or elongated parts, indication distance sometimes may provide misleading guidance. Fortunately, the impact of these issues is negligible with a large amount of data training. These scenarios can be covered in the training set by data simulation and iterative training, which will be discussed in Sections 2.2.3 and 2.2.5, respectively. In addition, to further reduce the impact of these circumstances, we force the indication distance to keep decreasing during the inference phase. Specifically, at each step, the current indication distance is determined by $curdist = min(lastdist, curdist)$, where $lastdist$ denotes the indication distance of the previous step.

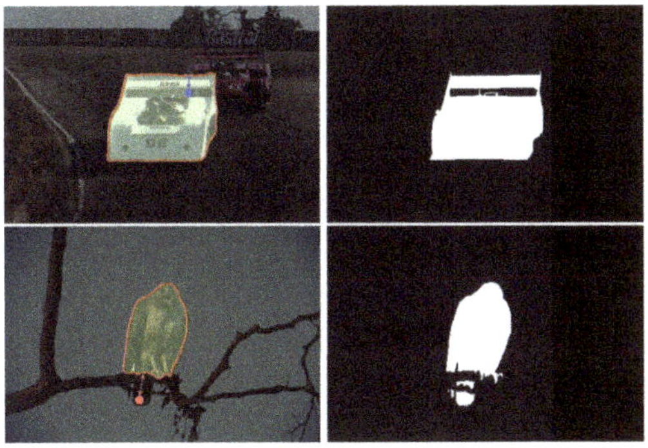

Figure 5. Some special cases of indication distance, such as objects with holes or multi-parts.

2.2.2. Guidance Representation and Network Structure

With humans in the loop, the interactive segmentation can be viewed as a sequential decision problem, which contains abundant information, such as previous segmentation results, history clicks, and newly added clicks, etc. There are various ways to encode the user input. Most of the existing methods follow [12] to simulate positive and negative clicks and transform them into guidance maps by Euclidean distance [14,16,17,19] or Gaussian masks [18]. Such representation treats all clicks indiscriminately. Furthermore, we argue that the click maps or their corresponding transformed distance-based maps are "weak" guidances, which is not conducive to the stability of the network prediction. Considering that the previous segmentation mask can provide "strong" guidance of the existence of target objects, we experimentally combine it with the newly added click map to feed into an interactive segmentation network and found out that the output tends to be consistent in the final detail refinement stage. We consider it is because this guidance representation lacks information of history clicks.

Therefore, we combine all the information mentioned above to form a new guidance representation and feed them to the interactive segmentation network. Figure 6 shows an overview of our method. Our guidance map consists of four parts, positive and negative click maps, newly added click map and the previous segmentation map. We utilize the same transform strategy for all click-related map generation. Given a set of click points $p_{ij} \in \mathcal{A}$, where (i,j) is the point location, then for any point p_{mn} in the 2D matrix with the same sized input image, its corresponding value $V(p_{mn}\mathcal{A})$ is computed as follows:

$$V(p_{mn}\mathcal{A}) = \min_{\forall p_{ij} \in \mathcal{A}} \sqrt{(i-m)^2 + (j-n)^2} \tag{1}$$

The 2D matrix is normalized into [0, 1] as the final guidance map.

$$V'(p_{mn}\mathcal{A}) = 1 - \frac{\min(V(p_{mn}\mathcal{A}), d)}{d} \tag{2}$$

For positive and negative click maps, we chose d as 10. Notably, we set d to 30 for the first click, which means there is only one positive click and no other negative clicks. The newly added click map is a single channel, d is set equal to its corresponding indication distance, and we set the matrix negative if the newly added click falls into the background.

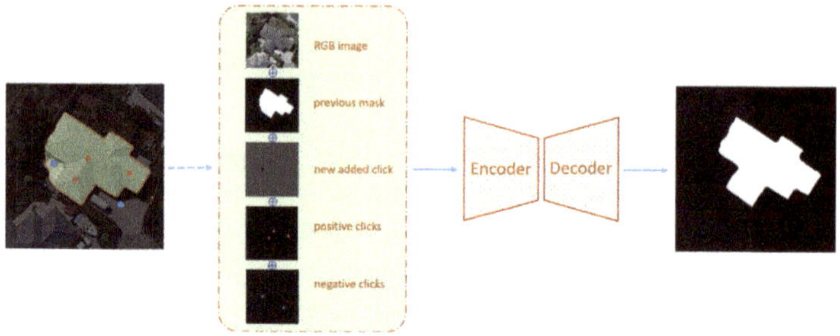

Figure 6. Overview of our method. The input to the network consists of the RGB image, the previous segmentation mask, the newly added click map (single channel), and the positive and negative click maps.

In this paper, we do not focus on the network architecture design. In Figure 7, we present the network structure of the PGR-Net. We follow [26] and utilize a modified ResNet [27] architecture as our backbone, in which the stride of the last two layers are reduced to one and dilation convolution is employed, which helps to increase the resolution of the output feature and maintain the receptive field of the network at the same time. In Appendix A, we present the detailed structure of the backbone network. Afterward, we add skip-connections in the encoder to aggregate both the low-level and high-level features. Specifically, the features "conv1", "res1", "res2" and "res4" of the backbone network are converted into a 128-d feature by a 3×3 convolutional layer, respectively. Subsequently, we upsample these features to the same resolution as "conv1" and concatenate these 128-d features to obtain a 512-d feature. Finally, we employ a PSP module [28] to obtain the prediction mask.

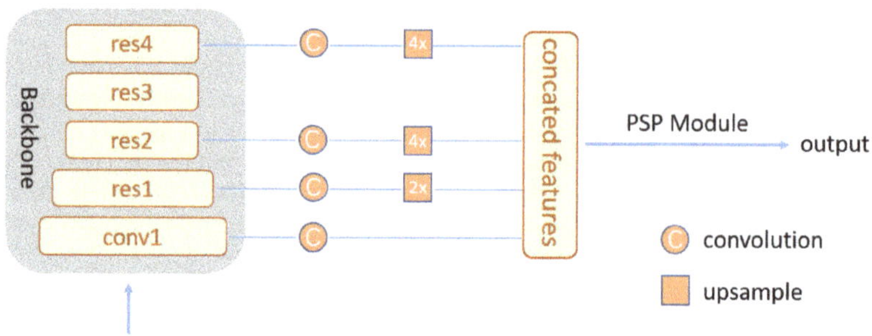

Figure 7. Architecture of our network.

2.2.3. Simulating User Input

To simulate the previous segmentation mask, we follow [29] to generate perturbed segmentations with various shapes. Specifically, we subsample the contour of the ground-truth mask, and then perform multiple random dilation and erosion operations. After the simulation of the previous segmentation mask, we set the center point of the largest error region as the newly added click for corresponding guidance map generation.

For the simulation of positive and negative clicks, we adopt the sampling strategy proposed in [12]. The numbers of positive and negative clicks are determined randomly within [1, 10] and [0, 10], respectively. The click points are generated in sequence, and for the positive clicks, the new point is sampled in the foreground with at least d_p^1 pixels from the object boundary and d_p^2 pixels from existing points. For negative clicks, the new point is sampled in the background with $d_n^1 \sim d_n^2$ pixels away from the object boundaries and d_n^3 pixels away from existing click points. We set $d_p^1 = 5$, $d_p^2 = 10$, $d_n^1 = 5$, $d_n^2 = 40$, and d_n^3 is set to 10.

Notably, the training samples are generated based on object instances, each object in the image is individually selected for training sample simulation. For each object, multiple training samples can be obtained by simulating different clicks and masks. In Figure 8, we present two examples of simulated training samples.

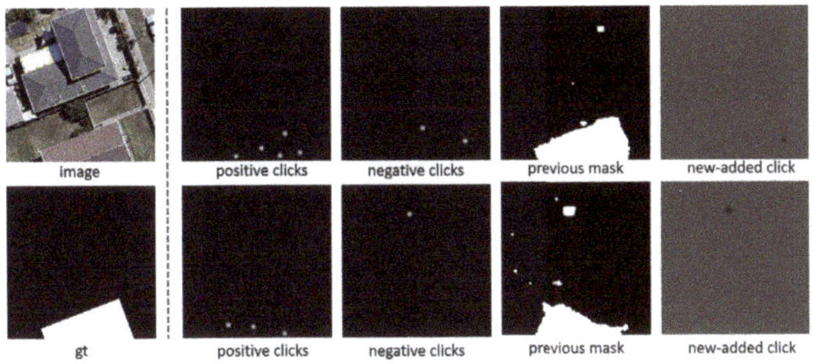

Figure 8. Two examples of simulated training samples.

2.2.4. Adaptive Zoom-In

Different from fully automatic image segmentation, in the interactive setting, objects are gradually refined and extracted in an iterative and interactive manner. Thus, cropping is a simple and effective manner for the detail refinement of the segmentation mask, especially for the small objects. Sofiiuk et al. [17] call this technique zoom-in and firstly introduce it to the interactive segmentation system.

The premise of applying zoom-in is that the network has predicted an approximate mask of the ground truth, inappropriate use may lead to degradation of the network output. Therefore, the evaluation of the quality of the current segmentation mask is of great importance. Sofiiuk et al. [17] notice that the first three clicks are sufficient for the network to obtain a rough mask of the target object. They empirically crop the image according to the bounding box of the predicted mask after the third click. Such a heuristic method is not applicable to all situations. In our work, the indication distance can accurately reflect the quality of the segmentation mask, and we exploit it to apply zoom-in prediction adaptively. Specifically, when the indication distance is smaller than a certain threshold $\mathcal{T}$, we crop an image according to the bounding box of the predicted object mask to obtain a zoom-in region, which will be resized to a target size $\mathcal{S}$ and fed into the network for the next prediction. For the bounding box, we extend $\mathcal{C}$ pixels along the longest side direction to preserve the context, and the shortest side will be extended adaptively to form a square box. Notably, we only apply zoom-in for small objects, which means the size of the extended bounding box should be smaller than $\mathcal{S}$, and the bounding box will be adjusted if a user provides a click outside the box.

By our adaptive zoom-in technique, the network is able to handle objects of different scales in a more flexible manner, which is helpful to promote the stability of the network and reduce the user interactions. We will further discuss the superiority of our adaptive zoom-in technique over the empirical way in Section 4.2.

2.2.5. Iterative Training Strategy

To facilitate the batch training of deep learning networks, most click-based interactive segmentation methods adopt a random sampling strategy for generating clicks. Such a sample generation strategy does not consider the sequential relationship of user-provided clicks during the inference stage. Thus, the iterative training strategy, where new clicks are added based on the prediction errors of the network during training, is often utilized to boost the performance, and this strategy is widely used and proved to be effective for interactive segmentation algorithms.

In our algorithm, there is always a gap between the simulated perturbed segmentations and the network predictions. Furthermore, our positive and negative clicks are also randomly generated. For the above considerations, we propose an iterative training strategy for the training of our model. Specifically, we incorporate the adaptive zoom-in technique into the standard iterative training procedure proposed by [13], and an example of the corresponding iterative training process is shown in Figure 9. The simulated guidance maps are fed into the network to obtain a prediction. Based on the misclassified region, we use the same clicking strategy as the inference stage to provide a new click. If the indication distance of the newly added click is smaller than $\mathcal{T}$, then we crop and resize the patch of the object to form new training data, and the related guidance maps will be transformed accordingly. In this way, we align the network training to the actual usage. This can also be regarded as a kind of data augmentation, which is helpful for improving the performance of the algorithm.

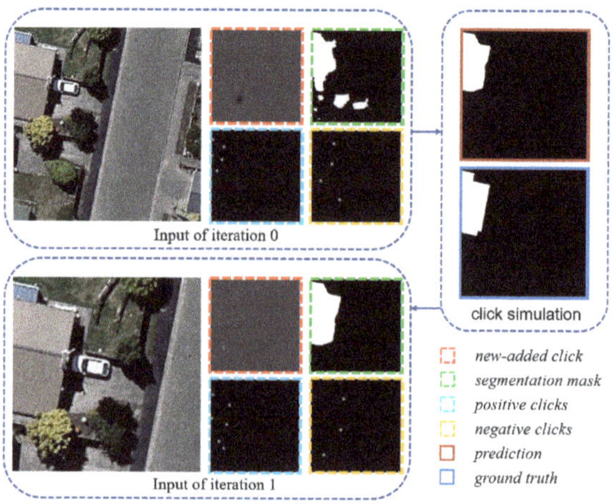

Figure 9. An example of training data generation by using the adaptive zoom-in technique during the iterative training process.

2.3. Implementation Details

We formulate the interactive segmentation problem as a binary segmentation task and use binary cross entropy loss for the network training. We use zero initialization for the extra channels of the first convolutional layer. We utilize Semantic Boundaries Dataset (SBD [21]) to train our model. The input images are randomly cropped into 384 × 384 pixels, and the dataset is augmented by a horizontal flip. We take ResNet-101 pre-trained on ImageNet [30] as the backbone. The batch size is 4. We set an initial learning rate of 3×10^{-5} for ResNet and 3×10^{-4} for other parts. We use the Adam [31] optimizer to train our network for 32 epochs. The learning rate decreases by a factor of 10 after every 10 epochs. The network is implemented in the PyTorch framework and trained on a single NVIDIA GeForce RTX 2080Ti GPU.

For the automatic evaluation of our algorithm, we use the same clicking strategy as the previous works [16–18] to simulate user interaction. At each step, we obtain a segmentation mask predicted by the network, and then the new click will be added at the center of the largest misclassified region. The first click is added in the same way, with the network prediction being regarded as zero. For the zoom-in prediction, we choose the indication distance threshold $\mathcal{T}$ as 20 pixels, and the extended size $\mathcal{C}$ is 30, the target size $\mathcal{S}$ is set to 480.

For the evaluation of related algorithms, the mask intersection over union (mask IoU) is adopted as a basic metric in our experiments. First, we follow [16,18] to utilize two performance measures to compare our algorithm with other state-of-the-art methods. One is the NoC metric, which indicates the average number of clicks to reach a certain IoU threshold on each sample of a dataset. We set the maximum number of clicks to 20 for each sample. The other is the plot of the mean IoU score according to the number of clicks. We also compute the area under curve (AuC) for each method, with each area normalized into [0, 1].

3. Experimental Results

Segmentation performance and generalization ability are two important indexes for the evaluation of an interactive segmentation method. To assess the effectiveness of the proposed PGR-Net, we devise two groups of experiments on natural scene images and high-resolution remote sensing datasets, respectively. We first follow References [14,16,17] to evaluate our algorithm on five widely used natural image datasets to demonstrate the generalization ability of PGR-Net. Afterward, to facilitate the detailed analysis of our method on the interactive extraction of buildings, we analyze and compare our algorithm with the latest state-of-the-art methods FCA [18] and f-BRS [17] on three building datasets of remote sensing images. In addition, we conduct detailed ablation experiments to verify the effectiveness of each component, and analyze the stability of our algorithm.

3.1. Evaluation on Natural Image Dataset

We compare our approach with other existing methods, including GrabCut (GC) [5], geodesic matting (GM) [32], random walk (RW) [11], Euclidean star convexity (ESC) [8], geodesic star convexity (GSC) [8], Growcut (GRC), deep object selection (DOS) [12], regional image segmentation (RIS) [19], latent diversity based segmentation (LD) [14], backpropagating refinement scheme (BRS) [16], content aware multi-level guidance (CMG) [15], feature backpropagating refinement scheme (f-BRS) [17], and first click attention network (FCA) [18].

Figure 10 illustrates the IoU scores of each method on the different number of clicks. In general, deep-learning-based methods (solid lines) have better performance than traditional interactive segmentation algorithms (dashed lines) in all datasets. The curve of our method is smooth, and we achieve the highest AuC scores across all five datasets, which means our algorithm has a better performance. Specifically, the curves of our method have obvious advantages on SBD and COCO datasets, and for the DAVIS, due to its difficulty, the curves of each algorithm are relatively close. In Table 2, we report the NoC results on five datasets. We achieve the best performance on four of them. As it can be seen, our algorithm requires fewer number of clicks to reach the same IoU score. Note that we do not utilize complicated architecture design. However, the improvement of performance is significant, which demonstrates the effectiveness of our algorithm.

Table 2. Comparison of the number of clicks (NoC) required to reach IoU 0.85 (NoC@85) and 0.9 (NoC@90) on GrabCut, Berkeley, DAVIS, SBD and COCO datasets. The best and the second-best results are boldfaced and underlined, respectively.

Method	GrabCut		Berkeley	DAVIS	SBD		COCO	
	NoC@85	NoC@90	NoC@90	NoC@90	NoC@85	NoC@90	NoC@85	NoC@90
GC [5]	7.98	10.00	14.22	17.41	13.6	15.96	15.23	17.61
GM [32]	13.32	14.57	15.96	19.50	15.36	17.60	16.91	19.63
RW [11]	11.36	13.77	14.02	18.31	12.22	15.04	13.62	16.74
ESC [8]	7.24	9.20	12.11	17.70	12.21	14.86	14.04	16.98
GSC [8]	7.10	9.12	12.57	17.52	12.69	15.31	14.39	16.89
DOS [12]	5.08	6.08	8.65	12.58	9.22	12.80	9.07	13.55
RIS [19]	-	5.00	6.03	-	-	-	-	-
LD [14]	3.20	4.79	-	9.57	7.41	10.78	7.86	12.45
BRS [16]	2.60	3.60	5.08	8.24	6.59	9.78	-	-
CMG [15]	-	3.58	5.60	-	-	-	5.92	-
FCA [18]	1.82	2.08	3.92	7.57	-	-	3.64	5.31
f-BRS [17]	2.30	2.72	4.57	7.41	4.81	7.73	4.11	5.91
Ours	1.99	2.26	3.66	7.05	3.70	5.67	3.25	4.26

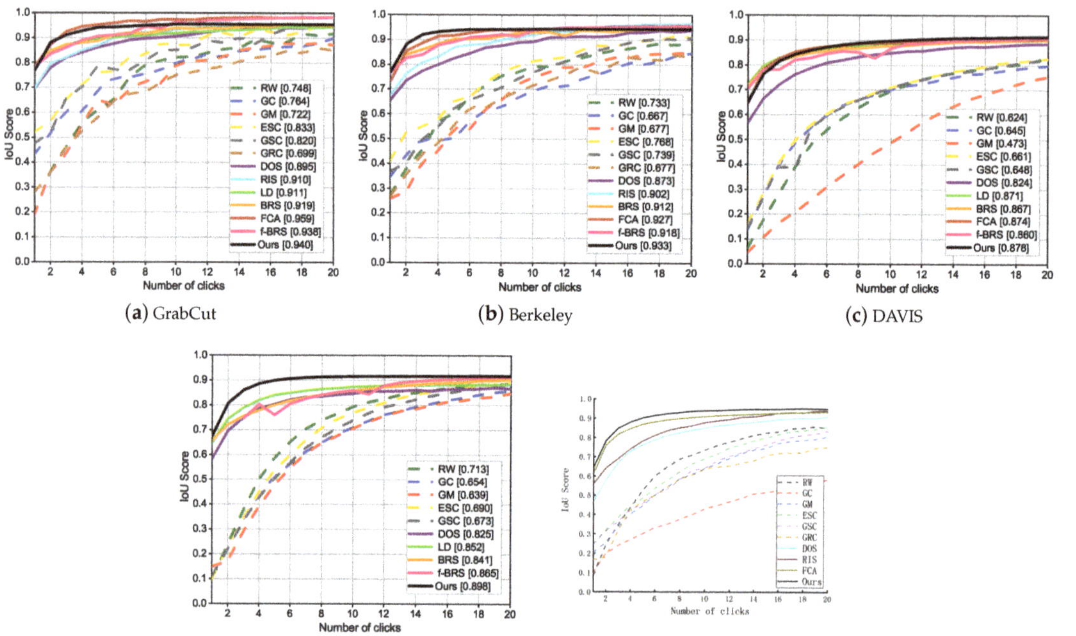

(**a**) GrabCut (**b**) Berkeley (**c**) DAVIS

(**d**) SBD (**e**) COCO

Figure 10. Comparison of the average IoU scores according to the number of clicks (NoC) on GrabCut, Berkeley, SBD, DAVIS and COCO datasets. The legend contains AuC scores for each algorithm.

3.2. Evaluation on Remote Sensing Dataset

In Section 3.1, we conduct a detailed comparison and analysis of related interactive segmentation algorithms, including traditional and deep-learning-based. In this section, we select the most recent state-of-the-art methods (FCA [18] and f-BRS [17]) for the comparison.

In Figure 11, we present the IoU scores of each algorithm under a different number of clicks on the three datasets. Compared with the other two methods, our algorithm achieves the highest AuC scores in all the three datasets. Concretely, the advantages of our method on AIRS and EVLabBuilding datasets are huge and significant, and on the CrowdAI dataset, the performance of the algorithms are close and worse; we consider that this is because there are many small buildings on the CrowdAI dataset. Basically, it only takes 3–4 clicks for our algorithm to achieve good results. In Tables 3–5, we report the quantitative NoC results of each algorithm on AIRS, CrowdAI and EVLabBuilding datasets, respectively. In fact, NoC results of small buildings (20×20) do not make much sense because it is always difficult for such small objects to reach 0.9 IoU. Nevertheless, our algorithm still performs significantly better than other methods in the NoC metric. For medium and large buildings, our algorithm basically only requires 1–2 fewer NoC than other methods to achieve the same IoU results, in some cases, it can reach 4–5 or more. Compared with the results of Table 2 in natural scenes, our results on buildings have more obvious advantages. We consider it is due to the better stability of our algorithm, and this advantage will be more prominent in "easy" buildings, of which we will make a further analysis in Section 3.4. In Figure 12, we also present some visualized comparisons of each algorithm on the three datasets.

Table 3. Comparison of the number of clicks (NoC) required to reach IoU 0.85 (NoC@85) and 0.9 (NoC@90) on the AIRS dataset.

Method	Small		Median		Large		All	
	NoC@85	NoC@90	NoC@85	NoC@90	NoC@85	NoC@90	NoC@85	NoC@90
f-BRS [17]	13.17	16.52	5.69	9.38	4.88	7.22	5.58	8.25
FCA [18]	16.62	18.83	5.05	8.51	3.26	4.80	4.48	6.45
Ours	15.94	17.45	4.27	7.05	2.57	3.33	3.78	4.99

Table 4. Comparison of the number of clicks (NoC) required to reach IoU 0.85 (NoC@85) and 0.9 (NoC@90) on the CrowdAI dataset.

Method	Small		Median		Large		All	
	NoC@85	NoC@90	NoC@85	NoC@90	NoC@85	NoC@90	NoC@85	NoC@90
f-BRS [17]	13.54	16.97	9.20	13.93	9.02	13.54	10.08	14.48
FCA [18]	14.48	17.98	6.34	10.26	5.50	8.76	7.86	11.52
Ours	13.61	16.39	5.16	8.33	4.47	6.63	6.78	9.61

Table 5. Comparison of the number of clicks (NoC) required to reach IoU 0.85 (NoC@85) and 0.9 (NoC@90) on the EVLabBuilding dataset.

Method	Small		Median		Large		All	
	NoC@85	NoC@90	NoC@85	NoC@90	NoC@85	NoC@90	NoC@85	NoC@90
f-BRS [17]	14.82	17.80	5.88	9.67	5.98	8.37	6.79	9.64
FCA [18]	18.22	19.72	6.19	11.06	4.53	6.55	6.31	9.11
Ours	10.98	13.64	3.67	5.25	3.73	4.65	4.40	5.68

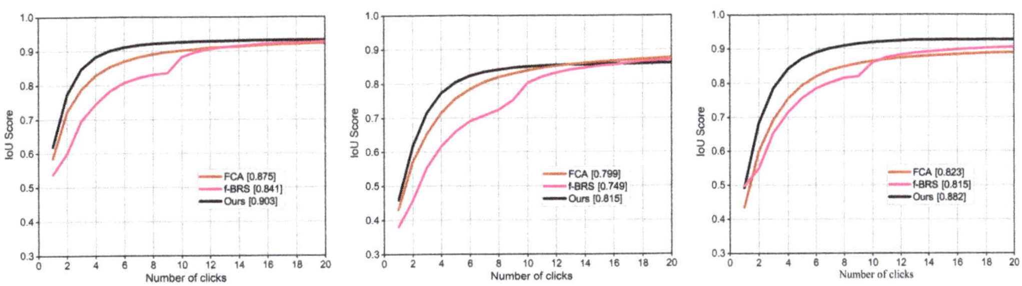

(**a**) AIRS (**b**) CrowdAI (**c**) EVLabBuilding

Figure 11. Comparison of the average IoU scores according to the number of clicks (NoC) on AIRS, CrowdAI and EVLabBuilding datasets. The legend contains AuC scores for each algorithm.

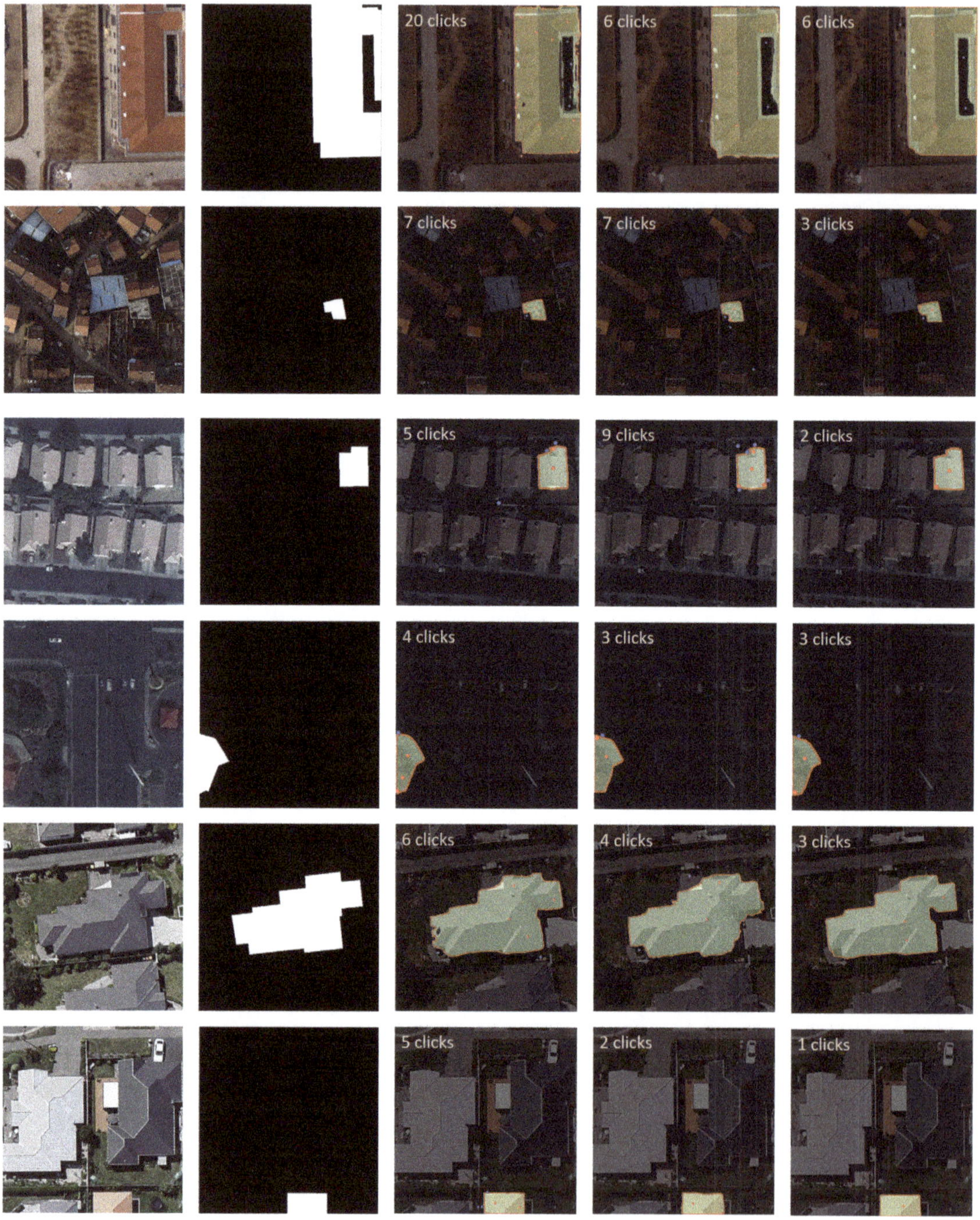

(a) image (b) gt (c) f-BRS (d) FCA (e) Ours

Figure 12. Visualized comparison of different methods on EVLabBuilding, AIRS and datasets. Red for positive clicks and blue for negative clicks.

3.3. Ablation Study

To further evaluate the efficacy of each component in our algorithm, we conduct an ablation study on Berkeley, COCO and EVLabBuilding datasets. We take the network that only uses the positive and negative clicks as input as the baseline, and we gradually add each component proposed in this paper for validation. In Table 6, we report the mean number of click (NoC) results of different settings. Overall, the iterative training (No.3) and segmentation guidance map (No.4) have brought significant performance improvement. By utilizing iterative training, the performance on the COCO dataset is improved with 0.94 and 1.48 alleviating of NoC; and for the EVLabBuilding dataset, the NoC value has been significantly reduced by 2.21 and 2.68. However, the improvement effect on the Berkeley dataset is very slight. We infer that it is because the iterative training and the data augmentation in it help the network to deal with some small or easy objects. However, for objects with complex shapes in the Berkeley dataset, the effect of improvement is limited. On the other hand, we consider that the previous segmentation mask can promote the stability of the mask refinement, which has more advantages in the detail refinement of complex objects. Thus, by adding the previous segmentation mask, the NoC on the Berkeley dataset is reduced from 4.51 to 3.66. We will compare the stability of No.3 and No.4 settings in Section 3.4 to further verify this point.

Table 6. Ablation study of the proposed method on Berkeley, COCO and EVLabBuilding datasets. BS: baseline; AZI: adaptive zoom-in technique; Iter: iterative training strategy; Seg: previous segmentation mask and the newly added click map.

Settings	Berkeley	COCO		EVLabBuilding	
	NoC@90	NoC@85	NoC@90	NoC@85	NoC@90
#1: BS	5.26	5.03	7.10	7.85	10.85
#2: BS + AZI	4.60	4.54	6.31	7.13	9.50
#3: BS + AZI + Iter	4.51	3.60	4.83	4.92	6.82
#4: BS + AZI + Iter + Seg	3.66	3.25	4.26	4.40	5.68

3.4. Stability Analysis

For the evaluation of the stability of interactive image segmentation algorithms, we use the frequency of "bad" clicks and the average IoU reduction caused by these "bad" clicks as a metric. The "bad" clicks here refer to clicks that cause the segmentation IoU to fall after being added to the network. Specifically, when calculating the NoC metric, we also count the frequency of "bad" clicks before the segmentation result reaches a certain IoU to evaluate the robustness of the interactive segmentation system. We compare our algorithm with f-BRS and FCA on the AIRS dataset. From the curves in Figures 10 and 11, we can see that the IoU is basically saturated after reaching 90% for all methods. Thus, here we set the target IoU to 90% for a more reasonable evaluation, and the comparison results are shown in Table 7. Our algorithm yields the lowest ratio of "bad" clicks, which means our algorithm is less prone to produce degraded results. Furthermore, the average IoU reduction of our method is 3.13%, and we consider it as a normal fluctuation during the final refinement stage of the segmentation masks. "Ours$^-$" denotes the No.3 setting in Table 6, and from this, we can see the importance of the previous segmentation mask and the newly added click map for improving the stability of the network. The stability of FCA-Net is better than that of the f-BRS, which is because the first click attention module is helpful to promote the stability of the network prediction. Furthermore, we notice that the average IoU drop of f-BRS is 11.83%, which means that each "bad" click will cause a drop in IoU of 0.12. From this, we can infer that the results of f-BRS fluctuate sharply.

Table 7. Comparison of the ratio of "bad" clicks and its corresponding average IoU reduction on the AIRS dataset. #ACs: number of all clicks; #BCs: number of bad clicks.

Method	#ACs	#BCs	Ratio	IoU Drop (avg.)
f-BRS [17]	28034	6132	21.87%	11.83%
FCA [18]	21925	3720	16.97%	7.46%
Ours$^-$	21646	4386	20.26%	11.65%
Ours	16952	1499	8.84%	3.13%

In Figure 13, we present a visualized interactive segmentation process of f-BRS, FCA and our algorithm on a test case of the AIRS dataset to further demonstrate the stability of our algorithm. We show the segmentation mask of each algorithm after each click is added before the IoU reaches 0.9. It only takes 4 clicks for our method to reach 0.9 IoU. In addition, our algorithm can continuously improve the segmentation results, while the segmentation masks of f-BRS (7th click) and FCA-Net (7th and 11th click) are easily degraded, which is harmful to the interactive segmentation system.

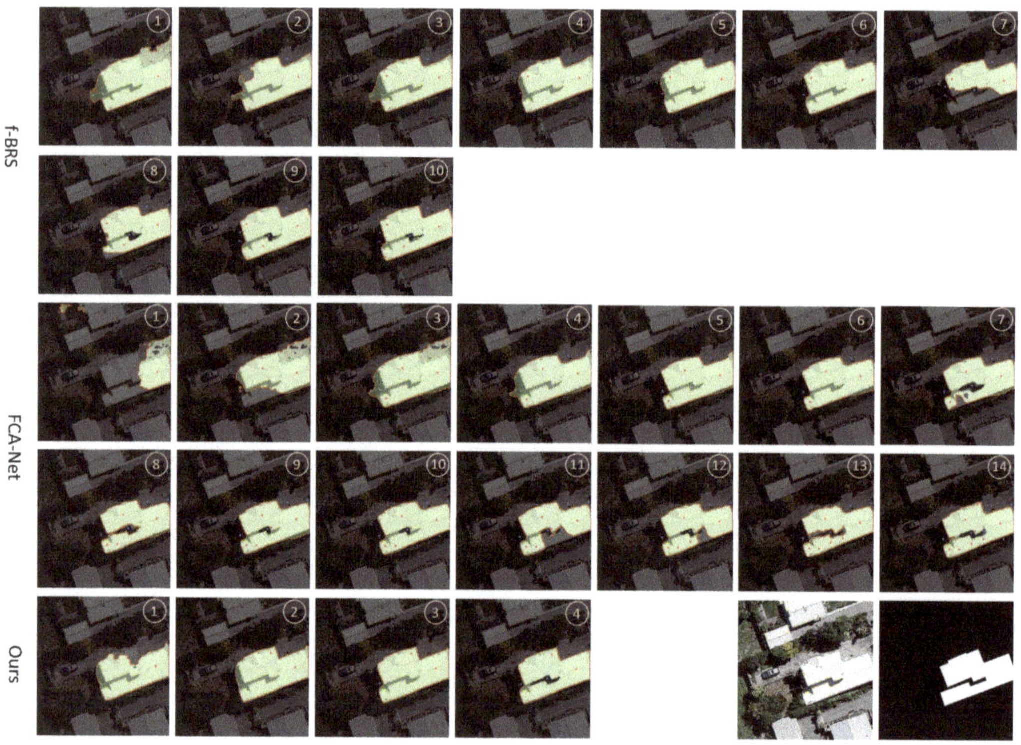

Figure 13. Visualized comparison of the segmentation process of f-BRS, FCA-Net and our algorithm on a test case of AIRS dataset. Red for positive clicks and blue for negative clicks.

4. Discussion

4.1. Building Extraction Analysis

From the results in Tables 2–5, we can see that our method surpasses FCA [18] and f-BRS [17] on both building and natural image datasets. Specifically, the advantage of our method is more obvious in dealing with buildings. We attribute this to the good stability of PGR-Net, which is essential for the interactive extraction of buildings. If an interactive segmentation algorithm is not stable, it will take a significant amount of clicks to extract even a very simple object. From the last row of Figure 12, we can see that for an easy building, f-BRS [17] takes 5 clicks to reach 0.9 IoU. This is because in the process of object extraction, the algorithm has been correcting the wrong prediction given by itself. We have provided an example in Figure 13 to demonstrate this point, and this is in line with our analysis in Section 1.

4.2. Adaptive Zoom-In

To demonstrate the superiority of our proposed adaptive zoom-in technique, we compare it with heuristic zoom-in strategies on Berkeley, DAVIS and EVLabBuilding datasets. We consider that the images in these datasets have different size, which is very suitable for such validation. Here we follow [17] to apply zoom-in prediction after 1, 3 and 5 clicks, and these three settings are denoted as "#1", "#3" and "#5", respectively. We compute and compare the NoC required for each setting to reach 0.9 IoU (NoC@90), and the results are reported in Table 8. For the empirical zoom-in strategies, we can see that "#1" is the best on Berkeley and EVLabBuilding datasets. However, on the DAVIS dataset, "#3" performs best. This indicates a single empirical setting does not always achieve optimal performance in the face of different scenarios. On the contrary, our adaptive zoom-in strategy can always achieve good performance on all of the datasets.

Table 8. Comparison of the different zoom-in settings on Berkeley, DAVIS and EVLabBuilding datasets.

Setting	#1	#3	#5	Adaptive
Berkeley(NoC@90)	3.93	4.02	4.18	3.66
DAVIS(NoC@90)	7.29	7.18	7.22	7.05
EVLabBuilding(NoC@90)	5.64	5.79	6.11	5.68

4.3. Clicking Strategy

To analyze the impact of the clicking strategy to our algorithm, we analyze the IoU score according to the number of clicks on the Berkeley dataset with three different clicking strategies, and the related results are shown in Figure 14. In strategy 1, we use the center of the largest error region (point p) as the new click. In clicking strategy 2, we first compute the minimum distance of point p to the boundary, denoted as d, and then we randomly select clicks from the error region within $0.5d$ from point p. In strategy 3, we randomly select clicks in the largest error region. We can see that strategy 2 has a similar performance to strategy 1. However, when we randomly select clicks by using strategy 3, the performance of the network is degraded to some degree. It is because the indication distance will provide misleading guidance in this situation. To sum up, our algorithm can perform well when users provide clicks around the center of the misclassified region. However, when users click near the boundary of the mislabeled region, the performance will decrease to a certain extent. Overall, by referring to the results of other similar methods [13], we believe such performance degradation is normal and acceptable. In Section 4.4, we will further verify our algorithm by human study.

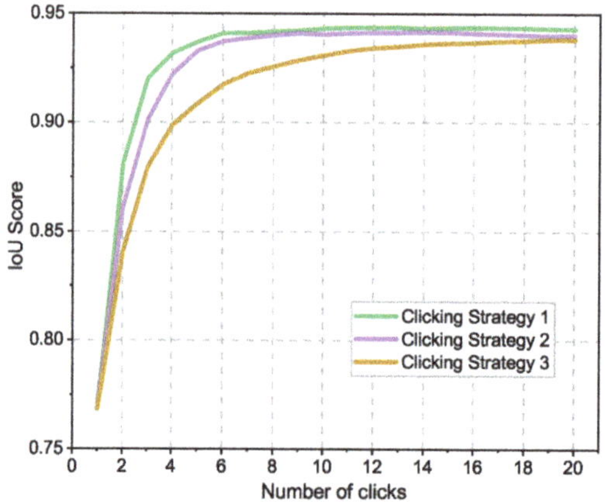

Figure 14. IoU scores according to the number of clicks of different clicking strategies.

4.4. Interaction with Human Annotators

Our algorithm is based on the assumption that the annotators are accustomed to click around the center of the largest error region. Notably, we just utilize this information roughly to obtain the progress information of the interactive segmentation process, and we do not require the annotators to follow this strategy strictly. To further validate the robustness of the proposed algorithm, we conduct a small experiment with real human annotators in the loop. Based on the proposed interactive segmentation model, we develop a tiny annotation tool for the evaluation. Taking into account the workload of the annotators, we asked two human subjects to annotate the objects in the GrabCut dataset (50 images). Notably, we only explained how to use this tool, and the annotators do not know the details of the algorithm behind it. For each object, the corresponding ground-truth image is presented, and the annotators can provide clicks according to their preferences. Once the IoU score reaches 0.9, the tool will give a prompt, which also means the annotators have completed the annotation of this object. In Table 9, we compare the NoC results of different annotators with an automatic clicking strategy. NoC@85 and NoC@90 denote the NoC required to reach IoU 0.85 and 0.90, respectively. In general, the results are very close. An interesting finding is that the NoC@85 results of human annotators are generally better than the results of the automatic clicking strategy, while for NoC@90, the situation is the opposite. This is because in the automatic clicking strategy, clicks are determined by calculating the maximum distance of the misclassified area, and sometimes these clicks are not the "real" center of the object, which can degrade the performance of the segmentation results; thus, it takes more clicks to achieve IoU 0.85 (NoC@85). As for NoC@90, the reason is that for small objects, it is sometimes difficult for human annotators to provide further clicks because the result is already pretty good, it just does not reach 0.9 IoU.

Table 9. Real human experiments on the GrabCut dataset.

	NoC@85	**NoC@90**
Automatic	1.99	2.26
Annotator#1	1.78	2.52
Annotator#2	1.80	2.56

5. Conclusions

In this work, we analyze the difference of objects in natural scenes and buildings in remote sensing images and realize that the stability is critical to an interactive segmentation system, especially for "easy" buildings. Focusing on the promotion of the robustness of the interactive segmentation, we utilize the distance of newly added clicks to the boundary of the previous segmentation mask as an indication of the interactive segmentation progress, and this information is employed with the previous segmentation mask and positive and negative clicks to form a progress guidance map. This progress guidance map is then fed into a CNN with the original RGB image. Furthermore, we propose an iterative training strategy for the training of the network. Moreover, we adopt an adaptive zoom-in technique during the inference stage for further performance promotion. Abundant experimental results show that our algorithm has good robustness and superiority. In particular, compared with the latest state-of-the-art methods, FCA [18] and f-BRS [17], the proposed PGR-Net basically requires 1-2 fewer clicks to achieve the same segmentation results on the three building datasets. Currently, our method is utilized for the extraction of building instances. In future research, we will try to improve our method for the interactive extraction of region objects.

Author Contributions: Conceptualization, Z.S. and X.H.; methodology, Z.S.; software and validation, Z.S. and H.D.; investigation, Z.S.; writing—original draft preparation, Z.S. and H.D.; writing—review and editing, Z.S., X.H. and H.D.; visualization, Z.S. and H.D.; supervision, X.H. All authors have read and agreed to the published version of the manuscript.

Funding: This research was funded by the Chinese National Natural Science Foundation with grant numbers 92038301 and 41771363.

Institutional Review Board Statement: Not applicable.

Informed Consent Statement: Not applicable.

Data Availability Statement: The data presented in this study are available on request from the first author.

Acknowledgments: The authors sincerely appreciate that academic editors and reviewers give their helpful comments and constructive suggestions.

Conflicts of Interest: The authors declare no conflict of interest.

Appendix A

The detail of the structure of the backbone network utilized in this study is depicted in Table A1.

Table A1. The detailed structure of the backbone network utilized in our work.

Block Group	Output Size	Channel	Parameters	Convolution Layout
input	384×384	7	-	-
conv1	192×192	64	stride 2, dilation 1	7×7
3×3 max pool, stride 2				
res1	96×96	256	stride 1, dilation 1	$\begin{bmatrix} 1 \times 1, 64 \\ 3 \times 3, 64 \\ 1 \times 1, 256 \end{bmatrix} \times 3$
res2	48×48	512	stride 2, dilation 1	$\begin{bmatrix} 1 \times 1, 128 \\ 3 \times 3, 128 \\ 1 \times 1, 512 \end{bmatrix} \times 4$
res3	48×48	1024	stride 1, dilation 2	$\begin{bmatrix} 1 \times 1, 256 \\ 3 \times 3, 256 \\ 1 \times 1, 1024 \end{bmatrix} \times 23$
res4	48×48	2048	stride 1, dilation 4	$\begin{bmatrix} 1 \times 1, 512 \\ 3 \times 3, 512 \\ 1 \times 1, 2048 \end{bmatrix} \times 3$

References

1. Ronneberger, O.; Fischer, P.; Brox, T. U-Net: Convolutional Networks for Biomedical Image Segmentation. In *Lecture Notes in Computer Science, Proceedings of the Medical Image Computing and Computer-Assisted Intervention—MICCAI 2015—18th International Conference, Munich, Germany, 5–9 October 2015*; Proceedings, Part III; Springer: Berlin/Heidelberg, Germany, 2015; Volume 9351, pp. 234–241. [CrossRef]
2. Badrinarayanan, V.; Handa, A.; Cipolla, R. Segnet: A deep convolutional encoder-decoder architecture for robust semantic pixel-wise labelling. *arXiv* **2015**, arXiv:1505.07293.
3. Chen, L.C.; Zhu, Y.; Papandreou, G.; Schroff, F.; Adam, H. Encoder-decoder with atrous separable convolution for semantic image segmentation. In Proceedings of the European Conference on Computer Vision (ECCV), Munich, Germany, 8–14 September 2018; pp. 801–818.
4. Mortensen, E.N.; Barrett, W.A. Intelligent scissors for image composition. In Proceedings of the 22nd Annual Conference on Computer Graphics and Interactive Techniques, SIGGRAPH 1995, Los Angeles, CA, USA, 6–11 August 1995; Mair, S.G., Cook, R., Eds.; ACM: New York, NY, USA, 1995; pp. 191–198. [CrossRef]
5. Boykov, Y.; Jolly, M. Interactive Graph Cuts for Optimal Boundary and Region Segmentation of Objects in N-D Images. In Proceedings of the Eighth International Conference On Computer Vision (ICCV-01), Vancouver, BC, Canada, 7–14 July 2001; Volume 1, pp. 105–112. [CrossRef]
6. Boykov, Y.; Kolmogorov, V. An experimental comparison of min-cut/max-flow algorithms for energy minimization in vision. *IEEE Trans. Pattern Anal. Mach. Intell.* **2004**, *26*, 1124–1137. [CrossRef] [PubMed]
7. Veksler, O. Star Shape Prior for Graph-Cut Image Segmentation. In *Lecture Notes in Computer Science, Proceedings of the Computer Vision—ECCV 2008, 10th European Conference on Computer Vision, Marseille, France, 12–18 October 2008*; Proceedings, Part III; Forsyth, D.A., Torr, P.H.S., Zisserman, A., Eds.; Springer: Berlin/Heidelberg, Germany, 2008; Volume 5304, pp. 454–467. [CrossRef]
8. Gulshan, V.; Rother, C.; Criminisi, A.; Blake, A.; Zisserman, A. Geodesic star convexity for interactive image segmentation. In Proceedings of the Twenty-Third IEEE Conference on Computer Vision and Pattern Recognition, CVPR 2010, San Francisco, CA, USA, 13–18 June 2010; pp. 3129–3136. [CrossRef]
9. Rother, C.; Kolmogorov, V.; Blake, A. "GrabCut": Interactive foreground extraction using iterated graph cuts. *ACM Trans. Graph.* **2004**, *23*, 309–314. [CrossRef]
10. Yu, H.; Zhou, Y.; Qian, H.; Xian, M.; Wang, S. Loosecut: Interactive image segmentation with loosely bounded boxes. In proceedings of the 2017 IEEE International Conference on Image Processing, ICIP 2017, Beijing, China, 17–20 September 2017; pp. 3335–3339. [CrossRef]
11. Grady, L. Random walks for image segmentation. *IEEE Trans. Pattern Anal. Mach. Intell.* **2006**, *28*, 1768–1783. [CrossRef] [PubMed]
12. Xu, N.; Price, B.; Cohen, S.; Yang, J.; Huang, T.S. Deep interactive object selection. In Proceedings of the IEEE Conference on Computer Vision and Pattern Recognition, Las Vegas, NV, USA, 27–30 June 2016; pp. 373–381.
13. Mahadevan, S.; Voigtlaender, P.; Leibe, B. Iteratively Trained Interactive Segmentation. In Proceedings of the British Machine Vision Conference 2018, BMVC 2018, Newcastle, UK, 3–6 September 2018; BMVA Press: Durham, UK, 2018; p. 212.
14. Li, Z.; Chen, Q.; Koltun, V. Interactive Image Segmentation with Latent Diversity. In Proceedings of the 2018 IEEE/CVF Conference on Computer Vision and Pattern Recognition (CVPR), Salt Lake City, UT, USA, 18–23 June 2018.
15. Majumder, S.; Yao, A. Content-Aware Multi-Level Guidance for Interactive Instance Segmentation. In Proceedings of the 2019 IEEE/CVF Conference on Computer Vision and Pattern Recognition (CVPR), Long Beach, CA, USA, 15–20 June 2019.
16. Jang, W.D.; Kim, C.S. Interactive Image Segmentation via Backpropagating Refinement Scheme. In Proceedings of the 2019 IEEE/CVF Conference on Computer Vision and Pattern Recognition (CVPR), Long Beach, CA, USA, 15–20 June 2019.
17. Sofiiuk, K.; Petrov, I.; Barinova, O.; Konushin, A. F-brs: Rethinking backpropagating refinement for interactive segmentation. In Proceedings of the IEEE/CVF Conference on Computer Vision and Pattern Recognition, Seattle, WA, USA, 14–19 June 2020; pp. 8623–8632.
18. Lin, Z.; Zhang, Z.; Chen, L.Z.; Cheng, M.M.; Lu, S.P. Interactive Image Segmentation With First Click Attention. In Proceedings of the 2020 IEEE/CVF Conference on Computer Vision and Pattern Recognition (CVPR), Seattle, WA, USA, 14–19 June 2020.
19. Liew, J.H.; Wei, Y.; Xiong, W.; Ong, S.H.; Feng, J. Regional Interactive Image Segmentation Networks. In Proceedings of the 2017 IEEE International Conference on Computer Vision (ICCV), Venice, Italy, 22–29 October 2017.
20. Mohanty, S.P. CrowdAI Dataset. Available online: https://www.crowdai.org/challenges/mapping-challenge/dataset_files (accessed on 12 June 2018).
21. Hariharan, B.; Arbelaez, P.; Bourdev, L.D.; Maji, S.; Malik, J. Semantic Contours from Inverse Detectors. In Proceedings of the International Conference on Computer Vision, Washington, DC, USA, 6–13 November 2011.
22. Mcguinness, K.; O'Connor, N.E. A comparative evaluation of interactive segmentation algorithms. *Pattern Recognit.* **2010**, *43*, 434–444. [CrossRef]
23. Perazzi, F.; Pont-Tuset, J.; McWilliams, B.; Gool, L.V.; Gross, M.H.; Sorkine-Hornung, A. A Benchmark Dataset and Evaluation Methodology for Video Object Segmentation. In Proceedings of the 2016 IEEE Conference on Computer Vision and Pattern Recognition, CVPR 2016, Las Vegas, NV, USA, 27–30 June 2016; pp. 724–732. [CrossRef]

24. Lin, T.; Maire, M.; Belongie, S.J.; Hays, J.; Perona, P.; Ramanan, D.; Dollár, P.; Zitnick, C.L. Microsoft COCO: Common Objects in Context. In *Lecture Notes in Computer Science, Proceedings of the Computer Vision—ECCV 2014—13th European Conference, Zurich, Switzerland, 6–12 September 2014*; Part V; Fleet, D.J., Pajdla, T., Schiele, B., Tuytelaars, T., Eds.; Springer: Berlin/Heidelberg, Germany, 2014; Volume 8693, pp. 740–755. [CrossRef]
25. Chen, Q.; Wang, L.; Wu, Y.; Wu, G.; Guo, Z.; Waslander, S.L. Aerial Imagery for Roof Segmentation: A Large-Scale Dataset towards Automatic Mapping of Buildings. *ISPRS J. Photogramm. Remote Sens.* **2019**, *147*, 42–55. [CrossRef]
26. Chen, L.; Papandreou, G.; Kokkinos, I.; Murphy, K.; Yuille, A.L. DeepLab: Semantic Image Segmentation with Deep Convolutional Nets, Atrous Convolution, and Fully Connected CRFs. *IEEE Trans. Pattern Anal. Mach. Intell.* **2018**, *40*, 834–848. [CrossRef] [PubMed]
27. He, K.; Zhang, X.; Ren, S.; Sun, J. Deep residual learning for image recognition. In Proceedings of the IEEE Conference on Computer Vision and Pattern Recognition, Las Vegas, NV, USA, 27–30 June 2016; pp. 770–778.
28. Zhao, H.; Shi, J.; Qi, X.; Wang, X.; Jia, J. Pyramid Scene Parsing Network. In Proceedings of the 2017 IEEE Conference on Computer Vision and Pattern Recognition, CVPR 2017, Honolulu, HI, USA, 21–26 July 2017; pp. 6230–6239. [CrossRef]
29. Cheng, H.K.; Chung, J.; Tai, Y.; Tang, C. CascadePSP: Toward Class-Agnostic and Very High-Resolution Segmentation via Global and Local Refinement. In Proceedings of the 2020 IEEE/CVF Conference on Computer Vision and Pattern Recognition, CVPR 2020, Seattle, WA, USA, 13–19 June 2020; pp. 8887–8896. [CrossRef]
30. Deng, J.; Dong, W.; Socher, R.; Li, L.J.; Li, K.; Fei-Fei, L. Imagenet: A large-scale hierarchical image database. In Proceedings of the 2009 IEEE Conference on Computer Vision and Pattern Recognition, Miami, FL, USA, 20–25 June 2009; pp. 248–255.
31. Kingma, D.P.; Ba, J. Adam: A method for stochastic optimization. *arXiv* **2014**, arXiv:1412.6980.
32. Bai, X.; Sapiro, G. Geodesic Matting: A Framework for Fast Interactive Image and Video Segmentation and Matting. *Int. J. Comput. Vis.* **2009**, *82*, 113–132. [CrossRef]

remote sensing

Article

B-FGC-Net: A Building Extraction Network from High Resolution Remote Sensing Imagery

Yong Wang [1,*], Xiangqiang Zeng [1,2], Xiaohan Liao [1] and Dafang Zhuang [1]

1 State Key Laboratory of Resources and Environmental Information System, Institute of Geographical Sciences and Natural Resources Research, Chinese Academy of Sciences, Beijing 100101, China; zengxiangqiang21@mails.ucas.ac.cn (X.Z.); liaoxh@igsnrr.ac.cn (X.L.); zhuangdf@igsnrr.ac.cn (D.Z.)
2 College of Resources and Environment, University of Chinese Academy of Sciences, Beijing 100049, China
* Correspondence: wangy@igsnrr.ac.cn; Tel.: +86-10-6488-8179

Abstract: Deep learning (DL) shows remarkable performance in extracting buildings from high resolution remote sensing images. However, how to improve the performance of DL based methods, especially the perception of spatial information, is worth further study. For this purpose, we proposed a building extraction network with feature highlighting, global awareness, and cross level information fusion (B-FGC-Net). The residual learning and spatial attention unit are introduced in the encoder of the B-FGC-Net, which simplifies the training of deep convolutional neural networks and highlights the spatial information representation of features. The global feature information awareness module is added to capture multiscale contextual information and integrate the global semantic information. The cross level feature recalibration module is used to bridge the semantic gap between low and high level features to complete the effective fusion of cross level information. The performance of the proposed method was tested on two public building datasets and compared with classical methods, such as UNet, LinkNet, and SegNet. Experimental results demonstrate that B-FGC-Net exhibits improved profitability of accurate extraction and information integration for both small and large scale buildings. The IoU scores of B-FGC-Net on WHU and INRIA Building datasets are 90.04% and 79.31%, respectively. B-FGC-Net is an effective and recommended method for extracting buildings from high resolution remote sensing images.

Keywords: deep learning; building extraction; spatial attention; global information awareness; cross level information fusion

Citation: Wang, Y.; Zeng, X.; Liao, X.; Zhuang, D. B-FGC-Net: A Building Extraction Network from High Resolution Remote Sensing Imagery. *Remote Sens.* **2022**, *14*, 269. https://doi.org/10.3390/rs14020269

Academic Editor: Gabriele Bitelli

Received: 7 December 2021
Accepted: 5 January 2022
Published: 7 January 2022

1. Introduction

Building extraction from high resolution remote sensing images plays a critical role in natural disaster emergency and management [1], land resource utilization and analysis [2], and intelligent city construction and planning [3], etc. With the ongoing development of earth observation technology, automatically extracting buildings from high resolution remote sensing imagery has gradually become one of the most vital research topics [4]. Despite the wealth of spectral information provided by high resolution remote sensing imagery [5], the spectral discrepancy among the various buildings coupled with complex background noise poses a significant challenge to automatic building extraction [6]. Therefore, a high precision and high performance extraction method for building extraction automation is urgently needed.

According to the different classification scales, there are two leading conventional approaches for the extraction of buildings from high resolution remote sensing imagery: pixel based and object based [7]. Pixel based thought regards a single pixel or its neighbouring pixels as a whole, which can extract building information by the spectral similarities principle [8]. Commonly used pixel based methods include maximum likelihood classification [9,10], decision tree, random forest, and support vector machine [11]. However, these methods may result in extremely serious salt and pepper noise [12] because of the characteristics of the same spectrum foreign matter and the same object heterogenic spectrum in remote sensing imagery. An object based approach normally takes the homogeneous pixels obtained by image segmentation [13] as basic units and classifies these homogeneous pixels on the basis of the variability of spectral, shadow, geometric, and other characteristics [14]. Although this method exploits the spatial information of buildings and effectively avoids the phenomenon of salt and pepper noise, the method is applicable only to the extraction of buildings with small areas and simple types; it is rather difficult to extract buildings with large ranges and high complexity because of the vulnerability to human factors [5]. The conventional methods may seem to have difficulty meeting the requirements of high precision, high performance, and automatic building extraction.

Recently, with the rapid advancement of artificial intelligence technology such as deep learning (DL), significant progress has been made in the extraction of various ground objects using convolutional neural networks (CNNs) [6]. CNNs have the potentiality to automatically learn the correlation features among ground objects from the input remote sensing imagery, avoiding the influence of human factors in conventional methods. Therefore, CNNs are widely applied in some files of feasibility prediction, classification extraction, and the automatic identification of ground objects [15], such as automatic mapping of cone karst [16], landslide susceptibility mapping [17] and automatic road extraction [18]. CNNs, which consist of multiple interconnected layers, including convolution layers, pooling layers, and activation functions [19], obtain hierarchical features of buildings by automatically encoding remote sensing imagery with the merits of local perception and parameter sharing [20]. CNNs have emerged as a building extraction method with high accuracy, great performance, and excellent automation capability. Simultaneously, the large amount of high resolution remote sensing imagery data provides sufficient training samples [21]. The performance of CNN based approaches is promoted in the data driven model, which dramatically enhances the generalization of building extraction. Notably, some studies showed that adding attention modules to CNNs can help the network pay more attention to and perceive contextual information and global features [22–25].

U-Net [26], as representative of CNN based approaches, has powerful feature extraction capability and superior recognition performance in the field of medical image segmentation. However, it is still extremely challenging to directly use U-Net to extract buildings from high resolution remote sensing images due to the spectral discrepancy, background, and complex noise interference of different buildings. Possible issues are as follows: (1) The difficulty of model training. U-Net acquires robust local information using continuous convolution; nevertheless, deep stacked convolutions tend to hinder model training and cause the degradation of the model performance [27,28]. (2) The lack of capacity for low level features (obtained by the U-Net encoder) representation. Due to the variety and complexity of buildings, the low level features acquired by the encoder convey less spatial detail information about the building features with much redundant information. Previous studies have shown that low level features may fail to convey the spatial detail information of ground objects in the face of high complexity ground objects [29,30]. (3) The insufficient integration of global information. U-Net aggregates the feature information extracted by the convolution layer through four max pooling steps, which not only reduces the computational complexity but increases the receptive fields of the feature maps. However, the standard convolution operation could capture only local neighborhood information and not effectively perceive global semantic information, for feature maps with large receptive fields [31]. (4) The inadequate cross level aggregation. Although U-Net employs skip connections to enhance the utilization of low level features, this method, with a simple concatenation operation, ignores the influence of redundant information and the semantic gap between low and high level features, which in turn limits the building extraction performance [6,21,32].

To solve the issues mentioned, a building extraction network (B-FGC-Net) based on residual learning, aggregated spatial attention (SA) units, global feature information awareness (GFIA) modules, and cross level feature recalibration (CLFR) modules is proposed in this work. The residual learning and SA unit is introduced in the encoder, which accelerates the convergence rate of gradient descent and highlights the features of spatial detail information of the buildings. The GFIA module captures the contextual information and improves the global awareness capability. The CLFR module, thoroughly considering the semantic gap between low and high level features, completes the effective fusion of cross level feature information from the channel dimension, suppresses the redundant information of low level features, and improves the building extraction performance of the model. Compared with the conventional building extraction methods, the B-FGC-Net, integrating residual learning, SA, GFIA, and CLFR, outperforms the capacity of feature highlighting, global awareness, and cross level information fusion, achieving superior performance in the building extraction from high resolution remote sensing imagery.

2. Related Work

Since fully convolutional neural network (FCN) [33] was proposed, the end to end deep convolutional neural network (DCNN) has received great attention. To solve the problem that spatially detailed information is difficult to recover in image segmentation, the low level features are mapped gradually by skip connection [26,34–36] and decoded in the decoder part. The methods based on skip connection allow the direct utilization of detailed low level features to restore the spatial resolution without additional parameters. However, using too much and stacked convolution in the encoder while obtaining more effective and sufficient low level features poses a risk of hindering the convergence speed and decreasing the prediction performance of the model. On this basis, residual learning was introduced into the end to end DCNN to alleviate the degradation problem due to multiple convolutional layers [31,37,38]. This scheme not only speeds up the training of the model but also effectively facilitates the utilization of low level features [39].

The DCNN with residual learning obtains rich low level features (e.g., semantic information) but the semantic information is less strong with significant redundant information [29]. The simple convolution operator, with the characteristic of focusing only on local regions, in addition to the difficulty of obtaining the spatial location relationship of each feature point, may fail to effectively capture detail rich spatial location information in low level features [40]. Therefore, it is urgent to design a new scheme in the encoder to capture the spatial relationship of feature points and highlight the expression of building features at the spatial level. The self attention mechanism [41], for example, was applied in the encoder of the GCB-Net [30] and the NL-LinkNet [42], which filtered the interference of noisy information and constructed the long range dependencies among each pixel. Furthermore, due to the semantic gap between low and high level features in the end to end DCNN, a simple cross level fusion method, such as channel concatenation in U-Net [26] and pixel addition in LinkNet [37], may cause the model to ignore the usefulness of all features and limit the propagation of spatial information between the encoder and decoder. For instance, LANet proposed an attention embedding module to bridge the gap in spatial distribution between high and low level features [43].

The encoder part of the end to end DCNN generates a feature map with small spatial resolution and large receptive fields. Actually, the standard convolution is weak in global information awareness for this feature map. A possible way to remedy the issues is to apply multiparallel dilated convolution or other submodules, which could capture the multiscale contextual and global semantic information, and enlarge the receptive fields to improve global information awareness. For instance, the pyramid pooling module (PPM) of PSPNet [44] captures multiscale information; DeepLabV3+ [45] constructs the atrous spatial pyramid pooling (ASPP) module based on dilated convolution to obtain contextual information; D-LinkNet [31] designs a specific cascaded operation of the dilated convolution unit (DCU) according to the spatial resolution of feature maps, which effectively obtains a larger range of feature information; HsgNet [46] proposes the high order spatial information global perception module to adaptively aggregate the long range relationships of feature points. However, the above methods have low extraction accuracy, excessive memory consumption, or computational complexity, which make it difficult to promote their application.

3. Methodology

In this section, we will describe the proposed method in detail. Firstly, the overall architecture of the model is described. Then, the spatial attention units, global feature information awareness modules, and cross level feature recalibration modules, and loss functions are elaborated.

3.1. Model Overview

The B-FGC-Net, consisting of the encoder, GFIA module, and decoder, is a standard end to end DCNN model, as shown in Figure 1. First, the method takes remote sensing images of buildings as the input to the encoder, which uses the residual learning block (Res-Block) and SA unit to obtain the feature information of the buildings automatically. Continuously, GFIA modules aggregate the contextual information by the self attention unit and the dilated convolution. Finally, the decoder uses multiple effective decoder blocks and CLFR modules to restore feature maps to the final building segmentation maps.

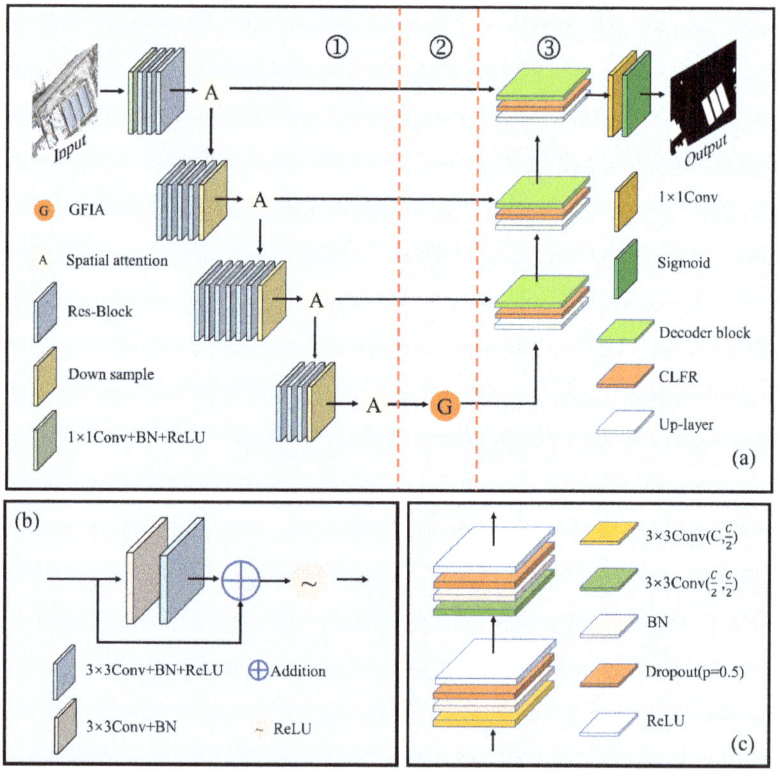

Figure 1. Overall of the proposed framework. (**a**) Structure of the B-FGC-Net, in which ①, ②, and ③ denote the encoder, the GFIA module, and the decoder, respectively, Down sample denotes downsampling, Up-layer denotes upsampling; (**b**) Res-Block; (**c**) Decoder-Block, where C denotes the number of channels of the feature map, and p is the probability of an element being zeroed. The addition and ReLU represent the pixel addition and the Rectified Linear Unit, respectively. The 1×1 and 3×3 denote the convolution kernel size.

The encoder takes ResNet-34 as the backbone network to extract low level features and removes the 7 × 7 convolution and max-pooling of the initial layer and the global average pooling and fully connected layer of the final layer. The input data is processed by four repeated groups of convolution layers, each of which contains multiple Res-Blocks (see Table 1) to generate different hierarchical low level features. At the end of each group of convolution layers, those low level features are delivered into the SA unit in four groups to further highlight potential information such as space, shape, and edge features of the building and to suppress backgrounds such as roads, trees, and farmland. A detailed description of the SA unit is provided in Section 3.2. Additionally, the stride of the convolution of downsampling is set to 2, achieving the goal of reducing the spatial resolution of feature maps by $\frac{1}{4}$ and doubling the number of channels. Although the receptive fields of feature maps are increased due to several downsampling operations, some rich spatial information is lost. It is rather difficult to recover the detailed and global semantic information by using only upsampling and standard convolution operations. In this work, we fuse the low level features generated in stages 1, 2 and 3 with high level features, expecting to recover the spatial information of feature maps. The GFIA module utilizes the low level features generated in stage 4 with the large receptive fields, which is helpful to obtain the semantic information of building features and improve the sensing ability of the global information. The encoder structure and the dimension variation of low level features are shown in Table 1.

Table 1. The encoder structure and the dimension variation of low level features. SA_1, SA_2, SA_3, and SA_4 denote the SA units of stages 1, 2, 3, and 4, respectively. Here, 3 × 256 × 256 represents the number of channels, height, and width, respectively. In addition, 3× Res-Block denotes three Res-Blocks.

Stage	Template	Size
Input	-	3 × 256 × 256
1	1 × 1 Conv + BN + ReLU 3 × Res-Block SA_1	64 × 256 × 256
2	4 × Res-Block SA_2	128 × 128 × 128
3	6 × Res-Block SA_3	256 × 64 × 64
4	3 × Res-Block SA_4	512 × 32 × 32

The GFIA module perceives a larger range of feature maps to capture the effective contextual information of the buildings by dilated convolution. Meanwhile, the self attention mechanism focuses on the spatial relationship of each feature point. The combination of the above two methods enables the high level features to enter the decoder to complete the decoding operation. The decoder perceives the global information and restores the spatial detail information of the features. Section 3.3 presents the GFIA module.

Bilinear interpolation and 1 × 1 convolution were adopted to recover the resolution of feature maps in the decoder. To overcome the semantic gap between low and high level features, we use the CLFR module described in Section 3.4 to focus on the complementary relationship between them, to diminish the interference of noise information and to improve the utilization of useful low level feature information. Thereafter, the decoder block decodes the fused feature maps through two convolution operations to output the final building extraction result. To prevent overfitting, dropout [47] and batch normalization (BN) [48] are introduced after each convolution operation of the decoder block to simplify the decoding structure and improve the training speed, respectively.

3.2. Spatial Attention

For the natural properties of buildings and the complexity of the background, such as roofs of various colors and shape features, the standard convolution operation focuses on neighborhood pixels and may fail to accurately obtain the distribution of each pixel and explore the spatial relationships on the overall space. Based on this observation, our study proposed an SA unit inspired by the convolutional block attention module (CBAM) [49], as shown in Figure 2. The SA unit aims to explore the spatial distribution regularity of pixels, highlight the building feature expression, and suppress the interference of background.

The SA consists of three major components: pooling, convolution, and excitation. Through three key steps, the SA automatically learns the feature expressions in spatial dimensions and adaptively acquires the spatial weights of each feature.

(1) Pooling: the feature map $x \in R^{C \times H \times W}$ is compressed in the channel dimension by the global average pooling and the global max pooling to optimize the spatial distribution information of each feature point. The pooling can be defined by Equation (1).

$$z = f_C(f_{GAP}(x), f_{GMP}(x)) \tag{1}$$

where $f_C(\cdot)$ represents the channel concatenate operation, $f_{GAP}(\cdot)$ and $f_{GMP}(\cdot)$ represent the global average pooling and global max pooling, respectively, and W *and* H are the width and height of the feature map, respectively.

(2) Convolution: 7×7 convolution and sigmoid activation function can autonomously learn the spatial distribution relationship of features and optimally assign weights to each feature point. The spatial attentional feature map $s \in R^{1 \times W \times H}$ is obtained by Equation (2).

$$s = f_{conv2d}(z) = \sigma_s(w(z)) \tag{2}$$

where $f_{conv2d}(\cdot)$ is a two-dimensional convolution operation, w denotes the convolution kernel parameters, and σ_s represents the sigmoid activation function.

(3) Excitation: the spatial attentional feature map s highly expresses the spatial distribution of feature points. Then, it performs point multiplication with the input feature map x. In this manner, the model focuses on learning building features and highlighting the spatial information expression during the training. The calculation process is as follows:

$$y = f_m(x, s) + x \tag{3}$$

where $f_m(\cdot)$ denotes the point multiplication. In summary, the SA successively completes the adaptive acquisition of spatial weights for each feature point by pooling, convolution and matrix dot product operations, which highlights the expression of building features in the spatial dimension and suppresses noise information interference.

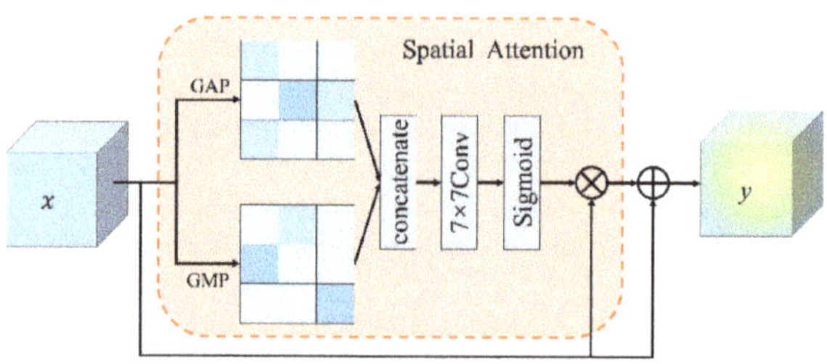

Figure 2. The structure of the spatial attention unit.

3.3. Global Feature Information Awareness

To capture multiscale contextual information and aggregate global information, we proposed the GFIA module, as illustrated in Figure 3, consisting of a dilated convolution (DC) unit and a self attention (also called nonlocal) unit. As shown in (b), compared with the standard convolution operation, the DC perceives a larger range of feature information by expanding the interval of convolution kernels. The DC unit uses five convolutions with different dilation rates to efficiently integrate the neighborhood information of the building features, which is calculated as follows:

$$F = \sum_i \sigma_r(w_i(L_{i-1})) \tag{4}$$

where $F \in R^{C\times W\times H}$ denotes the output of the DC unit, $i = \{0,1,2,3,4\}$ is the index of the values of the dilation rate, σ_r is the ReLU activation function, w_i is the parameters of the DC kernel and $L_{i-1} \in R^{C\times W\times H}$ represents the output of the previous DC. Specifically, L_{i-1} represents the input feature map x of the GFIA module when $i = 0$. In this work, the dilation rate was set to $dilation = \{1,2,3,4,8\}$, and the corresponding receptive fields of their convolutions were 3×3, 7×7, 11×11, 15×15, and 31×31, respectively. On the one hand, the DC with the continuous dilation rate avoids the omission extraction of feature information and effectively obtains multiscale contextual information. On the other hand, the convolution with a dilation rate of 8 can perceive a 31×31 feature area, which is basically able to cover the whole range of feature maps and complete the effective acquisition of global semantic information. In addition, depthwise separable convolution is introduced in the DC unit to reduce the complexity of the convolution operation. The nonlocal unit constructs three feature maps, $B \in R^{C\times H\times W}$, $C \in R^{C\times H\times W}$ and $D \in R^{C\times H\times W}$, with global information to capture the long range dependence between each feature point. The calculation process of the nonlocal unit is shown as Equations (5) and (6).

$$B = \sigma_r(w_b(x)), C = \sigma_r(w_c(x)), D = \sigma_r(w_d(x)) \tag{5}$$

$$N = f_m(D, f_m(C,B)) \tag{6}$$

where w_b, w_c and w_d denote the parameters of the convolution kernel, and $N \in R^{C\times H\times W}$ is the output of the nonlocal unit. As the model is continuously trained, the nonlocal unit automatically learns the correlation between arbitrary features and reweighs each feature to promote the concern of the model for the global information of the features.

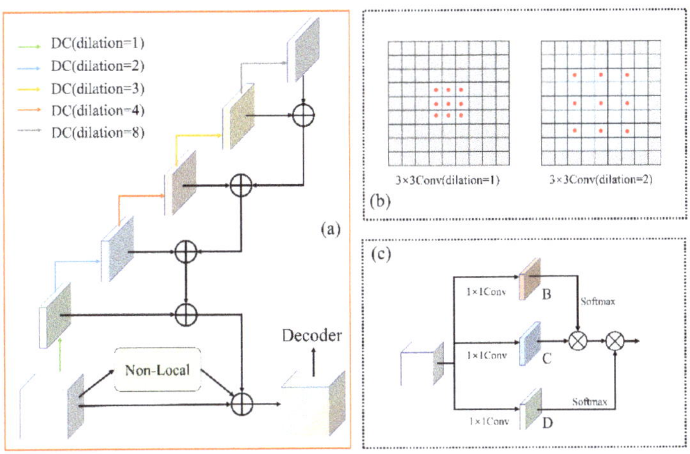

Figure 3. Overview of the GFIA module. (**a**) The structure of the GFIA module, (**b**) the comparison of standard convolution and dilated convolution, (**c**) the structure of the nonlocal units.

3.4. Cross Level Feature Recalibration

The direct feature fusion of low and high level features in the form of concatenated channels or pixel addition may cause the model to fail to learn effective complementary information among cross level features, and even inherent noise, as well as redundant information, which could affect the extraction performance of the model. Therefore, we were inspired by efficient channel attention (ECA) [50] and designed the CLFR module, as shown in Figure 4, to fuse low and high level features, which not only removes a large amount of redundant information but also eliminates the semantic gap between the pieces of redundant information.

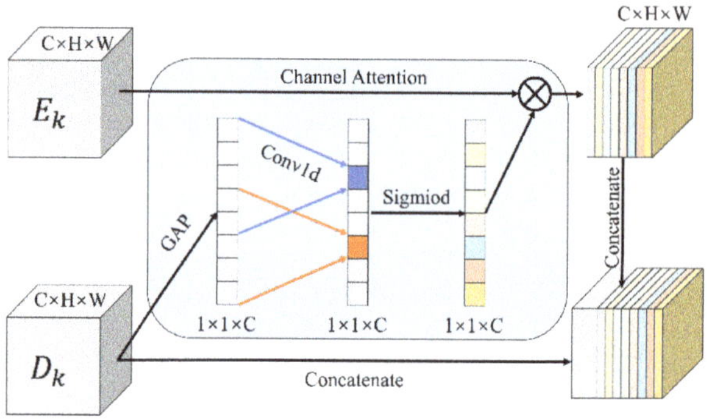

Figure 4. The structure of the CLFR module.

The CLFR module first compresses the high level features $D_k \in R^{C\times H\times W}$ in spatial dimensions by global average pooling to generate one-dimensional vectors and obtains the global semantic information of the channel dimension. Thereafter, a one-dimensional convolution is applied to obtain the weight parameters of feature points automatically. Then, the sigmoid activation function is used to highlight the correlation between the weights. In this manner, the building features in low level feature $E_k \in R^{C_k\times H_k\times W_k}$ are highlighted, and the semantic gap between D_k and E_k is eliminated. Finally, the fused feature map is fed into the decoder block for the decoding operation. The CLFR module is defined by Equations (7) and (8).

$$y_k = f_m(E_k, \sigma_s(w_k(f_{GAP}(D_k)))) \tag{7}$$

$$out_{CLFR} = [y_k, D_k] \tag{8}$$

in which $y_k \in R^{C\times H\times W}$ denotes the low level feature after channel recalibration, w_k is the parameter of the one-dimensional convolution, and $[\cdot]$ is the channel concatenate operation. The CLFR module adaptively acquires the channel weight parameters of the high level feature D_k and eliminates the large amount of redundant information in the channel dimension of the low level feature E_k by a dot product operation. Meanwhile, it also re-evaluates the degree of the contribution of each feature point, which makes the model learn the complementary information between D_k and E_k and overcome the semantic gap between them to maximize the effective information utilization of cross level features.

3.5. Loss Function

The binary cross entropy (BCE) loss, the boundary error (BE) loss [21], and the auxiliary loss were utilized to train the model, as shown in Figure 5.

BCE loss: given a couple of labels, y_{lab}, and prediction results, y_{pro}, the loss, l_{bce}, among them is calculated by Equation (9).

$$l_{bce} = -\frac{1}{HW}\sum_i^H \sum_j^W (y_{lab}\log y_{pro} + (1-y_{lab})\log(1-y_{pro})) \tag{9}$$

BE loss: while the BCE loss enables the model to focus on the correct classification of each pixel in the prediction results, there are still challenges in building boundary refinement. Thus, we use the BE loss to force the model to pay more attention to the boundary information of buildings. The boundary loss l_{be} is defined by Equation (10).

$$l_{be} = -\frac{1}{HW}\sum_i^H \sum_j^W \left(\frac{N}{P+N} z_{lab} \log z_{pro} + \frac{P}{P+N}(1 - z_{lab}) \log(1 - z_{pro}) \right) \quad (10)$$

where z_{lab} and z_{pro} denote the label and the prediction result after processing by the Laplacian operator, respectively, and P and N denote the number of positive and negative pixels in the label, respectively.

Auxiliary loss: To facilitate model training, the output of ResNet34 in stage 3 is upsampled to the same size as the label, and then the auxiliary loss, l_{aux}, between the label and prediction result is calculated by the BCE loss.

Thus, the final total loss of our network is:

$$l = \lambda_1 \times l_{bce} + \lambda_2 \times l_{be} + \lambda_3 \times l_{aux} \quad (11)$$

in which $\lambda_1 = \lambda_2 = 1$ and $\lambda_3 = 0.4$.

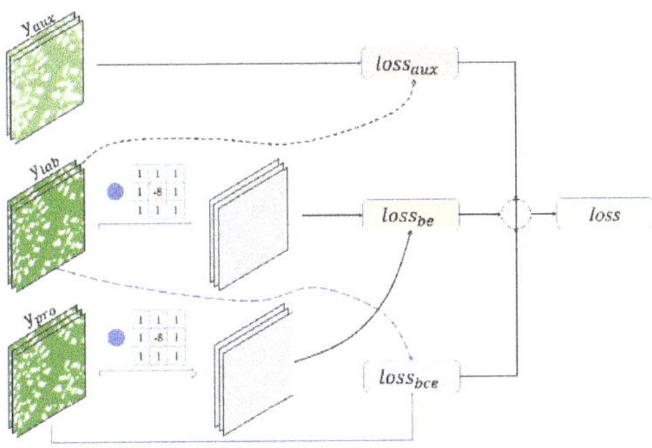

Figure 5. Flow chart of the loss function. The 3 × 3 matrix represents the Laplacian operator.

4. Experiments and Results

4.1. Datasets

In this work, the WHU building dataset and the INRIA aerial image labeling dataset were used to train and evaluate our proposed method.

The WHU building dataset, open source shared by Ji et al. [51], has become a very popular dataset in the field of remote sensing building extraction due to its wide coverage, high spatial resolution, and volume of data. This dataset covers 450 km^2 in Christchurch, New Zealand, with a spatial resolution of 7.5 cm and contains about 22,000 independent buildings with high image quality. The WHU building dataset consists of 4736, 1036 and 2416 images for training, validation and testing, respectively. Considering the limitation of computer graphics memory, we resized the original images and the ground truth from 512 × 512 pixels to 256 × 256 pixels. Figure 6 shows the processed training set, validation set, and test set data.

The INRIA aerial image labeling dataset [52] provides 360 remote sensing images with a size of 5000 × 5000 pixels and a spatial resolution of 0.3 m. The dataset contains various building types, such as dense residential areas in ten cities around the world. This dataset only provides ground truth in the training set but not in the testing set. Therefore, we selected the first five images of five cities in the training set for the testing set according to suggestions by the data organizers and [3]. Due to the large size of images and the limitation of the computer GPU memory, we cropped them into 500 × 500 pixels and resized them to 256 × 256 pixels to meet the input dimension requirements of the model. The original INRIA images and the preprocessed images are shown in Figure 7.

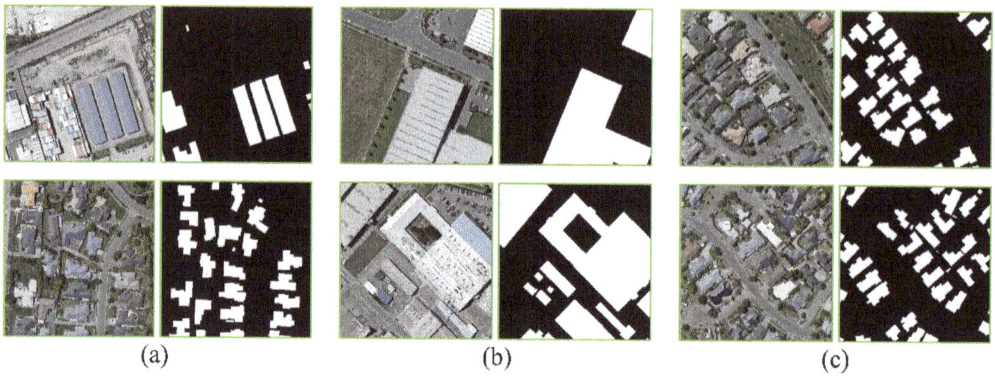

Figure 6. Examples of the original images and the ground truth of the WHU building dataset. (**a**–**c**) are training, validation, and testing samples, respectively.

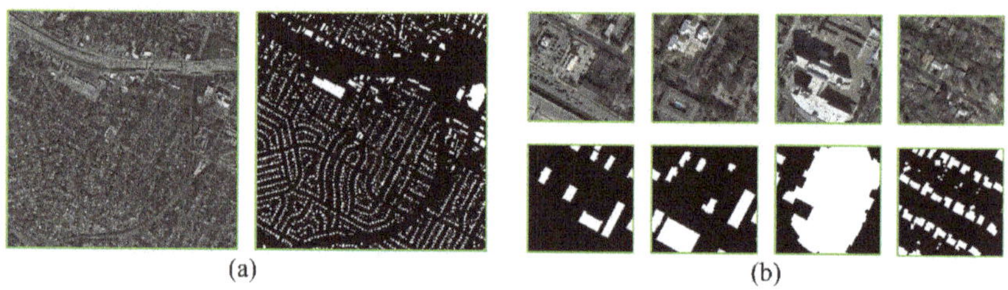

Figure 7. Examples of the images and the ground truth of the INRIA aerial image labeling dataset. (**a**,**b**) are the original dataset and the preprocessed image examples, respectively.

4.2. Experimental Settings

As shown in Table 2, the proposed B-FGC-Net was implemented based on Python-3.7 and PyTorch-1.7 in the CentOS 7 environment. We adopted an Adam optimizer [53] with an initial learning rate of 0.0001, which decayed at a rate of 0.85 after every five epochs. Additionally, we accelerated the training with two NVIDIA RTX 2080Ti GPUs. To avoid the risk of overfitting, data augmentation approaches were used during training, including random horizontal–vertical flipping and random rotation.

Table 2. Experimental environment and parameter settings.

Hardware Configuration		**Parameter Settings**	
Operating system	CentOS 7	Epoch	100
DL framework	Pytorch 1.7	Batch size	16
Language	Python 3.7	Optimizer	Adam
GPU	24G	Initial learning rate	1×10^{-4}

4.3. Evaluation Metrics

To objectively evaluate the performance of the proposed method, on the basis of [3,4,54,55], we use five evaluation metrics, including overall accuracy (OA), precision (P), recall (R), F1 score (F1), and intersection over union (IOU), to comprehensively evaluate the building extraction performance.

4.4. Result

4.4.1. Experiment Using the WHU Building Dataset

Figure 8 shows several extraction results of B-FGC-Net on the WHU building dataset. We randomly selected six typical images for testing, including both small scale buildings and large scale buildings, to verify the extraction performance of the proposed method. For the small scale buildings displayed in Columns 1 to 3 in Figure 7, the B-FGC-Net with SA introduced can accurately locate the spatial position of buildings and effectively identify the background as nonbuildings. Additionally, for the large scale buildings displayed in Columns 4 to 6 in Figure 7, B-FGC-Net with GFIA can extract the buildings quite completely and avoid building omission as much as possible. Comprehensively observing the labels and extraction results, although there are very few cases of building omission and error extraction, the B-FGC-Net proposed in this work can effectively and accurately extract most of the building information in both cases and shows superior building extraction performance.

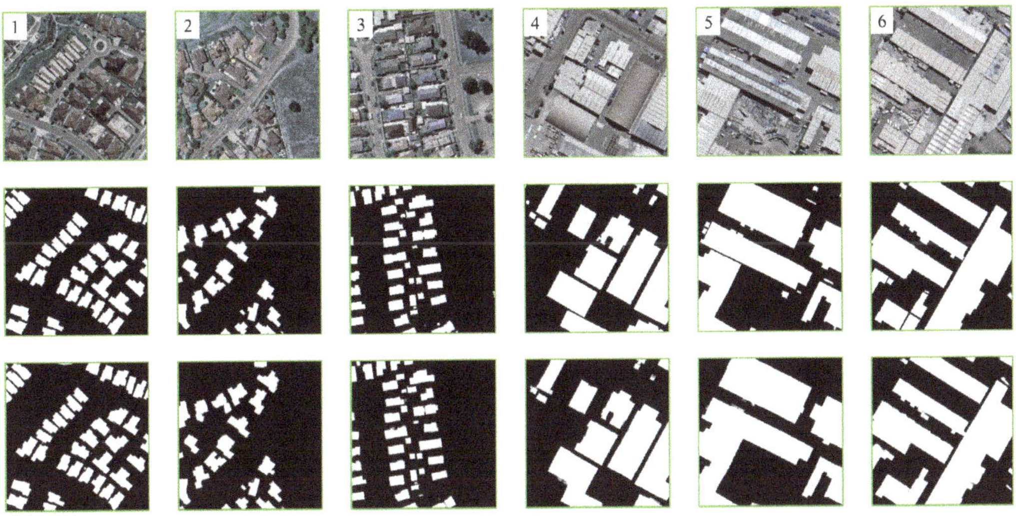

Figure 8. Building extraction results of the B-FGC-Net on the WHU building dataset. The first to third rows are the original images, labels, and results, respectively. The numbers 1–6 represent the index in which the image is located.

Figure 9 quantitatively evaluates the building extraction results of B-FGC-Net in Figure 8. According to Figure 9, the OA of B-FGC-Net is above 98.1% in both cases, indicating that B-FGC-Net can correctly distinguish between buildings and background. Extracting small scale buildings is still challenging because of their few building pixels. Nevertheless, the method proposed in this work achieves remarkable performance, with an F1 score above 96.7% and an IOU score above 93.6%. In addition, the F1 score and IOU of 97.6% and 95.4%, respectively, further demonstrate the high accuracy of the method for large scale building extraction. In short, B-FGC-Net possesses high accuracy for both small scale and large scale building extraction.

4.4.2. Experiment Using the INRIA Aerial Image Labeling Dataset

The building extraction results of randomly selected images from the INRIA aerial image labeling dataset are shown in Figure 10. From the results of Columns 1–3, B-FGC-Net is seen to show excellent recognition performance for small scale buildings and can accurately detect spatial location information. Similar results are observed in Figure 10 for large scale buildings, in which the proposed method can extract most of the buildings completely and avoid the phenomenon of missing extraction or incorrect extraction. In the extraction results of Column 4, B-FGC-Net exhibits excellent building extraction capability and avoids interference from noise information such as building shadows and trees. Particularly, in the case of complex urban building scenes (see Column 5), the B-FGC-Net model accurately extracts the vast majority of building information by fusing multiscale feature information.

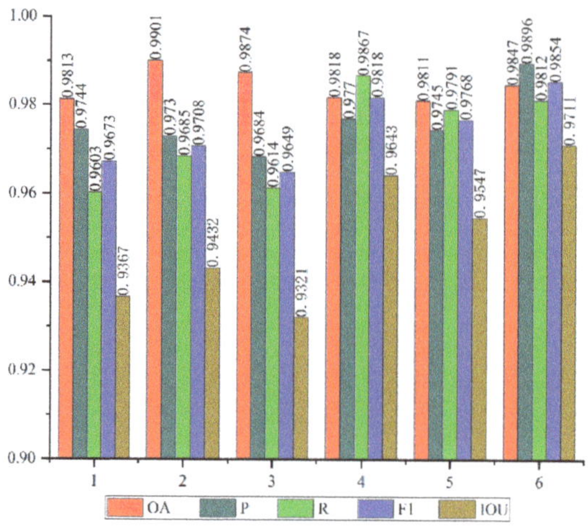

Figure 9. Evaluation results on the WHU building dataset.

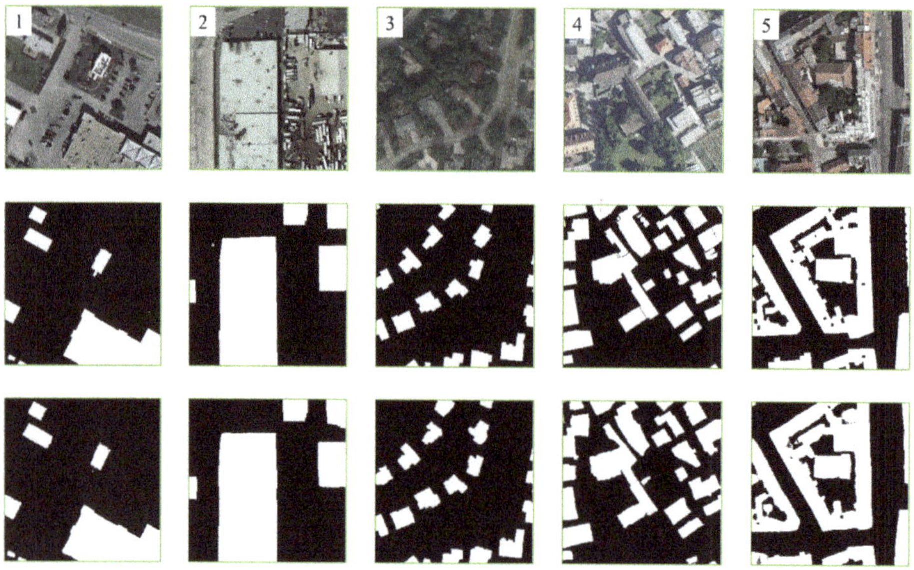

Figure 10. The extraction results of B-FGC-Net on the INRIA aerial image labeling dataset. The first to third rows are the test images, labels, and results, respectively. Numbers 1–5 represent Austin, Chicago, Kitsap County, West Tyrol, and Vienna, respectively.

Figure 11 presents the accuracy evaluation results of B-FGC-Net for five cities on the INRIA aerial image labeling. As shown in Figure 11, the OA score of B-FGC-Net exceeds 94% in all five cities, which indicates that the method proposed in this work can correctly distinguish between buildings and background. Since there are nonbuilding pixels of 97.89% and fewer building pixels of 2.11% in Kitsap County, this extreme imbalance among positive and negative sample numbers results in an OA of 99.19%, but is imprecise. In contrast, the F1 score of 80.44% and IOU of 67.28% in Kitsap County indicate that the method still achieves excellent extraction accuracy in this case. Observing the F1 score (90.5%) and IOU (82.65%) of Vienna thoroughly shows that the method performs well for buildings with high complexity. In sum, B-FGC-Net scored over 80.4% F1 on the five cities, with high extraction accuracy on small scale, large scale, and high complexity buildings.

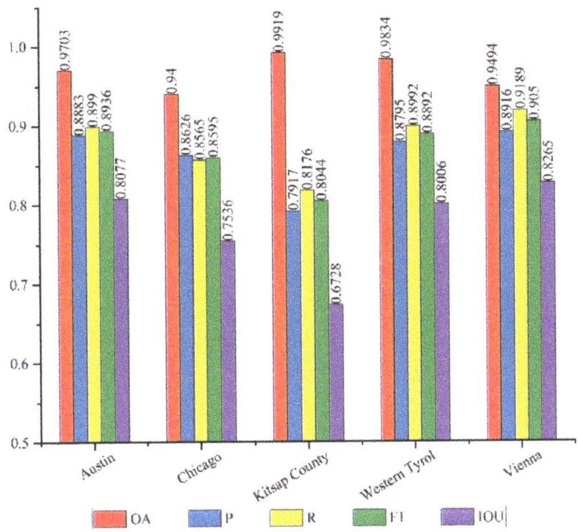

Figure 11. Evaluation results on the Inria Aerial Image Labeling dataset.

5. Discussion

5.1. Comparison of Different Classical Methods

To further examine the performance and accuracy of the proposed method, we used several different classical methods for semantic segmentation to compare and analyze, such as U-Net, LinkNet, SegNet, and DeepLabV3. These methods were trained at the same learning rate and optimized on two public building datasets. We also comprehensively analyzed the extraction accuracy of each method, and the experimental results were as follows.

5.1.1. On the WHU Building Dataset

Figure 12 exhibits the building extraction results of different methods on the WHU building dataset, including U-Net, Res-UNet, LinkNet, LinkNet*, and B-FGC-Net, where the encoder of Res-UNet is ResNet18 and LinkNet* removes the initial convolutional layer and max pooling in LinkNet.

As displayed in Figure 12, B-FGC-Net obtains superior visual results for building extraction compared with classical building extraction methods. Although UNet, Res-UNet, LinkNet, and LinkNet* can reasonably extract some building information, there is still a considerable number of results about building incorrect extraction and background error recognition. U-Net ignores the interference of building shadows in the fifth row in Figure 11 (see the blue rectangular box) and identifies the majority of building pixels. However, U-Net has a poor performance in locating small scale buildings and integrating large scale buildings, as shown in the red rectangular box in Figure 11. The extraction result of Res-UNet in the fourth row seems to be slightly better than the extraction result of UNet, but the majority of the buildings are misclassified as background, reflecting the poor extraction performance of Res-UNet. LinkNet, as a lightweight image segmentation network, greatly reduces the training time by reducing the image spatial resolution in the initial layer. From the extraction results, LinkNet identifies several building pixels in the fourth row, but too many holes occur. Therefore, we removed the LinkNet initial layer 7×7 convolution and max-pooling, called LinkNet*, to verify whether the excessive downsampling causes poor extraction performance and to reflect the rationality of the initial layer design of the B-FGC-Net. As displayed in Figure 12g, LinkNet* shows better integration ability for large scale buildings than the previous three methods but poorer capability for identifying small scale buildings and overcoming building shadows.

B-FGC-Net, with the merit of the SA, GFIA, and CLFR modules, effectively overcomes the interference of building shadows and performs favorably in extracting small scale and large scale buildings. From the yellow box, we find that the proposed method, with the support of SA, distinguishes the background and buildings properly and recognizes small scale buildings easily. Furthermore, almost all large scale building pixels are correctly and completely detected by B-FGC-Net, mainly because the CLFR module enhances the ability of global perception. Especially in the extraction results of the fourth row, compared with [4], B-FGC-Net extracts most of the buildings more completely. In the blue box, the proposed method can handle the interference of building shadows better, which makes the extraction results precise.

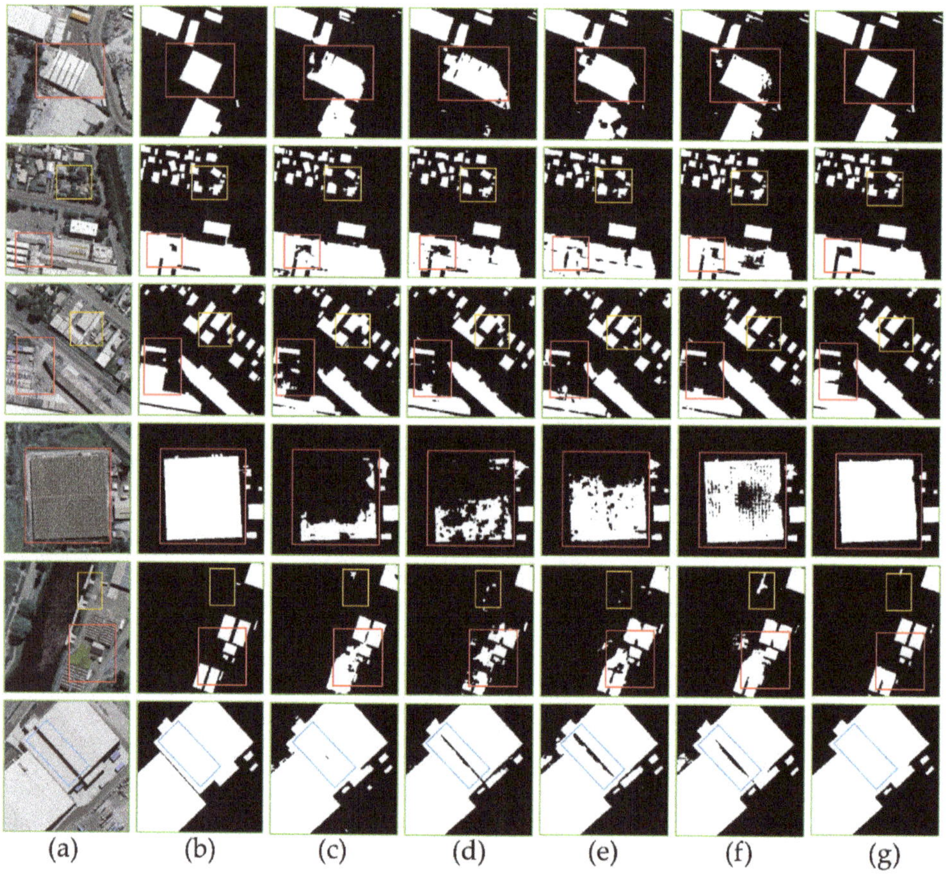

Figure 12. Extraction results of different models on the WHU Building Dataset. (**a**) Original image, (**b**) label, (**c**) U-Net, (**d**) Res-UNet, (**e**) LinkNet, (**f**) LinkNet*, (**g**) B-FGC-Net.

Table 3 quantifies the building extraction accuracy of several methods in the WHU building dataset. In contrast to other methods, B-FGC-Net achieved excellent accuracy in all evaluation metrics. In terms of OA score, the proposed method obtains 98.90%, which performs favorably against other methods and acquires the optimum extraction accuracy in distinguishing building and background. Compared with U-Net, the F1 score and IOU of B-FGC-Net were improved by 1.7% and 3.02%, respectively, indicating that the SA, GFIA, and CLFR can effectively improve the model precision. In particular, the result of the second best method (i.e., LinkNet*) proves that excessive downsampling can decrease the precision of the DL model and reflects the reasonableness of the B-FGC-Net design. Compared with LinkNet*, B-FGC-Net exhibited the best extraction performance on the test set with an increase in F1 score and IOU of 0.82% and 1.47%, respectively. Compared with recent work such as PISANet [56] and Chen's method [4], the evaluation results of this method are still optimal.

Table 3. Accuracy evaluation results of different methods on the WHU building dataset. PISANet and Chen's model are implemented by [56] and [4] respectively. '-' denotes that the paper did not provide relevant data.

Methods	OA (%)	P (%)	R (%)	F1 (%)	IOU (%)
U-Net	98.54	93.42	92.71	93.06	87.02
Res-UNet	98.49	91.44	94.00	92.70	86.40
LinkNet	97.99	92.16	89.09	90.60	82.82
LinkNet*	98.72	94.88	93.02	93.94	88.57
SegNet	97.15	85.90	86.78	86.33	75.95
DeeplabV3	97.82	88.93	90.16	89.54	81.06
PISANet	96.15	94.20	92.94	93.55	87.97
Chen's	-	93.25	95.56	94.40	89.39
B-FGC-Net	98.90	95.03	94.49	94.76	90.04

5.1.2. On the INRIA Aerial Image Labeling Dataset

Figure 13 exhibits the extraction results of B-FGC-Net and five other methods on the INRIA aerial image labeling dataset. From the results, we find that UNet, Res-UNet, LinkNet, SegNet, and DeepLabV3 identify most of the background, such as trees and roads, but suffer from error extraction and missing extraction compared with B-FGC-Net. Building extraction presents a great difficulty and challenge for classical methods due to the similar spectral features between buildings and backgrounds in the red rectangular box of Rows 1–3. Conversely, the proposed method extracts large scale buildings more completely and overcomes the interference of similar spectral features excellently. The extraction results of the classical methods can be seen in the red rectangular boxes in Row 4–5 of Figure 13, which are still unsatisfactory in terms of small and large scale buildings and serious building error extraction phenomena remain. However, B-FGC-Net almost perfectly eliminates the "sticking phenomenon" of small scale building extraction results by highlighting the building features in spatial and channel dimensions through the SA unit and the CLFR module. In other challenging building scenes, such as building shadows (the sixth row of Figure 13), tree shading (the seventh row of Figure 13) and complex urban architecture (the eighth row of Figure 13), the other five classical methods all present the disadvantages of incomplete extraction results and inaccurate location of the outer boundary of the building. Fortunately, B-FGC-Net achieved satisfactory visual performance through the SA unit, the GFLA module, and the CLFR module, to suppress the representation of noise information, to integrate multiscale contextual information, and to complete the effective fusion of cross level information.

The accurate results on the INRIA aerial image labeling dataset are shown in Table 4. We clearly found that the OA, F1 score, and IOU of all methods were above 95%, 83%, and 71%, respectively, further demonstrating the good performance of the end to end DCNN in the field of building extraction. Compared with other methods, the proposed method achieves the best performance in all metrics and obtains the highest OA, F1, and IOU, of 96.7%, 88.46%, and 79.31%, respectively. Furthermore, the IOU and F1 score of LinkNet* was increased by 5.65% and 3.67%, respectively, on this dataset compared to LinkNet, again showing that excessive downsampling in the initial layer may affect the extraction accuracy of the model and reflecting the rationality of removing downsampling in the initial layer in the proposed method. The F1 score and IOU of B-FGC-Net improved by 0.58% and 0.93%, respectively, over LinkNet*. In detail, when compared with U-Net, B-FGC-Net achieves a large increase in IOU and F1 scores, of 3.51% and 2.22%, indicating that the attention mechanism and dilated convolution are effective. As described in Section 4.4.2, the excessive sample imbalance makes the OA of AMUNet [32] slightly better than our method, but it is not accurate. In terms of IOU score, B-FGC-Net is 2.35% and 2.11% higher than AMUNet and He's model [3], respectively. These improvements demonstrate that the B-FGC-Net is robust enough to handle sample imbalances and complex buildings.

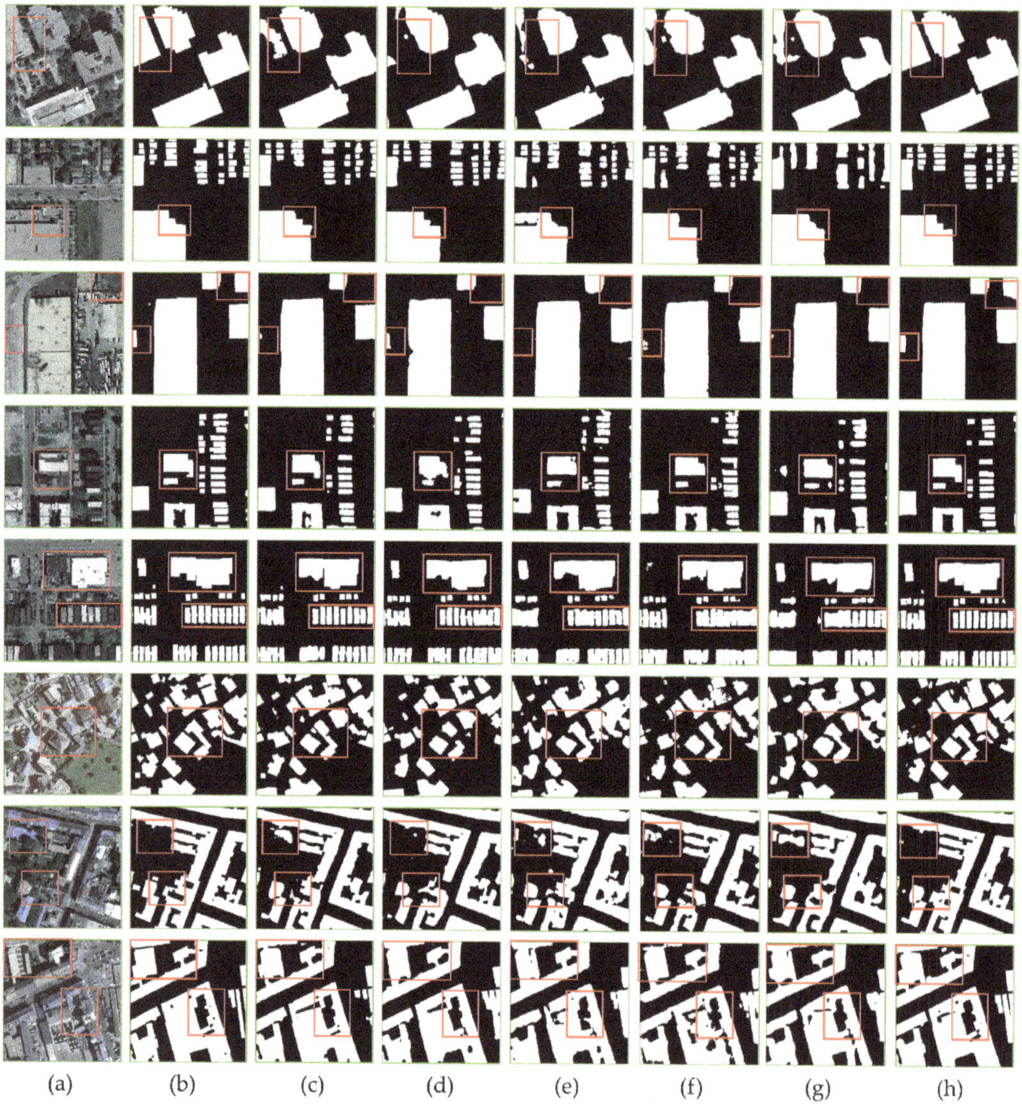

Figure 13. The extraction results of different methods on the INRIA aerial image labeling dataset. (**a**) Original images, (**b**) labels, (**c**) U-Net, (**d**) Res-UNet, (**e**) LinkNet, (**f**) SegNet, (**g**) DeepLabV3, (**h**) B-FGC-Net.

Table 4. Accuracy evaluation results of different methods on the INRIA aerial image labeling dataset. AMUNet and He's model are implemented by [32] and [3] respectively. Here, '-' denotes the unknown results that were not given by the authors.

Model	OA (%)	P (%)	R (%)	F1 (%)	IOU (%)
U-Net	96.10	84.76	87.76	86.24	75.80
Res-UNet	95.95	83.94	87.49	85.68	74.95
LinkNet	95.48	83.61	84.82	84.21	72.73
LinkNet*	96.55	86.85	88.93	87.88	78.38
SegNet	95.46	80.72	86.89	83.69	71.96
DeepLabV3	95.80	84.58	86.04	85.30	74.37
AMUNet	96.73	-	-	-	76.96
He's	-	83.50	91.10	87.10	77.20
B-FGC-Net	96.70	87.82	89.12	88.46	79.31

According to the visual results and the accuracy analysis above, we can conclude that B-FGC-Net highlights building features in the spatial dimension, aggregates multiscale contextual information and global semantic information, and effectively removes redundant information through SA, GFIA, and CLFR. Thus, B-FGC-Net achieved better visual extraction results in two datasets, especially in small scale, large scale, and complicated buildings, and overcame the noise information interference from building shadows and tree occlusions.

5.2. *Effectiveness Comparison of Different Levels of Spatial Attention*

To represent the effectiveness of different levels of spatial attention, we explored the mechanism and effects of spatial attention through contribution experiments and feature visualization operations on the WHU building dataset.

The evaluation results of different levels of SA units on the WHU Building Dataset are listed in Table 5. Compared with the No. 1 model, the No. 5 model (i.e., B-FGC-Net) achieved the best performance, with IOU and F1 score improving by 0.64% and 0.34%, respectively, indicating that the SA can increase the classification accuracy of the model. Comparing models No. 1–5 with each other, their IOU variations are 0.32%, 0.03%, 0.07% and 0.32%, respectively, demonstrating that the SA unit in layers 4 and 1 brings the most significant improvement but the importance of spatial attention in layers 2–3 cannot be neglected because Experiments 1–5 were performed gradually as the SA was added at different levels. As the SA unit is added gradually to the encoder, the F1 score and IOU gradually increase, further indicating that SA can highlight the relevant features of buildings in the spatial dimension and ignore the interference of other information.

Table 5. Evaluation results of different levels of SA units on the WHU building dataset. Note: the No. 1 model has no SA unit, and the No. 5 model is the B-FGC-Net.

No.	SA_4	SA_3	SA_2	SA_1	F1 (%)	IOU (%)
1					94.38	89.30
2	✓				94.52	89.62
3	✓	✓			94.54	89.65
4	✓	✓	✓		94.58	89.72
5	✓	✓	✓	✓	94.76	90.04

Figure 14 displays the feature visualization comparison of the B-FGC-Net model, where different brightnesses indicate different levels of attention to building features by the model. According to Figure 14, after adding the SA unit, the feature maps all appear to have different degrees of variation in brightness. The brightness of the building area is significantly increased after adding the SA unit, as shown in Figure 14b,c, suggesting that the SA unit in the first layer effectively ameliorates the overseeking of building boundary information, forcing the model to focus on building features and ignore other backgrounds.

Especially in the fourth row of visualization results, the SA highlights the representation of building features in the spatial dimension, more importantly, attenuates the brightness of building shadows, and effectively suppresses the interference of background. With the addition of the SA unit, the spatial semantic information of building features is gradually abstracted. However, the SA unit can easily be seen to increase the brightness contrast between buildings and nonbuildings, and make B-FGC-Net concentrate on learning building features. From the feature maps in Columns (h)–(j), we find that the features in the fourth layer are the most abstract, and the SA identifies buildings as red color, which enhances the ability of the B-FGC-Net to perceive the spatial information of the building features.

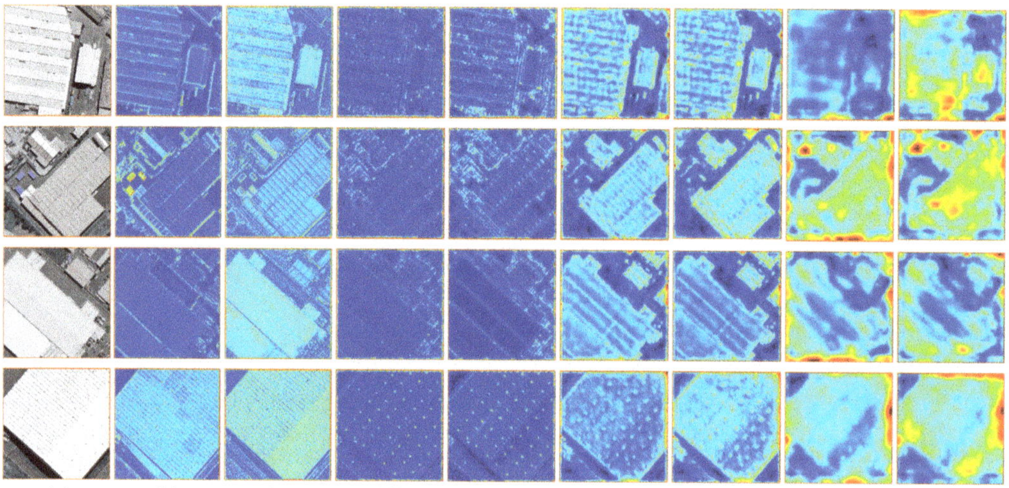

(a) (b) (c) (d) (e) (f) (g) (h) (i)

Figure 14. Visualization results of different levels of SA: (**a**) original images, (**b**,**c**) before and after SA_1, (**d**,**e**) before and after SA_2, (**f**,**g**) before and after SA_3, (**h**,**i**) before and after SA_4.

5.3. Comparison of Different Global Feature Information Awareness Schemes

To verify the performance of the proposed GFIA module, we compared it with several well verified global feature information awareness schemes, i.e., the PPM in PSPNet, the ASPP in DeepLabV3+, and the DCU in D-LinkNet. The giga floating-point operations per second (GFLOPs), parameters, and the speed (i.e., the image throughput per second) [57] are also reported, to analyze their computational complexity. According to Table 6, the GFIA module, although slightly slower than PPM, outperforms other global feature information awareness schemes in terms of GFLOPs, parameters, F1 scores and IOU. While PPM and ASPP can effectively improve the accuracy of the model in maintaining lower GFLOPs and parameters, the accuracy increments seem far from adequate compared to GFIA. Despite DCU aggregating the global information by dilated convolution, its GFLOPs and parameters are much larger and speed is much slower, which brings a greater computational complexity and reduces inference speed. On the basis of DCU, GFIA adds the depthwise separable convolution, greatly reducing GFLOPs and parameters and alleviating the model training complexity, despite the reduced inference speed. In addition, GFIA uses the nonlocal unit to enhance the spatial relationships between global semantic information and effectively aggregates building features. In comparison, GFIA obtained the best accuracy while maintaining a lower complexity, demonstrating that the GFIA module captures the multiscale contextual information of building features by dilated convolution and nonlocal units and accomplishes the effective aggregation of global semantic information.

Table 6. Evaluation results of different global feature information awareness schemes on the WHU Building Dataset. Note: GFLOPs and parameters are computed from a tensor with a size of $1 \times 512 \times 32 \times 32$. The speed is tested with a batch size of 2, full precision (fp32), input resolution of $3 \times 256 \times 256$, and measured in examples/second.

No.	Methods	GFLOPs	Parameters (M)	Speed	F1 (%)	IOU (%)
1	PPM	0.5417	0.7895	18.90	94.32	89.24
2	ASPP	4.0969	4.1318	17.19	94.39	89.37
3	DCU	12.082	11.799	16.93	94.60	89.75
4	GFIA	0.3036	0.2939	18.61	94.76	90.04

5.4. Comparison of Different Cross Level Feature Fusion Schemes

Figure 15 displays the comparison of different cross level feature fusion schemes based on B-FGC-Net, including the concatenate channel, pixel addition, CLFR-SE module, and proposed CLFR module. The CLFR-SE module replaces channel attention in the CLFR proposed in this paper with the squeeze and excitation (SE) module [58]. According to the results, the F1 and IOU of the concatenated channel and pixel addition are significantly lower than the F1 and IOU of the CLFR-SE and CLFR modules, mainly because of the large semantic gap between low and high level features and the extensive redundant noise information contained in the low level features. Considering the semantic gaps of low level features and the redundancy characteristics, our study designed a cross level feature recalibration scheme. The CLFR module can automatically pick up the complementary information from channel dimensions, completing the effective utilization of low level features and significantly enhancing the model performance. To choose superior channel attention in the CLFR module, we compared the learning ability of SE and ECA. The experimental results show that the latter achieves significant performance gains with only a few additional parameters. The comprehensive comparison of the four different cross level feature fusion schemes demonstrates that the ECA based CLFR completes the recalibration of the channel information of low level features and aggregates the cross level feature information by learning the channel semantic information of high level features.

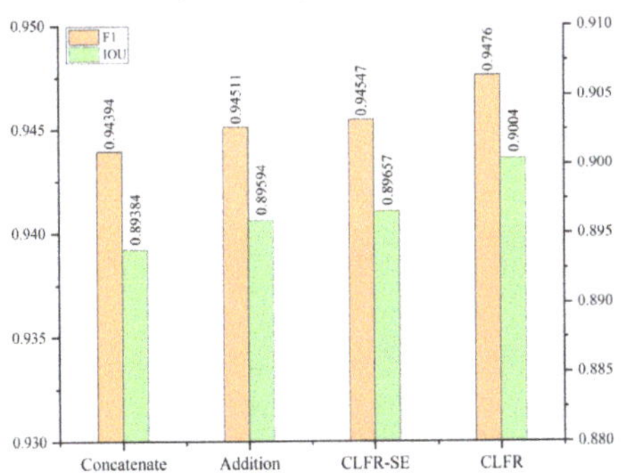

Figure 15. F1 scores and IOU of different cross level feature fusion schemes on the WHU building dataset.

5.5. Ablation Study

Ablation experiments were performed to verify the rationality and validity of each component of the B-FGC-Net on the WHU Building Dataset. U-Net with ResNet-34 was chosen as the baseline model, and the F1 score and IOU were adopted to quantitatively assess the effectiveness. The detailed results are shown in Table 7. The F1 and IOU are improved by 0.96% and 1.69% after ResNet34 was introduced in U-Net, demonstrating the robust feature extraction capability of ResNet34 as the encoder. The addition of the SA unit improves the baseline from 94.02% and 88.71% to 94.44% and 89.46% in terms of F1 and IOU, respectively, implying that the SA unit concentrates on building features in the spatial dimension and ignores other irrelevant backgrounds, such as building shadows.

After inserting the GFIA module with the DC and nonlocal units, the F1 score and IOU are improved by 0.54% and 0.97% compared with the baseline, indicating that larger scale building features are effectively captured and that global features are usefully integrated. By adding the CLFR module, the F1 score and IOU are improved by 0.74% and 1.33% compared with the basic model, meaning that the CLFR module eliminates the semantic gap between low and high level features and makes full use of the detailed spatial information of low level features. In summary, the SA, the GFIA, and the CLFR are proven to be able to effectively improve the performance through the ablation experiments of each module. Most importantly, to obtain the best building extraction results, each component of the proposed method is required.

Table 7. Ablation study with different component combinations on the WHU Building Dataset.

No.	Baseline	SA	GFIA	CLFR	F1 (%)	IOU (%)
1	✓				94.02	88.71
2	✓	✓			94.44	89.46
3	✓	✓	✓		94.56	89.68
4	✓	✓	✓	✓	94.76	90.04

5.6. Limitations and Future Research Work

Although the proposed method has achieved excellent extraction performance with superior extraction capability for small and large scale buildings on WHU and INRIA building datasets, there are still some difficulties in data dependence and the characteristics of the same spectrum foreign matter that should not be ignored.

Figure 16 displays examples of error extraction for U-Net and B-FGC-Net. According to the results, both methods suffer from partial building misidentification, which may be attributed to two main reasons: (1) Some nonbuildings (e.g., light gray concrete plots, containers, etc.) are similar to buildings in terms of spectral features and geometric features. End to end DCNN methods have extreme difficulty learning the potential difference in features between them from limited RGB image data, which is prone to misclassification. Thus, future work should use auxiliary information such as digital surface models (DSMs) [59] or multispectral images for building extraction to improve the extraction precision. (2) Some of the labels are mistaken, making it rather difficult for the model to learn all the information about buildings, resulting in the possible underfitting of the model. For this reason, semisupervised or unsupervised learning methods are suggested for future research to reduce the reliance on labeled data.

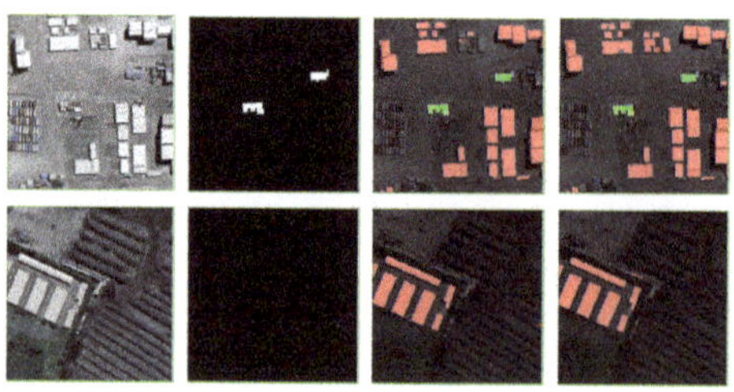

(a) (b) (c) (d)

Figure 16. Examples of error extraction. (**a**) Original images, (**b**) labels, (**c**) U-Net, (**d**) B-FGC-Net. The green and red indicate the correct and incorrect, respectively.

The comparison of the GFLOPs, parameters of several methods, and inference speed is illustrated in Figure 17. The B-FGC-Net model has larger GFLOPs (98.75) and model parameters (24M) and lower inference speed (18.61). Therefore, DL based DCNN models need to make a good trade off between computational complexity and precision in future work. For instance, smaller models can be used to extract buildings quickly in the deployment stage of various intelligent terminals (e.g., UAV identification terminals, handheld information collection terminals); larger models can be used to extract buildings accurately in the field of precision mapping. Furthermore, further work can pay more attention to the knowledge distillation scheme [60] that reduces the parameters of the model with good accuracy and high computational complexity and facilitates the deployment of the model.

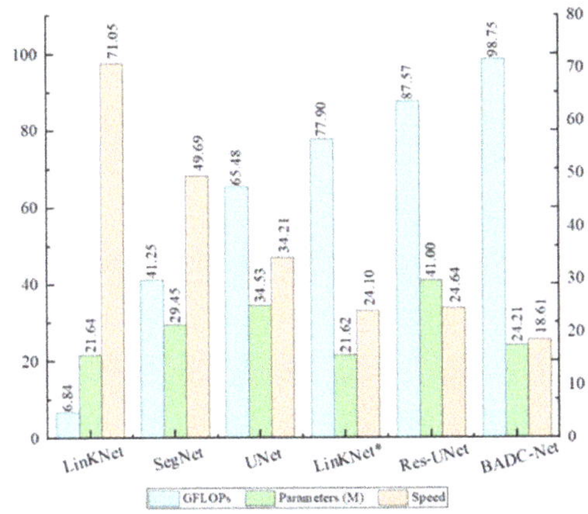

Figure 17. Comparison of GFLOPs and parameters for different methods. GFLOPs and parameters are computed from a tensor of dimension 1 × 3 × 256 × 256. The speed is tested with a batch size of 2, full precision (fp32), input resolution of 3 × 256 × 256, and measured in examples/second.

6. Conclusions

This study proposed a building extraction network (B-FGC-Net) for high resolution remote sensing imagery. The encoder combined the SA unit to highlight the spatial level of building feature representation, the GFIA module was applied to capture the multiscale contextual information and global semantic information, and the decoder used the CLFR module to achieve the effective fusion of cross level information. The proposed method was implemented and evaluated on two public datasets. The experimental results indicate that: (1) B-FGC-Net is a building extraction model with an outstanding extraction effect and high accuracy, especially in small and large scale buildings, and overcomes the influence of building shadows and tree shading. (2) Comparison from different perspectives reveals that the SA, GFIA, and CLFR can highlight building features, perceive global semantic information and recalibrate cross layer channel information, respectively. SA is able to autonomously learn the spatial distribution relationship of feature points, significantly improving the attention on building features in the form of weight assignment and weakening the representation of background noise such as building shadows; GFIA perceives a wider range of feature information with superior contextual information aggregation capability and brings greater accuracy gain through dilated convolution and self attention mechanisms; CLFR eliminates the semantic gap in low level features through adaptively acquiring channel information contributions from high level features and achieves significant performance gains by the effective fusion of different hierarchical features. (3) Future research should pay more attention to auxiliary information and semi supervised learning methods to improve extraction accuracy and reduce the dependence on labeled data.

Author Contributions: Conceptualization, Y.W. and X.Z.; methodology, Y.W. and X.Z.; software, X.Z.; validation, Y.W. and X.Z.; formal analysis, Y.W., X.L. and D.Z.; writing—original draft preparation, Y.W. and X.Z.; writing—review and editing, Y.W., X.Z., X.L. and D.Z.; visualization, X.Z.; project administration, Y.W.; funding acquisition, Y.W. All authors have read and agreed to the published version of the manuscript.

Funding: This research was partly supported by the Strategic Priority Research Program of Chinese Academy of Sciences (Grant No. XDA28050200), the Third Xinjiang Scientific Expedition Program

(Grant No. 2021xjkk1402), the Major Special Project—the China High resolution Earth Observation System (Grant No. 30-Y30F06-9003-20/22) and Fujian Province Highway Science and Technology Project: Key technology of Intelligent Inspection of Highway UAV Network by Remote Sensing (Grant No. GS 202101).

Institutional Review Board Statement: Not applicable.

Informed Consent Statement: Not applicable.

Data Availability Statement: WHU building data set can be downloaded in http://gpcv.whu.edu.cn/data/building_dataset.html (accessed on 5 December 2021) and INRIA building data set can be downloaded in https://project.inria.fr/aerialimagelabeling/download/ (accessed on 5 December 2021).

Acknowledgments: The authors appreciate Wuhan University and INRIA for sharing the building datasets for free.

Conflicts of Interest: The authors declare no conflict of interest.

References

1. Rashidian, V.; Baise, L.G.; Koch, M. Detecting Collapsed Buildings after a Natural Hazard on Vhr Optical Satellite Imagery Using U-Net Convolutional Neural Networks. *Int. Geosci. Remote Sens. Symp.* **2019**, 9394–9397. [CrossRef]
2. Zhang, Z.; Guo, W.; Li, M.; Yu, W. GIS-Supervised Building Extraction With Label Noise-Adaptive Fully Convolutional Neural Network. *IEEE Geosci. Remote Sens. Lett.* **2020**, *17*, 2135–2139. [CrossRef]
3. He, S.; Jiang, W. Boundary-Assisted Learning for Building Extraction from Optical Remote Sensing Imagery. *Remote Sens.* **2021**, *13*, 760. [CrossRef]
4. Chen, Z.; Li, D.; Fan, W.; Guan, H.; Wang, C.; Li, J. Self-attention in reconstruction bias U-net for semantic segmentation of building rooftops in optical remote sensing images. *Remote Sens.* **2021**, *13*, 2524. [CrossRef]
5. Liu, P.; Liu, X.; Liu, M.; Shi, Q.; Yang, J.; Xu, X.; Zhang, Y. Building footprint extraction from high-resolution images via spatial residual inception convolutional neural network. *Remote Sens.* **2019**, *11*, 830. [CrossRef]
6. He, N.; Fang, L.; Plaza, A. Hybrid first and second order attention Unet for building segmentation in remote sensing images. *Sci. China Inf. Sci.* **2020**, *63*, 140305. [CrossRef]
7. Zerrouki, N.; Bouchaffra, D. Pixel-based or Object-based: Which approach is more appropriate for remote sensing image classification? In Proceedings of the 2014 IEEE International Conference on Systems, Man, and Cybernetics (SMC), San Diego, CA, USA, 5–8 October 2014; pp. 864–869.
8. Maggiori, E.; Tarabalka, Y.; Charpiat, G.; Alliez, P. Convolutional Neural Networks for Large-Scale Remote-Sensing Image Classification. *IEEE Trans. Geosci. Remote Sens.* **2017**, *55*, 645–657. [CrossRef]
9. Myint, S.W.; Gober, P.; Brazel, A.; Grossman-Clarke, S.; Weng, Q. Per-pixel vs. object-based classification of urban land cover extraction using high spatial resolution imagery. *Remote Sens. Environ.* **2011**, *115*, 1145–1161. [CrossRef]
10. Dean, A.M.; Smith, G.M. An evaluation of per-parcel land cover mapping using maximum likelihood class probabilities. *Int. J. Remote Sens.* **2003**, *24*, 2905–2920. [CrossRef]
11. Duro, D.C.; Franklin, S.E.; Dubé, M.G. A comparison of pixel-based and object-based image analysis with selected machine learning algorithms for the classification of agricultural landscapes using SPOT-5 HRG imagery. *Remote Sens. Environ.* **2012**, *118*, 259–272. [CrossRef]
12. Blaschke, T.; Lang, S.; Lorup, E.; Strobl, J.; Zeil, P. Object-oriented image processing in an integrated GIS/remote sensing environment and perspectives for environmental applications. *Environ. Inf. Plan. Polit. Public* **2000**, *2*, 555–570.
13. Ding, Z.; Wang, X.Q.; Li, Y.L.; Zhang, S.S. Study on building extraction from high-resolution images using MBI. *Int. Arch. Photogramm. Remote Sens. Spat. Inf. Sci.* **2018**, *XLII-3*, 283–287. [CrossRef]
14. Sirmacek, B.; Unsalan, C. Building detection from aerial images using invariant color features and shadow information. In Proceedings of the 2008 23rd International Symposium on Computer and Information Sciences, Istanbul, Turkey, 27–29 October 2008; pp. 1–5.
15. Lin, Y.; Xu, D.; Wang, N.; Shi, Z.; Chen, Q. Road Extraction from Very-High-Resolution Remote Sensing Images via a Nested SE-Deeplab Model. *Remote Sens.* **2020**, *12*, 2985. [CrossRef]
16. Fu, H.; Fu, B.; Shi, P. An improved segmentation method for automatic mapping of cone karst from remote sensing data based on deeplab V3+ model. *Remote Sens.* **2021**, *13*, 441. [CrossRef]
17. Yang, X.; Liu, R.; Yang, M.; Chen, J.; Liu, T.; Yang, Y.; Chen, W.; Wang, Y. Incorporating landslide spatial information and correlated features among conditioning factors for landslide susceptibility mapping. *Remote Sens.* **2021**, *13*, 2166. [CrossRef]
18. Alshehhi, R.; Marpu, P.R.; Woon, W.L.; Mura, M.D. Simultaneous extraction of roads and buildings in remote sensing imagery with convolutional neural networks. *ISPRS J. Photogramm. Remote Sens.* **2017**, *130*, 139–149. [CrossRef]
19. Lecun, Y.; Bengio, Y.; Hinton, G. Deep learning. *Nature* **2015**, *521*, 436–444. [CrossRef]
20. Krizhevsky, A.; Sutskever, I.; Hinton, G.E. ImageNet classification with deep convolutional neural networks. *Commun. ACM* **2017**, *60*, 84–90. [CrossRef]

21. Jin, Y.; Xu, W.; Zhang, C.; Luo, X.; Jia, H. Boundary-aware refined network for automatic building extraction in very high-resolution urban aerial images. *Remote Sens.* **2021**, *13*, 692. [CrossRef]
22. Lan, Z.; Huang, Q.; Chen, F.; Meng, Y. Aerial Image Semantic Segmentation Using Spatial and Channel Attention. In Proceedings of the 2019 IEEE 4th International Conference on Image, Vision and Computing (ICIVC), Xiamen, China, 5–7 July 2019; pp. 316–320.
23. Seong, S.; Choi, J. Semantic segmentation of urban buildings using a high-resolution network (Hrnet) with channel and spatial attention gates. *Remote Sens.* **2021**, *13*, 3087. [CrossRef]
24. Chen, L.; Zhang, H.; Xiao, J.; Nie, L.; Shao, J.; Liu, W.; Chua, T.S. SCA-CNN: Spatial and channel-wise attention in convolutional networks for image captioning. In Proceedings of the 2017 IEEE Conference on Computer Vision and Pattern Recognition (CVPR), Honolulu, HI, USA, 21–26 July 2017; pp. 6298–6306.
25. Qi, X.; Li, K.; Liu, P.; Zhou, X.; Sun, M. Deep Attention and Multi-Scale Networks for Accurate Remote Sensing Image Segmentation. *IEEE Access* **2020**, *8*, 146627–146639. [CrossRef]
26. Ronneberger, O.; Fischer, P.; Brox, T. U-Net: Convolutional Networks for Biomedical Image Segmentation. In *Medical Image Computing and Computer-Assisted Intervention—MICCAI*; Navab, N., Hornegger, J., Wells, W.M., Frangi, A., Eds.; Springer International Publishing: Cham, Switzerland, 2015; Volume 9351, pp. 234–241.
27. He, K.; Zhang, X.; Ren, S.; Sun, J. Deep Residual Learning for Image Recognition. In Proceedings of the 2016 IEEE Conference on Computer Vision and Pattern Recognition (CVPR), Las Vegas, NV, USA, 27–30 June 2016; pp. 770–778.
28. Yi, Y.; Zhang, Z.; Zhang, W.; Zhang, C.; Li, W.; Zhao, T. Semantic segmentation of urban buildings from VHR remote sensing imagery using a deep convolutional neural network. *Remote Sens.* **2019**, *11*, 1774. [CrossRef]
29. Luo, H.; Chen, C.; Fang, L.; Zhu, X.; Lu, L. High-Resolution Aerial Images Semantic Segmentation Using Deep Fully Convolutional Network with Channel Attention Mechanism. *IEEE J. Sel. Top. Appl. Earth Obs. Remote Sens.* **2019**, *12*, 3492–3507. [CrossRef]
30. Zhu, Q.; Zhang, Y.; Wang, L.; Zhong, Y.; Guan, Q.; Lu, X.; Zhang, L.; Li, D. A Global Context-aware and Batch-independent Network for road extraction from VHR satellite imagery. *ISPRS J. Photogramm. Remote Sens.* **2021**, *175*, 353–365. [CrossRef]
31. Zhou, L.; Zhang, C.; Wu, M. D-linknet: Linknet with pretrained encoder and dilated convolution for high resolution satellite imagery road extraction. In Proceedings of the 2018 IEEE/CVF Conference on Computer Vision and Pattern Recognition Workshops (CVPRW), Salt Lake City, UT, USA, 18–22 June 2018; pp. 192–1924.
32. Guo, M.; Liu, H.; Xu, Y.; Huang, Y. Building Extraction Based on U-Net with an Attention Block and Multiple Losses. *Remote Sens.* **2020**, *12*, 1400. [CrossRef]
33. Long, J.; Shelhamer, E.; Darrell, T. Fully convolutional networks for semantic segmentation. In Proceedings of the 2015 IEEE Conference on Computer Vision and Pattern Recognition (CVPR), Boston, MA, USA, 7–12 June 2015; pp. 3431–3440.
34. Badrinarayanan, V.; Kendall, A.; Cipolla, R. SegNet: A Deep Convolutional Encoder-Decoder Architecture for Image Segmentation. *IEEE Trans. Pattern Anal. Mach. Intell.* **2017**, *39*, 2481–2495. [CrossRef] [PubMed]
35. Zhou, Z.; Rahman Siddiquee, M.M.; Tajbakhsh, N.; Liang, J. Unet++: A nested u-net architecture for medical image segmentation. In *Deep Learning in Medical Image Analysis and Multimodal Learning for Clinical Decision Support*; Springer International Publishing: Cham, Switzerland, 2018; pp. 3–11.
36. Kestur, R.; Farooq, S.; Abdal, R.; Mehraj, E.; Narasipura, O.; Mudigere, M. UFCN: A fully convolutional neural network for road extraction in RGB imagery acquired by remote sensing from an unmanned aerial vehicle. *J. Appl. Remote Sens.* **2018**, *12*, 1. [CrossRef]
37. Chaurasia, A.; Culurciello, E. LinkNet: Exploiting encoder representations for efficient semantic segmentation. In Proceedings of the 2017 IEEE Visual Communications and Image Processing (VCIP), St. Petersburg, FL, USA, 10–13 December 2017; pp. 1–4.
38. Chen, L.-C.; Papandreou, G.; Schroff, F.; Adam, H. Rethinking Atrous Convolution for Semantic Image Segmentation. *arXiv* **2017**, arXiv:1706.05587.
39. Zhang, Z.; Liu, Q.; Wang, Y. Road Extraction by Deep Residual U-Net. *IEEE Geosci. Remote Sens. Lett.* **2018**, *15*, 749–753. [CrossRef]
40. Das, P.; Chand, S. AttentionBuildNet for building extraction from aerial imagery. In Proceedings of the 2021 International Conference on Computing, Communication, and Intelligent Systems (ICCCIS), Greater Noida, India, 19–20 February 2021; pp. 576–580.
41. Wang, X.; Girshick, R.; Gupta, A.; He, K. Non-local Neural Networks. In Proceedings of the 2018 IEEE/CVF Conference on Computer Vision and Pattern Recognition, Salt Lake City, UT, USA, 18–23 June 2018; pp. 7794–7803.
42. Wang, Y.; Seo, J.; Jeon, T. NL-LinkNet: Toward Lighter But More Accurate Road Extraction With Nonlocal Operations. *IEEE Geosci. Remote Sens. Lett.* **2021**, *19*, 1–5. [CrossRef]
43. Ding, L.; Tang, H.; Bruzzone, L. LANet: Local Attention Embedding to Improve the Semantic Segmentation of Remote Sensing Images. *IEEE Trans. Geosci. Remote Sens.* **2021**, *59*, 426–435. [CrossRef]
44. Zhao, H.; Shi, J.; Qi, X.; Wang, X.; Jia, J. Pyramid scene parsing network. In Proceedings of the 2017 IEEE Conference on Computer Vision and Pattern Recognition (CVPR), Honolulu, HI, USA, 21–26 July 2017; pp. 6230–6239.
45. Chen, L.-C.; Zhu, Y.; Papandreou, G.; Schroff, F.; Adam, H. Encoder-Decoder with Atrous Separable Convolution for Semantic Image Segmentation. In Proceedings of the European Conference on Computer Vision (ECCV), Munich, Germany, 8–14 September 2018; pp. 801–818.
46. Xie, Y.; Miao, F.; Zhou, K.; Peng, J. HSGNet: A road extraction network based on global perception of high-order spatial information. *ISPRS Int. J. Geo-Inf.* **2019**, *8*, 571. [CrossRef]

47. Nitish, S.; Hinton, G.; Krizhevsky, A.; Sutskever, I.; Salakhutdinov, R. Dropout: A Simple Way to Prevent Neural Networks from Overfitting. *J. Mach. Learn. Res.* **2014**, *15*, 1929–1958.
48. Ioffe, S.; Szegedy, C. Batch Normalization: Accelerating Deep Network Training by Reducing Internal Covariate Shift. In Proceedings of the 32nd International Conference on Machine Learning, Lille, France, 7–9 July 2015; pp. 448–456.
49. Woo, S.; Park, J.; Lee, J.Y.; Kweon, I.S. CBAM: Convolutional block attention module. In *Proceedings of the European Conference on Computer Vision (ECCV), Munich, Germany, 8–14 September 2018*; Springer International Publishing: Cham, Switzerland, 2018; pp. 3–19.
50. Wang, Q.; Wu, B.; Zhu, P.; Li, P.; Zuo, W.; Hu, Q. ECA-Net: Efficient channel attention for deep convolutional neural networks. In Proceedings of the 2020 IEEE/CVF Conference on Computer Vision and Pattern Recognition (CVPR), Seattle, WA, USA, 13–19 June 2020; pp. 11531–11539.
51. Ji, S.; Wei, S.; Lu, M. Fully Convolutional Networks for Multisource Building Extraction from an Open Aerial and Satellite Imagery Data Set. *IEEE Trans. Geosci. Remote Sens.* **2019**, *57*, 574–586. [CrossRef]
52. Maggiori, E.; Tarabalka, Y.; Charpiat, G.; Alliez, P. Can semantic labeling methods generalize to any city? the inria aerial image labeling benchmark. *Int. Geosci. Remote Sens. Symp.* **2017**, *2017*, 3226–3229. [CrossRef]
53. Kingma, D.P.; Ba, J.L. Adam: A method for stochastic optimization. *arXiv* **2014**, arXiv:1412.6980.
54. Pan, X.; Yang, F.; Gao, L.; Chen, Z.; Zhang, B.; Fan, H.; Ren, J. Building extraction from high-resolution aerial imagery using a generative adversarial network with spatial and channel attention mechanisms. *Remote Sens.* **2019**, *11*, 917. [CrossRef]
55. Cao, Z.; Diao, W.; Sun, X.; Lyu, X.; Yan, M.; Fu, K. C3Net: Cross-Modal Feature Recalibrated, Cross-Scale Semantic Aggregated and Compact Network for Semantic Segmentation of Multi-Modal High-Resolution Aerial Images. *Remote Sens.* **2021**, *13*, 528. [CrossRef]
56. Zhou, D.; Wang, G.; He, G.; Long, T.; Yin, R.; Zhang, Z.; Chen, S.; Luo, B. Robust building extraction for high spatial resolution remote sensing images with self-attention network. *Sensors* **2020**, *20*, 7241. [CrossRef] [PubMed]
57. Ding, X.; Zhang, X.; Ma, N.; Han, J.; Ding, G.; Sun, J. RepVGG: Making VGG-style ConvNets Great Again. In Proceedings of the 2021 IEEE/CVF Conference on Computer Vision and Pattern Recognition (CVPR), Nashville, TN, USA, 20–25 June 2021; pp. 13728–13737.
58. Hu, J.; Shen, L.; Albanie, S.; Sun, G.; Wu, E. Squeeze-and-Excitation Networks. *IEEE Trans. Pattern Anal. Mach. Intell.* **2020**, *42*, 2011–2023. [CrossRef]
59. Cao, Z.; Fu, K.; Lu, X.; Diao, W.; Sun, H.; Yan, M.; Yu, H.; Sun, X. End-to-End DSM Fusion Networks for Semantic Segmentation in High-Resolution Aerial Images. *IEEE Geosci. Remote Sens. Lett.* **2019**, *16*, 1766–1770. [CrossRef]
60. Li, X.; Yu, L.; Chen, H.; Fu, C.W.; Xing, L.; Heng, P.A. Transformation-Consistent Self-Ensembling Model for Semisupervised Medical Image Segmentation. *IEEE Trans. Neural Networks Learn. Syst.* **2021**, *32*, 523–534. [CrossRef] [PubMed]

remote sensing

MDPI

Article

GA-Net-Pyramid: An Efficient End-to-End Network for Dense Matching

Yuanxin Xia [1,*], Pablo d'Angelo [1], Friedrich Fraundorfer [1,2], Jiaojiao Tian [1], Mario Fuentes Reyes [1] and Peter Reinartz [1]

[1] Department of Photogrammetry and Image Analysis, Remote Sensing Technology Institute, German Aerospace Center (DLR), 82234 Wessling, Germany; pablo.angelo@dlr.de (P.d.); fraundorfer@icg.tugraz.at (F.F.); jiaojiao.tian@dlr.de (J.T.); mario.fuentesReyes@dlr.de (M.F.R.); peter.reinartz@dlr.de (P.R.)

[2] Institute of Computer Graphics and Vision, Graz University of Technology (TU Graz), 8010 Graz, Austria

* Correspondence: yuanxin.xia@dlr.de; Tel.: +49-8153-2816-37

Abstract: Dense matching plays a crucial role in computer vision and remote sensing, to rapidly provide stereo products using inexpensive hardware. Along with the development of deep learning, the Guided Aggregation Network (GA-Net) achieves state-of-the-art performance via the proposed Semi-Global Guided Aggregation layers and reduces the use of costly 3D convolutional layers. To solve the problem of GA-Net requiring large GPU memory consumption, we design a pyramid architecture to modify the model. Starting from a downsampled stereo input, the disparity is estimated and continuously refined through the pyramid levels. Thus, the disparity search is only applied for a small size of stereo pair and then confined within a short residual range for minor correction, leading to highly reduced memory usage and runtime. Tests on close-range, aerial, and satellite data demonstrate that the proposed algorithm achieves significantly higher efficiency (around eight times faster consuming only 20–40% GPU memory) and comparable results with GA-Net on remote sensing data. Thanks to this coarse-to-fine estimation, we successfully process remote sensing datasets with very large disparity ranges, which could not be processed with GA-Net due to GPU memory limitations.

Keywords: dense matching; deep learning; convolutional neural networks; end-to-end; pyramid architecture

Citation: Xia, Y.; d'Angelo, P.; Fraundorfer, F.; Tian, J.; Fuentes Reyes, M.; Reinartz, P. GA-Net-Pyramid: An Efficient End-to-End Network for Dense Matching. *Remote Sens.* **2022**, *14*, 1942. https://doi.org/10.3390/rs14081942

Academic Editor: Sander Oude Elberink

Received: 8 March 2022
Accepted: 13 April 2022
Published: 17 April 2022

1. Introduction

Over the years, dense stereo matching has been studied persistently in the field of computer vision, remote sensing, and photogrammetry, as the corresponding applications keep promoting the development of self-driving, urban digitization, topographic survey, forest management, etc. [1–4]. Given a pair of images with the camera parameters and the relative distance (baseline) in between, the object depth is computed which extends 2D image information to 3D knowledge of the scene [5]. In stereo matching, the depth is obtained in the form of disparity which presents the (horizontal) displacement of two corresponding pixels from each of the (rectified) stereo pair, respectively. A disparity map allows each pixel to be triangulated to its location in the 3D space. Stereo vision methods define two terms for locating correspondences, the data term and smoothness term. The former searches pixels with similar intensity as potential matches, while the latter requires close disparity predictions between neighboring points for spatial smoothness. Semi-Global Matching (SGM) is a representative method in stereo matching [6]. The algorithm acquires dense correspondences via a simple pixel-wise cost comparison under a disparity searching range, and guarantees the (piece-wise) smoothness of the reconstructed surface simultaneously. For each target pixel, the previous point along a certain path is also considered to avoid neighboring disparity inconsistency. By repeatedly applying the strategy through multiple (normally 8 or 16) symmetric paths, 2D regularization is performed while keeping the algorithm computationally feasible.

As more high-quality, high-resolution data become available, the computational cost of dense matching rises exponentially, especially in the field of remote sensing. To limit the memory usage and runtime, Rothermel [7] proposed tSGM. Images are firstly downsampled to several scales constituting a pyramid structure, in which the dense matching is applied from the lowest resolution to the highest, level by level. On the pyramid top, the disparity range is downscaled accordingly together with the image size, leading to reduced workload. The matching result is then passed to the next higher resolution level as an initial prediction, from which a small disparity buffer is set as a new search range to locally refine the estimation. The coarse-to-fine scheme thus greatly reduces the demand for memory and runtime. Moreover, the influence of ambiguous disparity candidates is limited. Additionally, this strategy enables the use of deep learning-based algorithms, which typically only support small search ranges due to memory limits on datasets with large disparity ranges of sometime several thousand pixels, as typically occurring in extreme mountainous regions, such as the Himalayas.

Recently, Zhang et al. [8] introduced their GA-Net, which approximates SGM as a differentiable Semi-Global Guided Aggregation (SGA) layer, to construct an end-to-end neural network for stereo matching. All the user-defined parameters in SGM can be learned; thus, the smoothness requirement is satisfied in a smarter way depending on the specific scene situation. With SGA and only a few 3D convolutional layers to regularize the cost volume, GA-Net is more efficient than other networks, e.g., GC-Net [9], PSMNet [10], etc., and achieves state-of-the-art performance. For processing high-resolution remote sensing data, however, the training and prediction are still memory- and time-consuming (days are needed for training on patches of 384×576, with [0, 192] as the disparity search range, consuming around 15 GB GPU memory for each batch).

Inspired by tSGM and some corresponding pyramid networks [11–13], we adjust GA-Net to a pyramid architecture, and propose our GA-Net-Pyramid. The disparity is initially estimated for the full depth range at the coarsest resolution, then refined through the pyramid. Thus, we enhance the efficiency of the algorithm significantly, with moderately decreased accuracy especially for remote sensing data. To summarize our contributions:

- Firstly, we propose a hierarchical strategy for GA-Net stereo matching to estimate the depth from coarse to fine, for which two pyramid models are introduced with explicit or implicit image downsampling, respectively. A trainable Spatial Propagation Network (SPN) [14] is tested as a post-processing step to sharpen the depth boundaries. It is shown that the effect from SPN varies depending on the target data domain.
- Secondly, the proposed methods are tested on cross-domain datasets, from close-range benchmarks, Scene Flow [15] and KITTI-2012 [16], to large-scale aerial/satellite stereo data. We prove that our algorithm is robust and consistently more efficient in all cases. We also build a stereo dataset, consisting of simultaneously acquired 30-cm satellite and 6-cm aerial imagery which are co-registered to sub-pixel precision. This is particularly important for remote sensing scenarios, considering that the currently published data, such as [17], cannot provide reliable ground truth disparity maps, due to different sensing modalities or scene changes caused by temporal inconsistency.
- At last, we successfully solve a satellite stereo task on stereo pairs with very large disparity ranges, which cannot be handled by the baseline model GA-Net.

The rest of the paper is organized as follows: In Section 2, traditional stereo methods, SGM and its variants are recapped, which enlighten the main idea of GA-Net and our GA-Net-Pyramid. We also describe representative learning-based algorithms, from hybrid approaches replacing certain traditional components with deep learning-based ones, to full end-to-end stereo networks. Afterwards, we state the principle of our method, GA-Net-Pyramid, with a review of its prototype GA-Net in Section 3. In Section 4, we present a detailed comparison between GA-Net and our GA-Net-Pyramid on various datasets. At last, we discuss the strengths and limitations of the method in Section 5, and conclude the paper in Section 6.

2. Related Work

2.1. Traditional Stereo Methods

Conventional stereo matching algorithms define two terms to find dense correspondences from a stereo pair, data term and smoothness term [5]. The data term measures the photo consistency between potentially matched pixels through a pre-defined disparity range. The smoothness term guarantees a smooth reconstructed surface by limiting neighboring points' disparity differences. SGM well balanced the two terms via a scanline optimization strategy, which was widely applied thanks to the good compromise between accuracy and efficiency [6,18,19]. The strategy was further improved with a dynamic searching range for correspondences through a pyramid structure, leading to tSGM which consumed less memory and runtime [7]. As More Global Matching (MGM) was proposed, the support from neighboring pixels was increased without extra overhead, by additionally considering the previous scanline visited already [19,20]. Compared with other traditional stereo methods [21–25], which may solely rely on the cost function and winner-takes-all (WTA) strategy resulting in limited accuracy, or struggle to find the minimum global energy under certain runtime or memory budget, the SGM variants achieve robust stereo estimation consuming reasonable computational resource.

2.2. Learning-Assisted Stereo Methods

2.2.1. Integration of Conventional Stereo Methods and Machine Learning

Recent advances in machine/deep learning and convolutional neural networks (CNNs) enable the learning of data representation [26], and promote the development of stereo matching with a series of state-of-the-art algorithms. Deep learning could be exploited to extract features from images, in order to better measure the similarity for matching cost calculation. Zbontar and LeCun [27] used a Siamese network [28] to extract features from two patches symmetrically, after which a cost volume was constructed and regularized by SGM. The idea was adjusted by Luo et al. [29] based on multi-class classification, achieving faster estimation. Regarding the cost aggregation and disparity computation, Seki and Pollefeys [30] proposed their SGM-Net to learn the penalty terms for conflicting disparity predictions from neighboring points. Michael et al. [31] considered a specific weight for each scanline in SGM to achieve a weighted 2D scanline optimization, since varying performance could be obtained via each scanline depending on the scene structure. Poggi and Mattoccia [32] constructed a feature vector for each pixel according to the disparity estimation via a single scanline. The feature represented the statistical dispersion of surrounding disparities, which could be analyzed by a random forest to predict a confidence measure of the scanline for a weighted scanline summation. Similar work was accomplished in [33,34]. The disparity predicted by each scanline and the corresponding costs were fed to a random forest, so that the better performed scanlines were adaptively selected. The corresponding disparity estimation could serve as a reference to guide the further stereo prediction.

2.2.2. End-to-End Stereo Networks

The above methods mainly integrated deep learning with traditional stereo matching techniques for better performance, which were then followed by encoder-decoder structures for depth prediction as an end-to-end system. Dosovitskiy et al. [35] firstly presented a network, FlowNet, to estimate optical flow directly from a stereo pair. They used a correlation layer to measure the similarity between corresponding patches. Mayer et al. [15] then designed a large synthetic dataset, Scene Flow, allowing an initial training of deep neural networks before adjusting to specific scenarios. They also proposed DispNet and DispNet-Corr, as one of the first end-to-end stereo matching networks. Kendall et al. [9] proposed GC-Net, which applied 3D convolutions to regularize the cost volume, with both geometry and context information incorporated. Chang and Chen [10] introduced a pyramid pooling module in their PSMNet to aggregate multi-scale features. Thus, the global context and local details were simultaneously contained within the cost volume. Guo et al. [36] improved PSMNet by proposing the group-wise correlation stereo network (GwcNet). They constructed a group-wise correlation-based cost volume which required less parameters for the cost aggregation, achieving similar performance as PSMNet. Zhu et al. [37] proposed a multi-scale pyramid aggregation module to handle the cost volume, leading to MPANet with significantly better disparity estimation for foreground objects. Xu and Zhang [38] proposed AAnet, utilizing intra- and cross-scale cost aggregation, which delivered better results for depth discontinuities and large textureless area. Wang et al. [39] applied a recurrent unit to iteratively refine the stereo estimation, and designed a pyramid voting module to produce a semi-dense disparity map for self-supervision. Confident disparity prediction was achieved via seeking consistent estimation across scales. Inspired by SGM, Zhang et al. [8] proposed the GA-Net using so-called SGA layer for cost aggregation, to replace 3D convolution which was computationally expensive. They achieved great performance on multiple benchmark datasets, which coincided with the idea from [40] that classical stereo matching methods could serve as a robust guideline to develop deep learning-based algorithms, rather than designing a pure learning architecture. Semantic information could also be involved for stereo matching problems [41,42] as the object boundaries mostly corresponded to the depth discontinuities. The two tasks supported each other leading to a win-win situation. Other works included cost distribution study, disparity refinement, cross-domain prediction, stereo neural architecture search, etc., which boosted the state-of-the-art constantly [40,43–47].

Recently, the pyramid architecture was tested in a learning-based stereo framework, since the efficiency could be largely enhanced via a coarse-to-fine estimation [11–13,48,49]. Regarding the architecture in [11–13] as a baseline model, the stereo correspondences were firstly located on the pyramid top using downsampled features. Then, the disparity was iteratively refined through the network towards the pyramid bottom in full resolution, which considerably reduced the computational effort and GPU memory consumption. Chang et al. [48] benefited from the architecture to achieve real-time performance, with an attention-aware feature aggregation module for better representative ability of the feature. Compared with these methods, our contributions are different. At first, we additionally test our model on airborne and spaceborne images. We fill the application gap of the previous research, considering the very limited test cases applying newly proposed computer vision algorithms in the field of remote sensing. The proposed model is proven effective to process stereo imagery with large disparity range (thousand pixels) over mountain areas. It should be noted that our model acquires no supervision in training phase on stereo data with large baselines, with no need to normalize/denormalize the disparity measurement in test phase as [50]. This is, to the best of our knowledge, a novel showcase of adapting well-performed computer vision models to deliver high-quality geographical products in extreme regions. In addition, our baseline is the up-to-date model GA-Net-deep from [8], rather than the shallower and less accurate version GA-Net-11 used in [49].

3. Methodology

In this section, we recap GA-Net by presenting the proposed SGA and LGA (Local Guided Aggregation) layers, which approximate SGM for cost regularization and protect thin structures, respectively. SGM applies the scanline optimization strategy to efficiently locate stereo correspondences and avoids the streaking problem. For a detailed description of SGM, we encourage readers to follow the papers [6,51]. Afterwards, we describe our pyramidal extension of GA-Net, GA-Net-Pyramid. Two architectures are proposed. The first model explicitly downsamples the input stereo pair according to the pyramid level, and simply applies GA-Net on each level to regress disparity. The second model applies a different feature extraction strategy via a U-Net [52] structure to generate multi-scale features implicitly.

3.1. GA-Net

In traditional SGM, the scanline optimization technique [53] is applied to satisfy the spatial smoothness, by limiting the depth difference between neighboring pixels. To avoid the streaking problem, a pixel is accessed through multiple scanlines simultaneously along several canonical directions, typically 8 or 16, to consider the disparity estimation from its neighbor. Along a certain scanline traversing in direction r, the cost for a pixel located at the image position p assuming d as the disparity, is calculated as:

$$\begin{gathered} L_r(p,d) = C(p,d) + \min(\, L_r(p-r,d),\, L_r(p-r,d-1) + P_1, \\ L_r(p-r,d+1) + P_1,\, \min_i L_r(p-r,i) + P_2\,). \end{gathered} \tag{1}$$

In the above equation, the photo inconsistency is measured by $C(p,d)$, while P_1 and P_2 are defined for penalizing the prediction when the previous neighboring point $p-r$ prefers a different disparity value. In practice, however, two problems exist. Firstly, the users need expertise to determine appropriate P_1 and P_2 to punish neighboring disparity inconsistency. Tuning of P_1 and P_2 additionally depends on scene structure and the used similarity measure. Moreover, the values of P_1 and P_2 are fixed throughout the stereo processing or simply adapted according to, e.g., pixel gradients, which are not optimal for all the pixels within the image, especially under a varied scene structure, e.g., from plains to mountains.

GA-Net addresses these issues by introducing the SGA layer, a differentiable approximation of Equation (1) that is suitable for an end-to-end trainable network. Specifically, the master epipolar image provides guiding information through a sub-network to better penalize depth discontinuity, and enable a self-adaptive parameter setting. Thus, the penalty terms for conflicting neighboring disparities are determined according to the pixel location and scanline direction, which is more reasonable for smoothness regularization. Via the guidance sub-network, a weight is supplied for each term in Equation (1) to simulate the scanline optimization in SGM, leading to the following equation:

$$\begin{gathered} L_r(p,d) = C(p,d) + \mathrm{sum}(\, w_1(p,r)\cdot L_r(p-r,d), \\ w_2(p,r)\cdot L_r(p-r,d-1),\, w_3(p,r)\cdot L_r(p-r,d+1), \\ w_4(p,r)\cdot \max_i L_r(p-r,i)\,). \end{gathered} \tag{2}$$

Compared with Equation (1), the punishment from P_1 and P_2 is replaced by the relative importance (weight) w_i of each term, which is predicted separately for each pixel along a directed scanline. Moreover, there are two differences with SGM, one of which is that the first/external minimum operation is substituted by a weighted sum. This can be regarded as a replacement from a max-pooling layer to a convolution with strides, which is proven effective without accuracy loss [54]. In addition, the second/internal minimum search is changed to a maximum, which embodies the learning target to maximize the probability at the ground truth disparity rather than minimizing the cost. To avoid the exploding accumulation of $L_r(p,d)$ along the scanline, $C(p,d)$ is also included within the weighted summation, with the sum of all the weights equal to 1. Thus, SGA is finally formulated as:

$$\begin{gathered} L_r(p,d) = \mathrm{sum}(\, w_0(p,r)\cdot C(p,d),\, w_1(p,r)\cdot L_r(p-r,d), \\ w_2(p,r)\cdot L_r(p-r,d-1),\, w_3(p,r)\cdot L_r(p-r,d+1), \\ w_4(p,r)\cdot \max_i L_r(p-r,i)\,, \\ \textstyle\sum_{i=0,1,2,3,4} w_i(p,r) = 1. \end{gathered} \tag{3}$$

In SGM, the cost $L_r(p,d)$ from each scanline is simply summed up to approximate 2D smoothness, which was demonstrated to be not reasonable for incurring inferior scanlines [33,34]. Accordingly, GA-Net takes the maximum as $L(p,d) = \max_r L_r(p,d)$ to keep the best information.

The guidance sub-network also provides weights for another layer, LGA, to further filter the cost volume as below:

$$\begin{aligned} L_*(p,d) = \mathrm{sum}\Big(& \textstyle\sum_{q\in N_p} w_0(p,q)\cdot L(q,d), \\ & \textstyle\sum_{q\in N_p} w_1(p,q)\cdot L(q,d-1), \\ & \textstyle\sum_{q\in N_p} w_2(p,q)\cdot L(q,d+1)\Big), \end{aligned} \tag{4}$$

$$\textstyle\sum_{q\in N_p} w_0(p,q) + w_1(p,q) + w_2(p,q) = 1,$$

from which a 3D neighborhood (in both spatial and disparity dimensions) centered around each pixel within the cost volume is utilized for a weighted average to protect thin structures. Afterwards as suggested by [9], a softmax operation $\sigma(\cdot)$ is applied to the filtered cost volume in order to acquire a normalized probability for each disparity candidate (from $[0, D_{max}]$) and regress the final disparity value $\hat{d}$ as:

$$\hat{d} = \sum_{d=0}^{D_{max}} d \times \sigma(-L_{*d}). \tag{5}$$

3.2. *GA-Net-Pyramid with Explicit Downsampling*

GA-Net adapts the scanline optimization scheme to an end-to-end stereo matching system. Inspired by SGM, the disparity of each pixel can be estimated with the support from all the previous neighbors along multiple paths, instead of a pure convolution-based encoder-decoder to regularize the cost volume. Furthermore, the proposed SGA and LGA layers are computationally more efficient than 3D convolutions, which are used by most state-of-the-art methods [9,10]. However it can still take days to train a well performing model, when the computational power is limited. In our case, for example, the training on the Scene Flow dataset (patch size 384 × 576), which is normally used by most stereo matching networks for the initial learning phase, takes around 12 days to finish 8 epochs on two Quadro P6000 GPU cards. Hence, the employment of the network is hampered. In the field of remote sensing, it can be imagined that GA-Net would struggle to process high-resolution aerial or satellite stereo data, especially for wide baseline stereo pairs requiring larger disparity search ranges.

Rothermel [7] proposed an improved SGM, tSGM, which constructed a pyramid architecture to search correspondences between the stereo pair from coarse to fine. Based on this strategy, comparable quality was achieved with far less memory and runtime consumed. This inspires us to restructure GA-Net with a pyramid architecture as well, to regress the depth from coarse to fine. Figure 1 presents the schematic overview of our GA-Net-Pyramid. Three pyramid levels are depicted which could be extended. We use the same stacked hourglass module (a double U-Net structure) as GA-Net, which is essentially a Siamese network [28] for symmetric feature extraction from the left and right image, respectively. The input of the feature extraction module, however, is a stereo pair downsampled in accordance with the pyramid level. Afterwards, the cost volume is generated and then processed by SGA and LGA for disparity regression, in order to guide the subsequent level for the disparity refinement until the original resolution is recovered.

3.2.1. Pyramid Top

We start from the pyramid top with the original image downscaled by a factor of 4 along both row and column directions in our implementation (termed as 'Scale 1/4' in Figure 1). Then, the feature is extracted to construct a 4D cost volume by concatenating the left and right feature maps along the channel dimension, with a horizontal shift indicated by a disparity candidate within the search range. Assuming the cost volume on the original full-resolution image is of size $H \times W \times D_{max} \times 2C$, for the image height, width, the maximum disparity, and twice the channel number of the generated feature maps, respectively, our cost volume on the pyramid top reaches a highly reduced dimension as $H/4 \times W/4 \times D_{max}/4 \times 2C$. Thus, the memory consumption and computational complexity are decreased by a factor of 1/64.

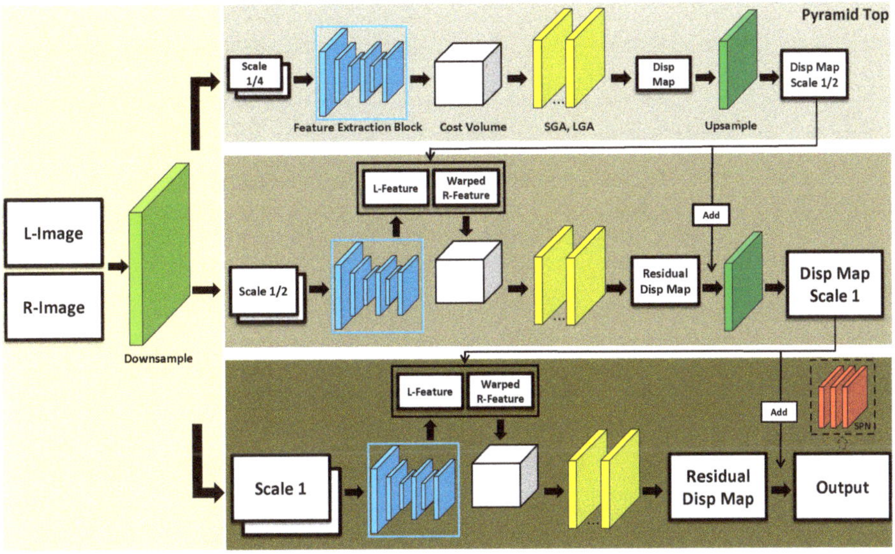

Figure 1. GA-Net-Pyramid with explicit downsampling. The input stereo pair is downsampled explicitly according to the resolution required by each pyramid level. At the pyramid top, the stereo correspondences are located within an absolute disparity range in low resolution. The following pyramid levels perform disparity refinement within a pre-defined residual disparity range until the original resolution is recovered at the pyramid bottom. SPN indicates the Spatial Propagation Network which is an optional module for depth boundary enhancement, as described in Section 3.3.

Afterwards, the cost volume enters the cost aggregation block containing SGA and LGA layers, for which the guiding information is obtained from the downscaled master epipolar image. At last, the filtered cost is used for the following disparity regression as GA-Net. Thus, a disparity map of the downsampled image 'Scale 1/4' is obtained for the pyramid top. From here, the depth of the scene is already roughly estimated and the large-scale context is perceived, which provides a good guidance for the following processing.

3.2.2. The Other Pyramid Levels

Based on the prediction of the pyramid top, the other levels thus only need to locally refine the disparity values. Therefore, the disparity map from 'Scale 1/4' level is upsampled by a factor of 2 via bilinear interpolation, to match the resolution of 'Scale 1/2' level as an initial estimation d_{ini}. Feature maps are computed for the left and right image of 'Scale 1/2' level as F_l and F_r, respectively. Assuming d_{ini} is accurate enough, we can warp F_r according to d_{ini} which would perfectly match F_l. However, considering the details lost through downsampling on the pyramid top and the corresponding matching error, a small shift would exist between the left and the warped right feature, which is named disparity residual and should be additionally considered for a perfect match. Accordingly, a cost volume CV is built in size of $H/2 \times W/2 \times (2disp\_resi + 1) \times 2C$ for 'Scale 1/2' level. Here, $disp\_resi$ is a pre-defined threshold, leading to a range $[d_{ini} - disp\_resi, d_{ini} + disp\_resi]$ around the initial disparity estimation d_{ini} for refinement. The cost volume is thus formed by concatenation of F_l and F_r as:

$$CV(x, y, d) = F_l(x, y) \oplus F_r(x + (d_{ini}(x, y) + d), y), \quad d \in [-disp\_resi, +disp\_resi]. \tag{6}$$

In Equation (6), x and y are the indices of a pixel along the width and height dimension. $\oplus$ represents the concatenation. Then, the cost volume is regularized by SGA and LGA, and a residual disparity map d_{resi} is calculated via multiplying each residual candidate to the corresponding probability and summing them up. The disparity estimation for the current level is obtained by adding the residual and the previously upscaled disparity map as: $d_{resi} + d_{ini}$.

The stereo pair on 'Scale 1/2' level is twice larger in height and width; however, the search for correspondences is restricted within a narrow range. Hence, only a small overhead is accumulated. We apply the same procedure for the remaining pyramid level, to continuously improve the disparity estimation until the original resolution is reached.

Each pyramid level only requires the input epipolar imagery at its level and the disparity image of the previous level. For an efficient and memory saving implementation during disparity estimation, computation of the levels could be decoupled to significantly lower the memory footprint while allowing large input image sizes. Compared to GA-Net, it is thus feasible to significantly increase both image size and disparity range, as only the pyramid top needs to process the full disparity search range, for example, processing of images with a four times larger width, height, and disparity range is possible without additional GPU memory requirements in this case. Note that the evaluation in Section 4 is recorded without adding these optimizations.

3.2.3. Loss

We train the model using the same smooth L_1 loss function as GA-Net in [8]. However, our pyramid architecture predicts more than one disparity map, which should all be considered to allow for intermediate supervision. Hence, a weight is assigned to each pyramid level for a weighted loss summation as:

$$L = \sum_{i=1}^{N} l(|\hat{d}_i - \bar{d}|) \cdot \omega_i, \tag{7}$$

in which $\hat{d}_i$ denotes the disparity predicted by the pyramid level i (starting from 1 at the pyramid top), and $\bar{d}$ is the corresponding ground truth. l computes the smooth L_1 loss from the disparity difference. A weight ω_i is assigned to the level i for a weighted summation through all N pyramid levels. The disparity map from each level is upscaled to the original full resolution before computing the loss. As the estimation is improved from the pyramid top to the bottom, the corresponding weight is also increased (details for parameter setting are in Section 4).

3.3. *GA-Net-Pyramid with Implicit Downsampling*

The paper focuses on presenting a more efficient model based on the structure of GA-Net, in order to achieve robust estimation on datasets from multiple domains. Thus, we design different feature extractors and observe the corresponding performance, so that an appropriate model could be used to handle specific data types. The architecture in Figure 1 simply applies GA-Net in a pyramidal manner, which takes the linearly downsampled stereo pair as input to extract features for further processing. Therefore, we propose another architecture to implicitly learn the downsampled feature, as displayed Figure 2, such that both explicit and implicit image downsampling strategies are tested.

Instead of downsampling the input stereo pair level by level, we only use the stacked hourglass once to extract the feature from the original (full-resolution) images for feeding all the pyramid levels. The input images are firstly downsampled via convolutions with stride two, and then deconvolved to gradually recover the resolution, in which a skip connection is exerted between corresponding feature maps of the encoder and decoder at the same resolution. Before reaching the original size, we directly extract the intermediate feature maps from the decoder to feed each level, as long as the expected resolution is acquired. To differentiate the GA-Net-Pyramid with explicit and implicit downsampling, in the following sections we name the two variants as GA-Net-PyramidED and GA-Net-PyramidID, respectively.

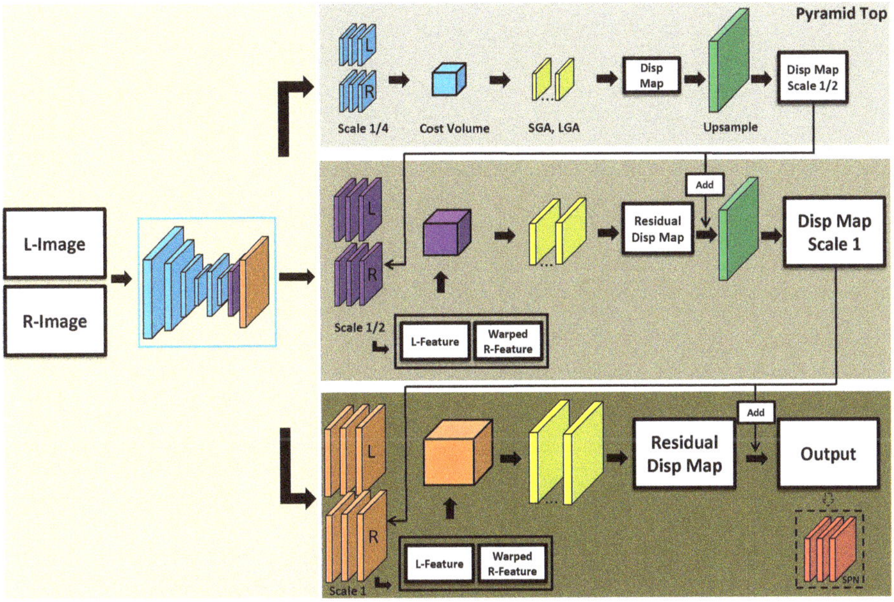

Figure 2. GA-Net-Pyramid with implicit downsampling. The feature extractor is applied on the stereo pair in original resolution, with the intermediate feature maps from its decoder to feed each pyramid level according to the expected resolution. SPN indicates the Spatial Propagation Network which is an optional module for depth boundary enhancement, as described in Section 3.3.

As the disparity is estimated and refined through the pyramid, we add a Spatial Propagation Network (SPN) as a post-processing step to explore its influence on the matching results. SPN is capable of sharpening the object boundaries, by learning from the source image (in our case, the master epipolar image) in a data-driven mode, which is appropriate as a further refinement in our pyramid architecture, especially for close-range data with rich details. Hence, four models are finally proposed including GA-Net-PyramidED and GA-Net-PyramidID, respectively, with or without SPN added at the end of the pyramid bottom.

4. Experiments

In this section, we compare our GA-Net-Pyramid with GA-Net through a series of experiments on close-range, including Scene Flow and KITTI-2012, aerial, and satellite stereo datasets. For a fair comparison, the implementation details are rigidly controlled between the two algorithms. Regarding the training, we use the same patch size with a pre-defined disparity search range, to train the networks for certain epochs, based on Adam optimization strategy [55]. Each stereo pair is normalized, according to the mean and standard deviation of the pixel values from each channel, before feeding to the network. SGA is applied along four directions (horizontally and vertically) for both GA-Net-Pyramid and GA-Net.

For GA-Net-Pyramid specifically, the number of pyramid levels is 3 and the search range for the disparity residual after the pyramid top is set as $[-6, +6]$ to refine the matching results. Details about the pyramid setting are discussed in Section 4.2.3. We apply 3 SGA and 2 LGA layers to regularize the cost volume on our pyramid top, which is the same as GA-Net. With regard to the other pyramid levels, only 1 SGA layer (with 2 LGA layers) is utilized due to the small disparity search range. The weight is set as 0.25, 0.5, and 1, to the pyramid level 1 (top), 2 and 3 (bottom), respectively, to calculate the final loss in Equation (7). The implementation of the methods is based on Python and Pytorch.

4.1. Experiments on Close-Range Stereo Data

We firstly test the networks on Scene Flow and KITTI-2012 datasets, in which the scene structure is relatively complicated with rich details. Referring to most learning-based dense matching algorithms, we train the models on Scene Flow data from scratch, and utilize real data, KITTI-2012 in our case, for finetuning. Both the pre-trained and finetuned models are tested on the corresponding dataset. Regarding the former, the whole Scene Flow training dataset is used for training (8 epochs), while only 1000 stereo pairs from its validation set are selected for test to save time. On the other hand, 170 images from KITTI-2012's training data are exploited to finetune the models for 800 epochs, with the remaining 24 images for test. All the data selection is random, so that a fair evaluation is achieved. In training, we use the same patch size (384 $\times$ 576) with the maximum disparity set to 192. The networks are trained with a batch size of two on two Quadro P6000 GPU cards.

4.1.1. Close-Range Stereo Data

Scene Flow is a synthetic dataset via randomly combining human-made objects with backgrounds from real images, which is used by most stereo networks for initial training. Afterwards, only a small dataset from a specific field is sufficient to adjust the model into practical scenarios. The dataset contains three subsets, namely FlyingThings3D, Monkaa and Driving, including around 35,000 images for training and 4370 images for validation. KITTI-2012 is a stereo dataset with a focus on outdoor street views, which is normally applied in the field of autonomous driving. The dataset includes 194 training and 195 test stereo pairs, with ground truth disparity maps based on LiDAR measurements provided or withheld.

4.1.2. Visualization and Evaluation on Close-Range Stereo Data

The pre-trained networks are firstly tested on the Scene Flow dataset. The quantitative and visual comparison between our pyramid models and GA-Net is shown in Table 1 and Figure 3. As indicated by the table, we calculate the percentage of pixels, for which the estimation error is smaller than 1, 2, and 3 pixels, respectively, and the end point error (EPE) for accuracy evaluation. Regarding the efficiency, the runtime and GPU memory consumption are reported. For all the experiments in this paper, the runtime in the test period is counted for processing the whole test dataset. Specifically, we generate a binary file to save the disparity value of each correspondence, and a png (Portable Network Graphics) file to visualize the result. In the tables, M denotes megabytes for the GPU memory consumed by each network, while the time spent in training and test is expressed in hours (h) or seconds (s). Better performance is highlighted in bold.

Table 1. Accuracy and efficiency comparison between GA-Net-Pyramid, including GA-Net-PyramidED and GA-Net-PyramidID, and GA-Net on Scene Flow data.

	Accuracy				Training Efficiency		Test Efficiency	
	1 pix	2 pix	3 pix	EPE	Memory	Runtime	Memory	Runtime
GA-Net-PyramidED	81.77%	88.59%	91.42%	1.61	**7052 M**	38.25h	**2761 M**	**0.39 h**
GA-Net-PyramidED+SPN	83.04%	89.97%	92.67%	1.44	7140 M	40.62 h	**2761 M**	**0.39 h**
GA-Net-PyramidID	81.26%	89.10%	92.05%	1.49	7264 M	**30.07 h**	3501 M	0.40 h
GA-Net-PyramidID+SPN	84.27%	91.09%	93.64%	1.23	7422 M	31.69 h	3501 M	**0.39 h**
GA-Net	**91.41%**	**95.35%**	**96.60%**	**0.86**	30,464 M	280.53 h	6983 M	2.10 h

Bold font means the best accuracy/efficiency in each group.

From the results, it is found that GA-Net outperforms the two pyramid models in accuracy; however, the latter consume much less memory and runtime in both training and test periods. In case of the close-range data, the objects are captured under an ideal viewing condition, thus very high resolution is achieved with plenty of details and texture information contained. Moreover, as Scene Flow is a synthetic dataset, the random arrangement of man-made objects makes the scene non-natural, non-logical, and highly complicated with many occlusions. Hence, our GA-Net-Pyramid is surpassed by GA-Net, considering the information loss due to a sequence of downsampling-upsampling through the pyramid levels. On the other hand, our hierarchical strategy highly simplifies the problem complexity, consuming far less computational source but at a much higher speed. Between the two pyramid models, GA-Net-PyramidED and GA-Net-PyramidID, similar accuracy is obtained. Regarding the SPN processing, a positive effect is achieved for both pyramid structures, while GA-Net-PyramidID could be improved by a larger extent. The experiments of this paper are implemented on a server open to multiple users; therefore, the runtime of each model could be slightly influenced by unknown processes. We recommend referring to the training time to evaluate the speed of the algorithms, especially for each pyramid model with similar efficiency, considering the relatively long training process compared with the test period. GA-Net-PyramidID is faster than GA-Net-PyramidED, since the feature extraction in the former case is applied only once on the full-resolution stereo pair, rather than repeatedly learning from the corresponding downsampled images level by level. In case of the GPU memory consumption, GA-Net-PyramidED performs better.

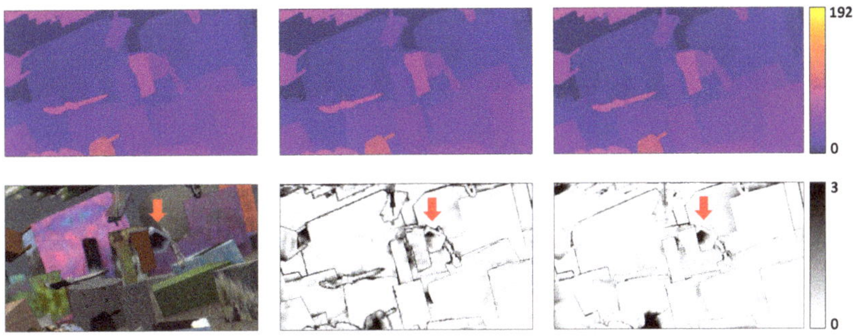

(**a**)

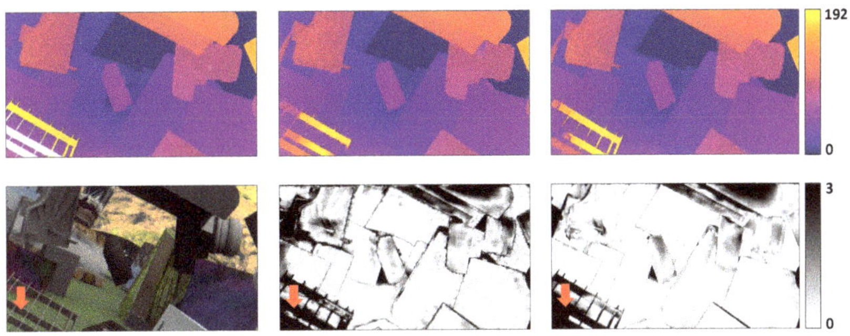

(**b**)

Figure 3. Visual comparison on Scene Flow data. Two test cases are displayed in subfigure (**a**,**b**). In each subfigure, the disparity maps from the ground truth, GA-Net-PyramidID+SPN and GA-Net are displayed from left to right in the first row. The second row provides the master epipolar image and the corresponding error map of each model. Regions where the proposed algorithm outperforms GA-Net are marked with red arrows.

As for the figures in the paper, only the best performed pyramid model is visually compared with GA-Net, e.g., GA-Net-PyramidID+SPN on Scene Flow dataset. Accordingly, we display the master epipolar image, where the guidance information is acquired for SGA and LGA, the ground truth, and the corresponding results from each algorithm. The color bar at the end shows the disparity and error changes. In Figure 3, it is found that GA-Net obtains a generally better disparity result than GA-Net-PyramidID+SPN, with clear edges and more details included. However, our pyramid model still produces a disparity map in good quality, even including superior depth results in certain regions. We discover that GA-Net-PyramidID+SPN is capable of better reconstructing hollow-shaped objects, e.g., the barrel and the shelf as indicated by the red arrows. The finding is also supported by the following experiments on the KITTI dataset.

The pre-trained networks are finetuned on part of KITTI-2012's training data and tested on the remaining stereo pairs. In Table 2 and Figure 4, the corresponding quantitative and qualitative results are provided. Regarding the training efficiency, only the time spent for finetuning is recorded. Similar to the previous experiment, GA-Net acquires the best accuracy, however, the pyramid models are faster and more memory friendly. SPN still improves the results of all the pyramid models, among which GA-Net-PyramidID+SPN achieves the highest accuracy. It should be noted that our GA-Net-Pyramid performs better for real data, leading to a further reduced accuracy gap compared with GA-Net. From the visual inspection, the depth result of each algorithm is barely distinguishable. Moreover as mentioned before, we obtain a better depth prediction for hollow-shaped structures (see the regions indicated by the red arrows). KITTI-2012 does not provide ground truth for the whole scene; nevertheless, according to the image content, it is obvious that our pyramid architecture gives a clean and more reasonable depth estimation.

Table 2. Accuracy and efficiency comparison between GA-Net-Pyramid, including GA-Net-PyramidED and GA-Net-PyramidID, and GA-Net on KITTI-2012 data.

	Accuracy				Training Efficiency		Test Efficiency	
	1 pix	2 pix	3 pix	EPE	Memory	Runtime	Memory	Runtime
GA-Net-PyramidED	86.54%	93.57%	95.76%	0.89	**7140 M**	17.81 h	**2641 M**	28.07 s
GA-Net-PyramidED+SPN	86.56%	93.53%	95.66%	0.88	7242 M	18.49 h	**2641 M**	29.29 s
GA-Net-PyramidID	83.20%	92.68%	95.12%	1.10	7546 M	**13.77 h**	3379 M	**27.02 s**
GA-Net-PyramidID+SPN	86.88%	94.13%	96.18%	0.83	7680 M	15.02 h	3379 M	29.89 s
GA-Net	**91.55%**	**96.64%**	**97.65%**	**0.60**	30,514 M	135.47h	6565 M	165.72 s

Bold font means the best accuracy/efficiency in each group.

4.2. Experiments on Aerial Stereo Data

In this section, the networks are tested using our aerial data. The airborne and satellite (discussed in the following section) stereo processing is the target domain of this research, since the corresponding data are usually large in size and own a much wider stereo baseline, which presents a higher demand on the algorithm's efficiency. The networks are trained on synthetic remote sensing data (854 stereo pairs) from scratch for 200 epochs, then finetuned on a subset (200 stereo pairs) of our aerial data for 100 epochs (data details are in Section 4.2.1). We randomly select another 20 aerial stereo pairs, possessing no overlap with the finetuning data, to test the trained models. Image patches in size of 384×576 are randomly cropped for training, and the test images are 1152×1152. The data may contain negative or very large disparity values; hence, we exclude the stereo pairs with large baselines in order to keep the disparity range processible by both GA-Net-Pyramid and GA-Net. Accordingly, the disparity range is also set as [0, 192]. The models are trained with a batch size of two on two Quadro P6000 GPU cards.

In addition, SGM is utilized as a baseline model in our aerial and satellite experiments, since the algorithm is widely used in the field of remote sensing for dense reconstruction. We exploit Census [56] to calculate the matching cost with a 7×7 window. The penalty terms P_1 and P_2 (see Equation (1)) are set to 19 and 33, respectively. The cost from 8 symmetric scanlines along horizontal, vertical, and diagonal directions are accumulated to compute the disparity based on the WTA strategy, which is then further refined using a left-right consistency check.

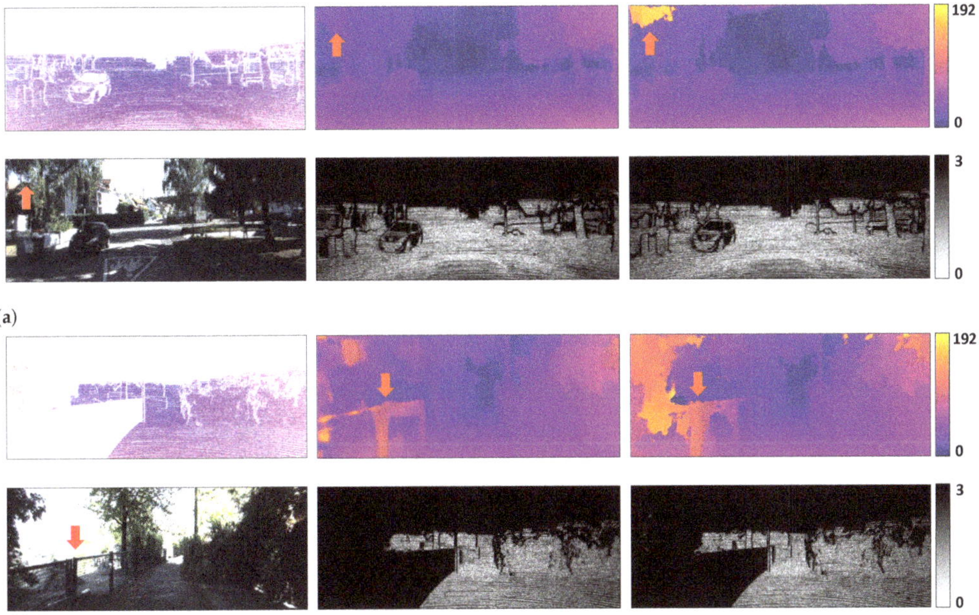

(a)

(b)

Figure 4. Visual comparison on KITTI-2012 data. Two test cases are displayed in subfigure (**a**,**b**). In each subfigure, the disparity maps from the ground truth, GA-Net-PyramidID+SPN and GA-Net are displayed from left to right in the first row. The second row provides the master epipolar image and the corresponding error map of each model. Regions where the proposed algorithm outperforms GA-Net are marked with red arrows.

4.2.1. Aerial Stereo Data

Nowadays, most state-of-the-art dense matching algorithms are data-driven deep neural networks [8–10,12,41–43]. The high performance usually originates from a thorough training, for which a synthetic dataset is preferred for an initial learning phase, to avoid time-consuming data collection and annotation. In the field of remote sensing, nevertheless, a well-annotated stereo dataset is scarce. For example, the aerial image matching benchmark [57,58] provides reference data using LiDAR measurement. However, each algorithm is finally evaluated by the median of the DSM estimation from all the evaluated approaches, due to the limited accuracy of the reference data. Therefore, we propose a synthetic dataset, which is designed specifically for airborne and satellite stereo tasks. The dataset focuses on urban regions via referring to six city models provided by the software CityEngine: Paris, Venice, New York, Philadelphia, and two small development scenes. The models were exported and processed in Blender to preserve the textures and relevant information. Afterwards, we used BlenderProc [59] to render the dataset according to the geometry of the model which included RGB images and the corresponding disparity maps. Considering both aerial and satellite platforms, the simulated camera for rendering was located at 200 m and 500 km above the cities, respectively. A total of 854 stereo pairs in size of 1024 $\times$ 1024 pixels were generated, with the ground sampling distance (GSD) ranging from 5 cm to 50 cm.

Regarding our real aerial data, we use the 4K sensor system mounted on a helicopter for the data collection [60]. Three off-the-shelf Canon EOS cameras (one 1D-C and two 1D-X) constitute the imaging unit. The data contain geo-referenced images with a size of 17.9 megapixels, acquired over Gilching in the southwest of Munich, Germany. Equipped with 50-mm lenses looking in varying view directions, a field of view (FOV) up to 104° is reached. The flight height was 500 m above ground, enabling 6.9-cm nadir GSD. A multi-view stereo matching based on SGM was applied, in which the calculated heights (depths) from multiple highly overlapped images were fused to achieve a high-quality digital surface model (DSM). The DSM was used to compute disparity maps for each stereo pair, which were utilized as reference data for finetuning and evaluation.

4.2.2. Visualization and Evaluation on Aerial Stereo Data

In Table 3, the performance of each algorithm is recorded. We firstly find that all the GA-Net models outperform the baseline SGM by a certain margin. Moreover, our pyramidal revision leads to a very small accuracy decrease compared with the original structure, but highly improves the efficiency. Our GA-Net-PyramidED (without SPN added) is the best performing pyramid model, which is only around 1% worse than GA-Net in accuracy. Nevertheless, the pyramid models are about 8 and 7 times faster than GA-Net, by only expending around 25% and 40% memory usage for training and prediction, respectively. It should be noted that for airborne data, SPN cannot improve the performance for either of the pyramid models, which is different from the close-range experiments. A visual comparison among the methods is provided in Figure 5.

Table 3. Accuracy and efficiency comparison between GA-Net-Pyramid, including GA-Net-PyramidED and GA-Net-PyramidID, and GA-Net on aerial data (baseline model: SGM).

	Accuracy			Training Efficiency		Test Efficiency	
	1 pix	**2 pix**	**3 pix**	**Memory**	**Runtime**	**Memory**	**Runtime**
GA-Net-PyramidED	77.28%	86.19%	89.70%	**7124 M**	25.18 h	**5623 M**	**83.60 s**
GA-Net-PyramidED+SPN	74.06%	86.08%	89.69%	7238 M	26.19 h	**5623 M**	89.08 s
GA-Net-PyramidID	76.35%	85.46%	89.14%	7544 M	**20.59 h**	6979 M	84.02 s
GA-Net-PyramidID+SPN	76.14%	84.82%	88.21%	7676 M	21.54 h	6979 M	86.19 s
GA-Net	**78.75%**	**86.99%**	**90.13%**	30,512 M	187.59 h	15,685 M	616.74 s
SGM	72.14%	75.89%	77.15%	—	—	—	—

Bold font means the best accuracy/efficiency in each group.

We select two regions, one vegetation and one building area from the test data for the visualization. It is shown that GA-Net-PyramidED archives good performance in airborne stereo matching. When the scene is relatively simple, containing fewer depth discontinuities and a smooth depth change, the hierarchical estimation and refinement of disparity is capable of highly enhancing the efficiency, without a noteworthy sacrifice of the result's quality.

4.2.3. Pyramid Setting

To further understand our GA-Net-Pyramid when applied in the field of remote sensing, we explore the impact of different pyramid architectures using our aerial data. Regarding the pyramid structure, two variants are the most important factors, the number of pyramid levels and the residual search range for disparity refinement. The main difference between GA-Net-PyramidED and GA-Net-PyramidID is the strategy to extract features, which is not directly related to the above two factors. In addition, our two pyramid models achieve similar accuracy. Therefore, we select GA-Net-PyramidED without SPN for post-processing to study the pyramid setting, since it is the more intuitive pyramidal modification of GA-Net. As for the number of pyramid levels, we start from 2, since a 1-level GA-Net-Pyramid will degenerate to GA-Net, to 4 levels, with a fixed residual range [−6, +6]. The model is trained on our synthetic dataset from scratch and evaluated on the same test data. We use the same hyperparameter setting as before, except that the size of the training patches changes to 384 × 768 to facilitate the downsampling when more levels are applied. We train the model on one GPU card due to the less memory requirement of GA-Net-Pyramid. The results are in Table 4.

According to the table, it is found that the architecture with 4 pyramid levels acquires the best efficiency. However, with slightly increased memory and runtime, the model with 3 pyramid levels achieves better results. Along with GA-Net-PyramidED regresses towards GA-Net (from 3 to 2 levels), the efficiency drastically deteriorates as expected, nevertheless, without a noticeable improvement of the accuracy. Therefore, we determine to use the number of pyramid levels as 3. Then, we adjust the residual search range to [−3, +3], [−6, +6] and [−12, +12], respectively. The model is also trained from scratch on our synthetic dataset using one GPU card, and tested on the same 20 aerial images. We keep the training setting unchanged, except that the patch size is set back to 384 × 576. In Table 5, the performance for different residual search ranges is recorded.

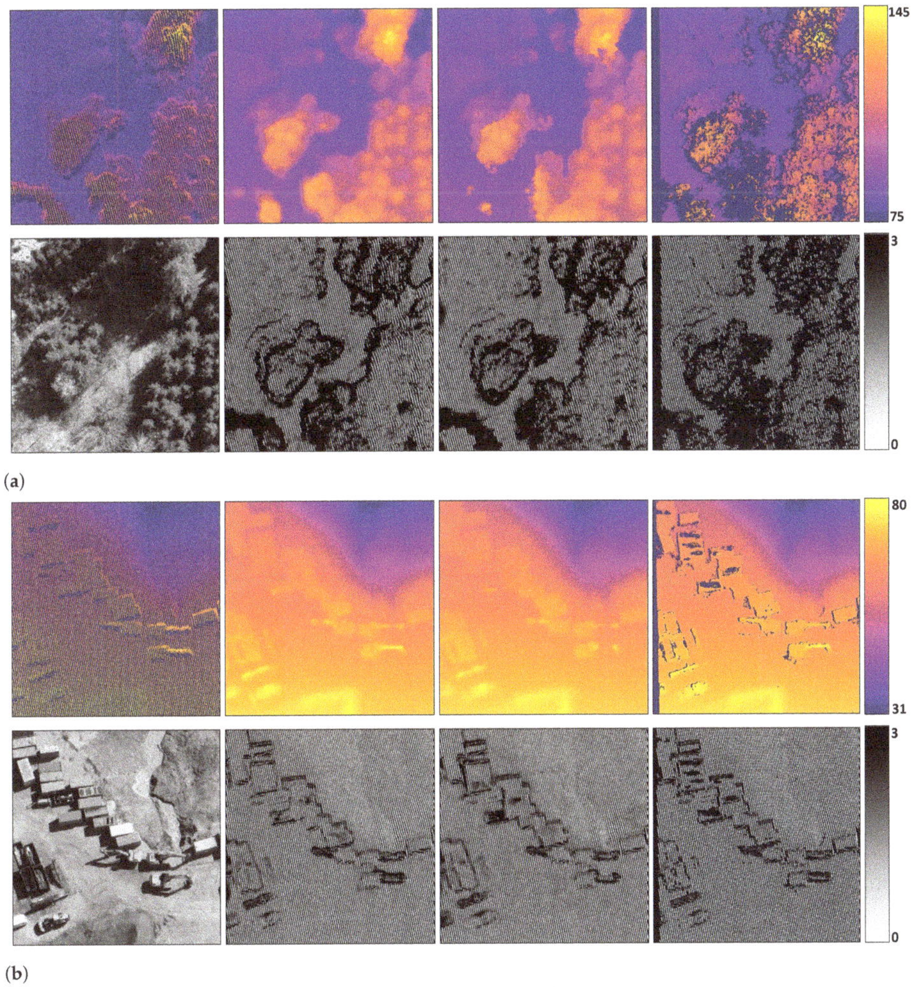

Figure 5. Visual comparison on aerial data. Two test cases regarding vegetation and building area are displayed in subfigure (**a**,**b**), respectively. In each subfigure, the reference disparity map and the stereo results from GA-Net-PyramidED, GA-Net and SGM are displayed from left to right in the first row. The second row provides the master epipolar image and the corresponding error map of each model.

Table 4. Accuracy and efficiency comparison for GA-Net-PyramidED with different pyramid levels.

	Accuracy			Training Efficiency		Test Efficiency	
Pyramid Levels	**1 pix**	**2 pix**	**3 pix**	**Memory**	**Runtime**	**Memory**	**Runtime**
2	**72.38%**	80.89%	85.14%	11521 M	70.25 h	5813 M	120.28 s
3	72.17%	**81.22%**	**85.69%**	8121 M	29.13 h	5623 M	82.11 s
4	72.08%	81.19%	85.57%	**7647 M**	**27.80 h**	**5589 M**	**63.92 s**

Bold font means the best accuracy/efficiency in each group.

Table 5. Accuracy and efficiency comparison for GA-Net-PyramidED with different residual search ranges.

	Accuracy			Training Efficiency		Test Efficiency	
Residual Range	**1 pix**	**2 pix**	**3 pix**	**Memory**	**Runtime**	**Memory**	**Runtime**
[−3, +3]	73.38%	81.95%	86.04%	**5941 M**	**23.49 h**	**5467 M**	**55.23 s**
[−6, +6]	**73.76%**	**82.21%**	**86.40%**	6283 M	26.35 h	5623 M	84.50 s
[−12, +12]	73.38%	82.11%	86.37%	7033 M	34.96 h	6489 M	123.09 s

Bold font means the best accuracy/efficiency in each group.

Table 5 indicates that as the residual range becomes larger, the efficiency naturally decreases. Moreover, when the residual buffer expands over [−6, +6], the accuracy cannot be further enhanced. Hence, the structure of our pyramid is determined as 3 levels, with the maximum/minimum residual set as 6/−6. To keep the experiments consistent, the pyramid structure is used for both GA-Net-PyramidED and GA-Net-PyramidID in this paper.

4.3. Experiments on Satellite Stereo Data

The flight campaign regarding our aerial 4K images was performed during a WorldView-3 stereo acquisition of the same area [61]. Due to the minimal time difference of less than 1 hour of each aerial image from the satellite images, the higher resolution airborne data are well suited as reference data for the satellite stereo matching to finetune the models and evaluate the results. This is a notable improvement over other satellite stereo datasets [17,62], which do not provide sub-pixel disparity accuracy due to different sensing modalities and scene changes due to time difference between the image and ground truth acquisition. In contrast, the data used in this article allow reliable evaluation for 1- and 2-pixel accuracy metrics. This is especially import for photogrammetry and remote sensing, as many applications require highly precise elevation measurements.

Similar to Section 4.2, the networks are pre-trained on our synthetic remote sensing data for 200 epochs, and finetuned on the generated satellite training data for 150 epochs. The training conditions stay the same, including the patch size (384 × 576), disparity range ([0, 192]), batch size (2), GPU usage (2 Quadro P6000 cards), etc. SGM is also tested for reference.

4.3.1. Satellite Stereo Data

WorldView-3 is a very-high-resolution imaging satellite currently offering the most detailed publicly available spaceborne imagery, at a resolution of 30 cm. After bundle-adjustment of the data with the 4K aerial imagery and DSM as reference, we generated an epipolar rectified stereo pair using the algorithm implemented by the CARS stereo pipeline [63]. Similar to the aerial imagery, a reference disparity map was calculated by projecting each point of the 4K DSM into the epipolar satellite stereo pair. The stereo pair has a dimension of 20,815 × 28,264 pixels, which was cut into 98 tiles (in size of 1152 × 1152) owning an overlap larger than 25% with the 4K data coverage. From them, 78 tiles were randomly selected for finetuning the pre-trained GA-Net models, with the other 20 image pairs as the test data.

As the airborne data were geo-referenced in two separate blocks using differential GPS and only few ground control points (GCPs), a slight height offset was found between the aerial and satellite data, yielding disparity differences between the aerial reference and the satellite stereo pair in the pixel range, but rising up to 4 pixels at the corner of one aerial block. Since these systematic differences strongly affected training and evaluation of the networks, a second-order offset surface was fitted to the difference of the airborne reference disparity map and the satellite disparity map estimated by SGM, on each of the 98 tiles. The offset was added to the reference disparity map to alleviate the systematic bias which was reduced from 0.97 to 0.51 pixels.

4.3.2. Visualization and Evaluation on Satellite Stereo Data

In Table 6, we record the performance of GA-Net-Pyramid, GA-Net and SGM. Similar to the results of airborne data, GA-Net achieves the highest accuracy, after which GA-Net-PyramidED still acquires the best performance among all the other models. The 1-pixel accuracy of our GA-Net-PyramidED, without SPN added for post-processing, is only surpassed by GA-Net by 0.08%. However, the former is around 8 and 13 times faster than the latter, consuming only 23% and 36% GPU memory in training and test, respectively. In addition, GA-Net-PyramidED performs better than GA-Net_PyrmaidID, with less GPU memory consumption but longer training time. SPN also impairs the performance of the pyramid models which is consistent with our experiments on aerial data. The visual comparison is in Figure 6, including a vegetation and a building area as well. It is found that both networks predict a smoother disparity map than SGM, with less erroneous estimation. Moreover, similar results are obtained between our GA-Net-PyramidED and GA-Net, considering the reconstruction density and quality.

Table 6. Accuracy and efficiency comparison between GA-Net-Pyramid, including GA-Net-PyramidED and GA-Net-PyramidID, and GA-Net on satellite data (baseline model: SGM).

	Accuracy			Training Efficiency		Test Efficiency	
	1 pix	**2 pix**	**3 pix**	**Memory**	**Runtime**	**Memory**	**Runtime**
GA-Net-PyramidED	83.76%	90.70%	93.00%	**7144 M**	23.77 h	**5623 M**	**31.53 s**
GA-Net-PyramidED+SPN	82.99%	91.05%	93.34%	7250 M	24.56 h	**5623 M**	35.93 s
GA-Net-PyramidID	81.45%	89.58%	92.40%	7558 M	**19.11 h**	6979 M	33.11 s
GA-Net-PyramidID+SPN	80.66%	89.10%	92.00%	7700 M	20.27 h	6979 M	32.87 s
GA-Net	**83.84%**	**91.42%**	**93.74%**	30,514 M	179.19 h	15,685 M	401.91 s
SGM	79.98%	82.74%	83.32%	—	—	—	—

Bold font means the best accuracy/efficiency in each group.

4.3.3. Stereo Processing over Mountain Area

In this section, we apply our pyramid network on a stereo pair with a large disparity range, in order to indicate the model's ability to process large-scale remote sensing data. The imagery is from WorldView-2 [64] at a resolution of 50 cm, covering the Matterhorn mountain, Switzerland. We select a stereo pair with 14° conversion angle for which the disparity varies in range of thousand pixels, due to the very large ground height difference from 1800 m to 4478 m. The best performing model finetuned in our previous satellite experiments, GA-Net-PyramidED, is directly used for disparity prediction in this test. Regarding the evaluation, we follow our processing chain in Section 4.3.1, using an aerial dataset with good stereo geometry to the same area to generate reference data. The test region, the reference disparity map, and our stereo results are displayed in Figure 7.

The mountain peak is located at the center of the image with a disparity up to around 1250 pixels; thus, we set the disparity range as [0, 1248]. Note that the model we use receives no supervision and knowledge regarding the mountain area with that large disparity difference. However, we achieve a 3-pixel accuracy of 87.34%. There are temporal inconsistencies between the satellite and reference data, leading to varying snow cover. Therefore, we use 3-pixel as the threshold. The visual comparison shows very similar results between our disparity prediction and the reference, considering the reconstruction density, smoothness, etc. Disparity holes are found from certain regions in our results. According to the image content, the regions are in shadow with limited texture information, where the network suffers from collecting enough information to locate the correspondences.

In the test period, the patch in size of 768 × 6912 is fed to the network for disparity prediction. Considering the disparity range [0, 1248], GA-Net will theoretically need more than 200 GB GPU memory to process the same data. Our GA-Net-PyramidED, however, consumes only around 20 GB.

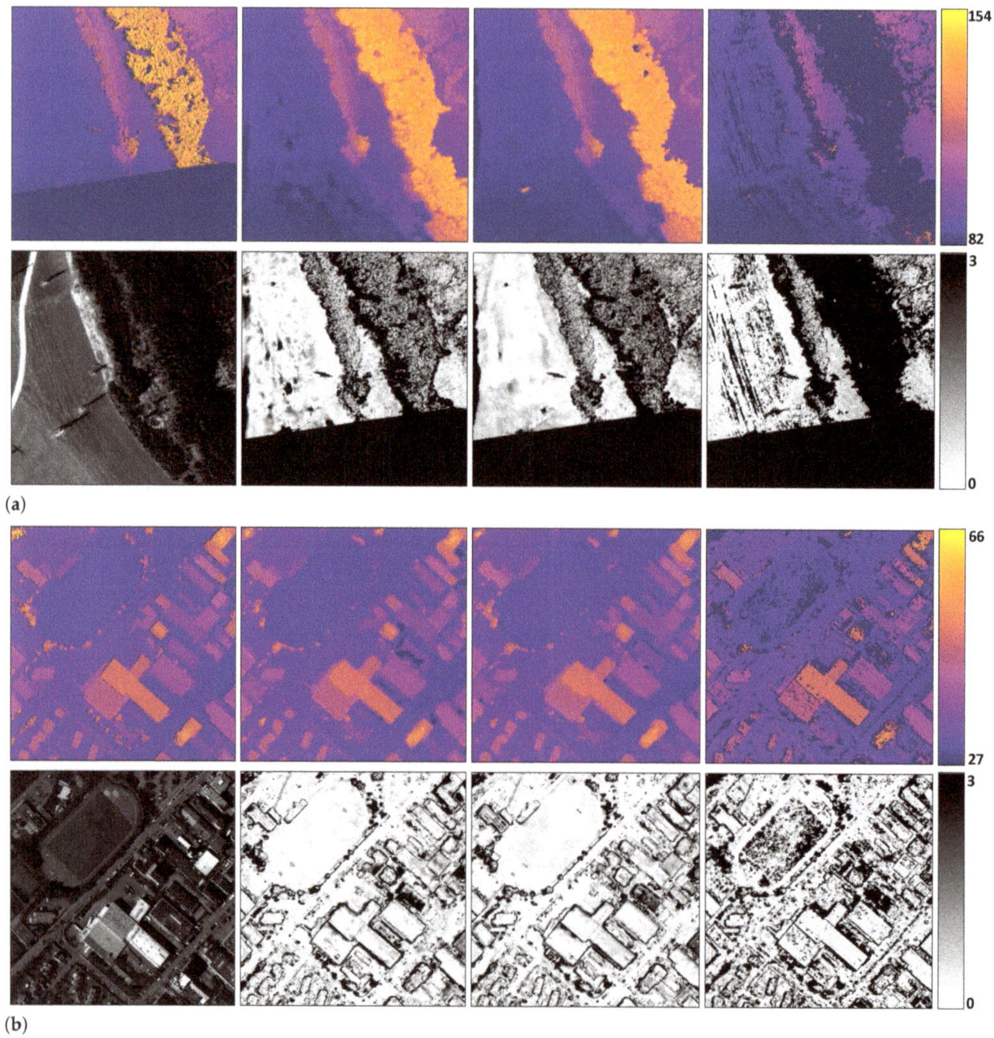

Figure 6. Visual comparison on satellite data. Two test cases regarding vegetation and building area are displayed in subfigure (**a**,**b**), respectively. In each subfigure, the reference disparity map and the stereo results from GA-Net-PyramidED, GA-Net and SGM are displayed from left to right in the first row. The second row provides the master epipolar image and the corresponding error map of each model.

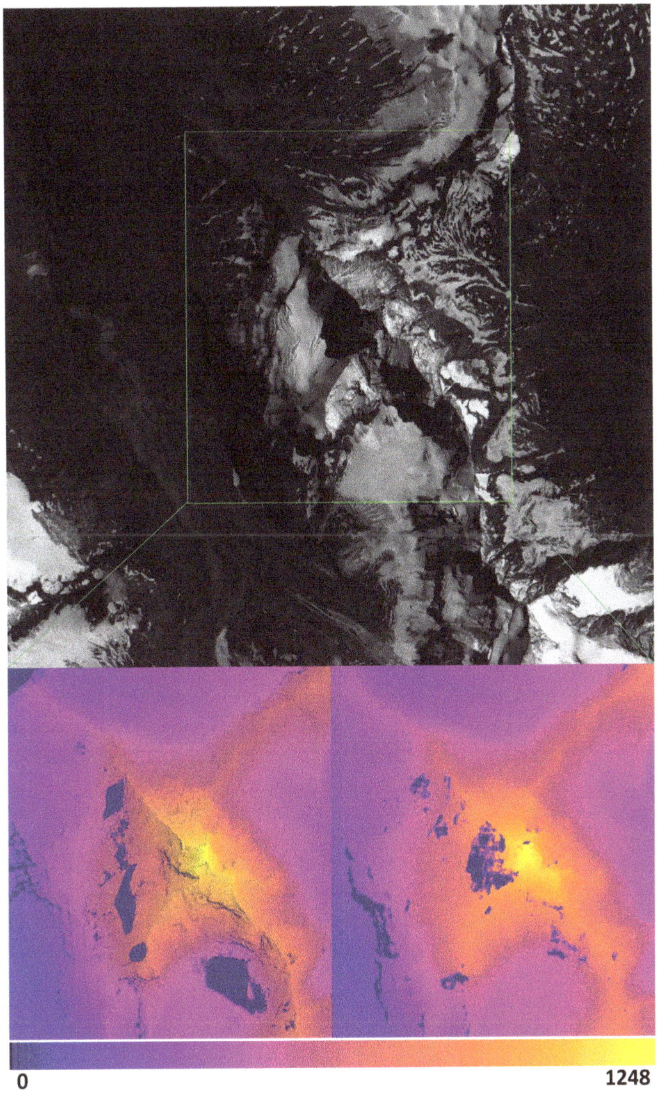

Figure 7. A showcase to indicate the ability of our pyramid network in processing remote sensing stereo pair with large baseline. The test image and the corresponding stereo reconstruction from the reference disparity map (**lower left**) and our pyramid model (**lower right**) are shown. The reconstructed region is highlighted by the green rectangle with a size of 19,791 × 15,639 pixels. Test region: Matterhorn mountain, Switzerland. Test model: GA-Net-PyramidED.

5. Discussion

Based on a pyramid architecture, our GA-Net-Pyramid is able to roughly estimate the depth from a downsampled feature, and then refine the prediction level by level until the original resolution is recovered. Thus, the efficiency is significantly enhanced with the accuracy maintained to be comparable with GA-Net on remote sensing datasets. Some technical details are found below.

We firstly propose GA-Net-PyramidED which applies the GA-Net model hierarchically. In our experiments on airborne and satellite data, it is demonstrated that GA-Net-

PyramidED is able to achieve similar results as GA-Net, nevertheless, consuming much less GPU memory and runtime for both training and prediction. Considering that only the pyramid top exploits the absolute disparity range in low resolution to locate the stereo correspondence, GA-Net-PyramidED is capable of processing stereo pairs with wider baselines if the same GPU memory for GA-Net is available. This is particularly suitable to process large stereo pairs with high-disparity search ranges in the field of remote sensing, which usually triggers the bottleneck of most memory-hungry deep neural networks. On the other hand, the aerial/satellite images mainly focus on large-scale landscapes such as city areas, for which the local object heights/depths are generally smoother and regular with fewer occlusions, depth discontinuities, fine structures, etc., compared with the close-range datasets. Thus, the results from the previous pyramid level can better guide the disparity estimation on the current level. When a large height variance exists within the scene, e.g., in mountain areas, a rough depth prediction from lower resolution pyramid level is effective to limit the search range and avoid influence from ambiguous disparity candidates for higher resolution level.

Another architecture is designed as GA-Net-PyramidID, which implicitly downsamples the input stereo pair via a U-Net feature extractor to feed each pyramid level using the intermediate feature map of its decoder. Concerning the close-range datasets, especially for Scene Flow that contains very complex and non-logical scene structures, both GA-Net-PyramidED and GA-Net-PyramidID are not competitive with GA-Net (GA-Net-PyramidID+SPN performs the best among all the pyramid models). The accuracy could be influenced when details are possibly omitted by the low-resolution level. Moreover, the residual search range may not support refinement for regions with rapid depth changes and discontinuities. Although GA-Net outperforms the proposed pyramid approaches on both close-range datasets, Scene Flow and KITTI, the performance difference is smaller for the real-world KITTI 2012 data.

SPN is applied on image segmentation to refine the object boundaries. In our experiments on close-range data, better depth estimation is achieved by our pyramid networks with SPN added, especially for GA-Net-PyramidID. However, it is found that negative influence from SPN occurs on airborne and satellite data, for both GA-Net-PyramidED and GA-Net-PyramidID. The reason is that the resolution of aerial/satellite data is relatively low, with fewer details and depth discontinuities included; thus, the strength of SPN is not embodied. More importantly, the training of SPN cannot be well supervised, considering that the number of valid training patches from airborne (987 millions) and satellite (934 millions) datasets is far less than the close-range datasets (18 billions). The condition to collect reference data is not as ideal as close-range scenarios using precise LiDAR scanning, structured light or synthetic labeling. In addition, SPN essentially refers to the input to improve the output, which are the master epipolar image and the disparity result in our case, respectively. The natural land texture and shadows, which are not necessarily related to ground height variation, may confuse SPN to locate the correct depth borders. The slightly changing and rolling ground height, e.g., in natural regions, could confuse the disparity post-processing as indicated by the lower 1-pixel accuracy.

6. Conclusions

Nowadays, the rapid development of deep learning and CNNs has made the technique dominate in the field of dense matching, leading to a sequence of high-rank algorithms in different close-range benchmarks. Compared to conventional approaches, the depth estimation for ill-posed areas, e.g., textureless regions, occlusions, etc., is better accomplished resulting in a considerable improvement. However, a large amount of well-annotated data and a time-consuming training are usually required before a network reaches high performance. In the field of remote sensing, a huge amount of high-definition data is supplied by unmanned aerial vehicles, helicopters, airplanes or satellites at all times. The data cover large areas with varying stereo baselines and image sizes of up to multiple gigapixels. Hence, a well-performed deep network from the field of computer vision would struggle to process the remote sensing data, under a certain time and memory budget. Since that stereo datasets with reliable ground truth are not available in remote sensing, we build a dataset consisting of simultaneously acquired 30-cm satellite and 6-cm aerial imagery which are co-registered to sub-pixel disparity precision. The experimental results demonstrate that our proposed model can largely enhance the efficiency in training and test, while maintaining a comparable accuracy. The test on a satellite stereo pair over Matterhorn specifically highlights the significance of our method for processing large baseline stereo data.

We suggest to use GA-Net-PyramidED for remote sensing stereo processing. With slightly increased runtime, GA-Net-PyramidED produces better depth results than GA-Net-PyramidID, while consuming less GPU memory. As for the close-range dataset, GA-Net-PyramidID with an SPN module to enhance the depth borders is preferred. Regarding the effect of SPN, it is demonstrated that a minor improvement is obtained on close-range data; nevertheless, the depth estimation could be impaired using SPN in case of remote sensing data, especially when the reference data own limited quantity or quality for training.

In future research, more reference data should be collected for urban, rural and mountainous scenarios for remote sensing, in order to better supervise a learning-based model in stereo prediction. Thus, we can better handle the ill-posed regions in shadows, depth boundaries, etc., and obtain high-quality geographical measurements for earth observation.

Author Contributions: Conceptualization, Y.X.; data curation, Y.X., P.d. and M.F.R.; funding acquisition, J.T. and P.R.; investigation, Y.X., P.d., F.F., J.T., M.F.R. and P.R.; methodology, Y.X.; supervision, P.d., F.F., J.T. and P.R.; validation, Y.X.; visualization, Y.X.; writing—original draft, Y.X.; writing—review and editing, P.d., F.F., J.T., M.F.R. and P.R. All authors have read and agreed to the published version of the manuscript.

Funding: This research was funded by "ForDroughtDet" project (FKZ: 22WB410602), from the Waldklimafonds, under joint leadership of Bundeslandwirtschafts (BMEL) and Bundesumweltministerium (BMU). Yuanxin Xia is funded by a DLR-DAAD Research Fellowship (No. 57265855).

Data Availability Statement: The Scene Flow dataset can be accessed in https://lmb.informatik.uni-freiburg.de/resources/datasets/SceneFlowDatasets.en.html/. The KITTI-2012 dataset can be accessed in http://www.cvlibs.net/datasets/kitti/eval_stereo_flow.php?benchmark=stereo/.

Acknowledgments: We are indebted to University of Freiburg, Karlsruhe Institute of Technology, and Toyota Technological Institute at Chicago for providing the close-range benchmark datasets. We would like to thank Franz Kurz from the German Aerospace Center (DLR) for providing the aerial data, and DigitalGlobe and European Space Imaging (EUSI) for providing the satellite data used in the research.

Conflicts of Interest: The authors declare no conflict of interest.

Abbreviations

The following abbreviations are used in this manuscript:

GA-Net	Guided Aggregation Network
GA-Net-Pyramid	GA-Net based on a pyramid architecture
GA-Net-PyramidED	GA-Net-Pyramid with Explicit Downsampling
GA-Net-PyramidID	GA-Net-Pyramid with Implicit Downsampling
SGM	Semi-Global Matching

References

1. Hirschmüller, H. Semi-global Matching—Motivation, Developments and Applications. In *Photogrammetric Week*; Wichmann Verlag: Heidelberg, Germany, 2011; Volume 11, pp. 173–184.
2. Kuschk, G.; d'Angelo, P.; Qin, R.; Poli, D.; Reinartz, P.; Cremers, D. DSM Accuracy Evaluation for the ISPRS Commission I Image Matching Benchmark. *Int. Arch. Photogramm. Remote Sens. Spat. Inf. Sci.* **2014**, *40*, 195–200. [CrossRef]
3. Qin, R.; Huang, X.; Gruen, A.; Schmitt, G. Object-based 3-D building change detection on multitemporal stereo images. *IEEE J. Sel. Top. Appl. Earth Observ. Remote Sens.* **2015**, *8*, 2125–2137. [CrossRef]

4. Xia, Y.; d'Angelo, P.; Tian, J.; Fraundorfer, F.; Reinartz, P. Self-supervised convolutional neural networks for plant reconstruction using stereo imagery. *Photogramm Eng. Remote. Sens.* **2019**, *85*, 389–399. [CrossRef]
5. Bleyer, M.; Breiteneder, C. Stereo matching—State-of-the-art and research challenges. In *Advanced Topics in Computer Vision*; Farinella, G.M., Battiato, S., Cipolla, R., Eds.; Springer: London, UK, 2013; pp. 143–179. [CrossRef]
6. Hirschmüller, H. Accurate and Efficient Stereo Processing by Semi-global Matching and Mutual Information. In Proceedings of the 2005 IEEE Computer Society Conference on Computer Vision and Pattern Recognition (CVPR), San Diego, CA, USA, 20–26 June 2005; Volume 2, pp. 807–814. [CrossRef]
7. Rothermel, M. Development of a SGM-Based Multi-View Reconstruction Framework for Aerial Imagery. Ph.D. Thesis, University of Stuttgart, Stuttgart, Germany, 2017.
8. Zhang, F.; Prisacariu, V.; Yang, R.; Torr, P.H. GA-Net: Guided Aggregation Net for End-to-End Stereo Matching. In Proceedings of the IEEE Conference on Computer Vision and Pattern Recognition (CVPR), Long Beach, CA, USA, 15–20 June 2019; pp. 185–194.
9. Kendall, A.; Martirosyan, H.; Dasgupta, S.; Henry, P.; Kennedy, R.; Bachrach, A.; Bry, A. End-to-End Learning of Geometry and Context for Deep Stereo Regression. In Proceedings of the 2017 IEEE International Conference on Computer Vision (ICCV), Venice, Italy, 22–29 October 2017; pp. 66–75. [CrossRef]
10. Chang, J.; Chen, Y. Pyramid Stereo Matching Network. In Proceedings of the 2018 IEEE/CVF Conference on Computer Vision and Pattern Recognition (CVPR), Salt Lake City, UT, USA, 18–23 June 2018; pp. 5410–5418. [CrossRef]
11. Tonioni, A.; Tosi, F.; Poggi, M.; Mattoccia, S.; Stefano, L.D. Real-Time Self-Adaptive Deep Stereo. In Proceedings of the 2019 IEEE/CVF Conference on Computer Vision and Pattern Recognition (CVPR), Long Beach, CA, USA, 15–20 June 2019; pp. 195–204. [CrossRef]
12. Wang, Y.; Lai, Z.; Huang, G.; Wang, B.H.; van der Maaten, L.; Campbell, M.; Weinberger, K.Q. Anytime Stereo Image Depth Estimation on Mobile Devices. In Proceedings of the 2019 International Conference on Robotics and Automation (ICRA), Montreal, QC, Canda, 20–24 May 2019; pp. 5893–5900.
13. Yang, G.; Manela, J.; Happold, M.; Ramanan, D. Hierarchical Deep Stereo Matching on High-Resolution Images. In Proceedings of the 2019 IEEE/CVF Conference on Computer Vision and Pattern Recognition (CVPR), Long Beach, CA, USA, 15–20 June 2019; pp. 5510–5519. [CrossRef]
14. Liu, S.; De Mello, S.; Gu, J.; Zhong, G.; Yang, M.H.; Kautz, J. Learning Affinity via Spatial Propagation Networks. In *Advances in Neural Information Processing Systems*; Guyon, I., Luxburg, U.V., Bengio, S., Wallach, H., Fergus, R., Vishwanathan, S., Garnett, R., Eds.; Curran Associates, Inc.: Red Hook, NY, USA, 2017; Volume 30, pp. 1520–1530.
15. Mayer, N.; Ilg, E.; Häusser, P.; Fischer, P.; Cremers, D.; Dosovitskiy, A.; Brox, T. A Large Dataset to Train Convolutional Networks for Disparity, Optical Flow, and Scene Flow Estimation. In Proceedings of the 2016 IEEE Conference on Computer Vision and Pattern Recognition (CVPR), Las Vegas, NV, USA, 27–30 June 2016; pp. 4040–4048.
16. Geiger, A.; Lenz, P.; Urtasun, R. Are We Ready for Autonomous Driving? The KITTI Vision Benchmark Suite. In Proceedings of the 2012 IEEE Conference on Computer Vision and Pattern Recognition (CVPR), Providence, RI, USA, 16–21 June 2012; pp. 3354–3361. [CrossRef]
17. Le Saux, B.; Yokoya, N.; Hansch, R.; Brown, M.; Hager, G.; Kim, H. 2019 IEEE GRSS data fusion contest: Semantic 3D reconstruction [Technical Committees]. *IEEE Geosci. Remote Sens. Mag.* **2019**, *7*, 103–105. [CrossRef]
18. d'Angelo, P.; Reinartz, P. Semiglobal Matching Results on the ISPRS Stereo Matching Benchmark. In *ISPRS—International Archives of the Photogrammetry, Remote Sensing and Spatial Information Sciences*; ISPRS: Hanover, Germany, 2011; Volume XXXVIII-4/W19, pp. 79–84. [CrossRef]
19. d'Angelo, P. Improving Semi-global Matching: Cost Aggregation and Confidence Measure. In *ISPRS—International Archives of the Photogrammetry, Remote Sensing and Spatial Information Sciences*; ISPRS: Prague, Czech Republic, 2016; Volume XLI-B1, pp. 299–304. [CrossRef]
20. Facciolo, G.; de Franchis, C.; Meinhardt, E. MGM: A Significantly More Global Matching for Stereovision. In Proceedings of the British Machine Vision Conference (BMVC), Swansea, UK, 7–10 September 2015; Xie, X., Tam, G.K.L., Eds.; BMVA Press: Swansea, UK, 2015; pp. 90.1–90.12. [CrossRef]
21. Geman, S.; Geman, D. Stochastic relaxation, gibbs distributions, and the bayesian restoration of images. *IEEE Trans. Pattern Anal. Mach. Intell.* **1984**, *PAMI-6*, 721–741. [CrossRef]
22. Pollard, S.B.; Mayhew, J.E.W.; Frisby, J.P. PMF: A stereo correspondence algorithm using a disparity gradient limit. *Perception* **1985**, *14*, 449–470. [CrossRef] [PubMed]
23. Barnard, S. Stochastic stereo matching over scale. *Int. J. Comput. Vis.* **1989**, *3*, 17–32. [CrossRef]
24. Kolmogorov, V.; Zabih, R. Computing Visual Correspondence with Occlusions using Graph Cuts. In Proceedings of the Proceedings Eighth IEEE International Conference on Computer Vision (ICCV), Vancouver, BC, Canada, 7–14 July 2001; Volume 2, pp. 508–515. [CrossRef]
25. Boykov, Y.; Kolmogorov, V. An experimental comparison of min-cut/max- flow algorithms for energy minimization in vision. *IEEE Trans. Pattern Anal. Mach. Intell.* **2004**, *26*, 1124–1137. [CrossRef]
26. Lecun, Y.; Bottou, L.; Bengio, Y.; Haffner, P. Gradient-based learning applied to document recognition. *Proc. IEEE* **1998**, *86*, 2278–2324. [CrossRef]
27. Zbontar, J.; LeCun, Y. Stereo matching by training a convolutional neural network to compare image patches. *J. Mach. Learn. Res.* **2016**, *17*, 1–32.

28. Bromley, J.; Bentz, J.; Bottou, L.; Guyon, I.; LeCun, Y.; Moore, C.; Sackinger, E.; Shah, R. Signature verification using a "siamese" time delay neural network. *Int. J. Pattern Recognit. Artif. Intell.* **1993**, *7*, 25. [CrossRef]
29. Luo, W.; Schwing, A.G.; Urtasun, R. Efficient Deep Learning for Stereo Matching. In Proceedings of the 2016 IEEE Conference on Computer Vision and Pattern Recognition (CVPR), Las Vegas, NV, USA, 27–30 June 2016; pp. 5695–5703. [CrossRef]
30. Seki, A.; Pollefeys, M. Sgm-nets: Semi-global Matching with Neural Networks. In Proceedings of the 2017 IEEE Conference on Computer Vision and Pattern Recognition (CVPR), Honolulu, HI, USA, 21–26 July 2017; pp. 6640–6649. [CrossRef]
31. Michael, M.; Salmen, J.; Stallkamp, J.; Schlipsing, M. Real-time Stereo Vision: Optimizing Semi-global Matching. In Proceedings of the 2013 IEEE Intelligent Vehicles Symposium (IV), Gold Coast City, Australia, 23 June 2013; pp. 1197–1202. [CrossRef]
32. Poggi, M.; Mattoccia, S. Learning a General-purpose Confidence Measure based on O(1) Features and a Smarter Aggregation Strategy for Semi Global Matching. In Proceedings of the 2016 Fourth International Conference on 3D Vision (3DV), Stanford, CA, USA, 25–28 October 2016; pp. 509–518. [CrossRef]
33. Schönberger, J.L.; Sinha, S.N.; Pollefeys, M. Learning to Fuse Proposals from Multiple Scanline Optimizations in Semi-global Matching. In *Computer Vision—ECCV 2018*; Ferrari, V., Hebert, M., Sminchisescu, C., Weiss, Y., Eds.; Springer International Publishing: Cham, Switzerland, 2018; pp. 758–775.
34. Xia, Y.; d'Angelo, P.; Tian, J.; Fraundorfer, F.; Reinartz, P. Multi-label learning based semi-global matching forest. *Remote Sens.* **2020**, *12*, 1069. [CrossRef]
35. Dosovitskiy, A.; Fischer, P.; Ilg, E.; Häusser, P.; Hazirbas, C.; Golkov, V.; van der Smagt, P.; Cremers, D.; Brox, T. FlowNet: Learning Optical Flow with Convolutional Networks. In Proceedings of the 2015 IEEE International Conference on Computer Vision (ICCV), Santiago, Chile, 7–13 December 2015; pp. 2758–2766.
36. Guo, X.; Yang, K.; Yang, W.; Wang, X.; Li, H. Group-Wise Correlation Stereo Network. In Proceedings of the 2019 IEEE/CVF Conference on Computer Vision and Pattern Recognition (CVPR), Long Beach, CA, USA, 15–20 June 2019; pp. 3268–3277. [CrossRef]
37. Zhu, Z.; Guo, W.; Chen, W.; Li, Q.; Zhao, Y. MPANet: Multi-Scale Pyramid Aggregation Network For Stereo Matching. In Proceedings of the 2021 IEEE International Conference on Image Processing (ICIP), Anchorage, AK, USA, 19–22 September 2021; pp. 2773–2777. [CrossRef]
38. Xu, H.; Zhang, J. AANet: Adaptive Aggregation Network for Efficient Stereo Matching. In Proceedings of the 2020 IEEE/CVF Conference on Computer Vision and Pattern Recognition (CVPR), Seattle, WA, USA, 13–19 June 2020; pp. 1956–1965. [CrossRef]
39. Wang, H.; Fan, R.; Cai, P.; Liu, M. PVStereo: Pyramid voting module for end-to-end self-supervised stereo matching. *IEEE Robot. Autom. Lett.* **2021**, *6*, 4353–4360. [CrossRef]
40. Stucker, C.; Schindler, K. ResDepth: Learned Residual Stereo Reconstruction. In Proceedings of the IEEE/CVF Conference on Computer Vision and Pattern Recognition Workshops, Seattle, WA, USA, 13–19 June 2020.
41. Wu, Z.; Wu, X.; Zhang, X.; Wang, S.; Ju, L. Semantic Stereo Matching with Pyramid Cost Volumes. In Proceedings of the 2019 IEEE/CVF International Conference on Computer Vision (ICCV), Seoul, Korea, 27 October–2 November 2019; pp. 7483–7492. [CrossRef]
42. Song, X.; Zhao, X.; Fang, L.; Hu, H.; Yu, Y. EdgeStereo: An effective multi-task learning network for stereo matching and edge detection. *Int. J. Comput. Vis.* **2020**, *128*, 910–930. [CrossRef]
43. Cheng, X.; Wang, P.; Yang, R. Learning depth with convolutional spatial propagation network. *IEEE Trans. Pattern Anal. Mach. Intell.* **2019**, 42, 2361–2379. [CrossRef]
44. Zhang, F.; Qi, X.; Yang, R.; Prisacariu, V.; Wah, B.; Torr, P. Domain-invariant Stereo Matching Networks. In *European Conference on Computer Vision*; Springer: Cham, Switzerland, 2019.
45. Zhang, Y.; Chen, Y.; Bai, X.; Yu, S.; Yu, K.; Li, Z.; Yang, K. Adaptive Unimodal Cost Volume Filtering for Deep Stereo Matching. In Proceedings of the Thirty-Fourth AAAI Conference on Artificial Intelligence, New York, NY, USA, 7 February–12 February 2020.
46. Cheng, X.; Zhong, Y.; Harandi, M.; Dai, Y.; Chang, X.; Li, H.; Drummond, T.; Ge, Z. Hierarchical Neural Architecture Search for Deep Stereo Matching. In Proceedings of the Advances in Neural Information Processing Systems 34: Annual Conference on Neural Information Processing Systems 2020, NeurIPS 2020, Vancouver, BC, Canada, 6–12 December 2020.
47. Song, X.; Yang, G.; Zhu, X.; Zhou, H.; Wang, Z.; Shi, J. AdaStereo: A Simple and Efficient Approach for Adaptive Stereo Matching. In Proceedings of the IEEE/CVF Conference on Computer Vision and Pattern Recognition 2021, Nashville, TN, USA, 19–25 June 2021.
48. Chang, J.R.; Chang, P.C.; Chen, Y.S. Attention-Aware Feature Aggregation for Real-time Stereo Matching on Edge Devices. In Proceedings of the Asian Conference on Computer Vision (ACCV), Kyoto, Japan, 30 November–4 December 2020.
49. Gu, X.; Fan, Z.; Zhu, S.; Dai, Z.; Tan, F.; Tan, P. Cascade Cost Volume for High-Resolution Multi-View Stereo and Stereo Matching. In Proceedings of the 2020 IEEE/CVF Conference on Computer Vision and Pattern Recognition (CVPR), Seattle, WA, USA, 13–19 June 2020; pp. 2492–2501. [CrossRef]
50. Hu, Y.; Wang, W.; Yu, H.; Zhen, W.; Scherer, S. ORStereo: Occlusion-Aware Recurrent Stereo Matching for 4K-Resolution Images. In Proceedings of the 2021 IEEE/RSJ International Conference on Intelligent Robots and Systems (IROS), Prague, Czech Republic, 27 September–1 October 2021.
51. Hirschmüller, H. Stereo processing by semiglobal matching and mutual information. *IEEE Trans. Pattern Anal. Mach. Intell.* **2008**, *30*, 328–341. [CrossRef]

52. Ronneberger, O.; Fischer, P.; Brox, T. U-Net: Convolutional Networks for Biomedical Image Segmentation. In *Medical Image Computing and Computer-Assisted Intervention— MICCAI 2015*; Navab, N., Hornegger, J., Wells, W.M., Frangi, A.F., Eds.; Springer International Publishing: Cham, Switzerland, 2015; pp. 234–241.
53. Scharstein, D.; Szeliski, R. A taxonomy and evaluation of dense two-frame stereo correspondence algorithms. *Int. J. Comput. Vis.* **2002**, *47*, 7–42. [CrossRef]
54. Springenberg, J.; Dosovitskiy, A.; Brox, T.; Riedmiller, M. Striving for Simplicity: The All Convolutional Net. *arXiv* **2016**, arXiv:1606.04038.
55. Kingma, D.P.; Ba, J. Adam: A Method for Stochastic Optimization. *arXiv* **2014**, arXiv:1412.6980.
56. Zabih, R.; Woodfill, J. Non-parametric Local Transforms for Computing Visual Correspondence. In *Computer Vision—ECCV'94*; Eklundh, J.O., Ed.; Springer: Berlin/Heidelberg, Germany, 1994; pp. 151–158.
57. Haala, N. *The Landscape of Dense Image Matching Algorithms*; Wichmann/VDE: Belin/Offenbach, Germany, 2013.
58. Haala, N. Dense image matching final report. *Eurosdr Publ. Ser. Off. Publ.* **2014**, *64*, 115–145.
59. Denninger, M.; Sundermeyer, M.; Winkelbauer, D.; Zidan, Y.; Olefir, D.; Elbadrawy, M.; Lodhi, A.; Katam, H. BlenderProc. *arXiv* **2019**, arXiv:1911.01911.
60. Kurz, F.; Rosenbaum, D.; Meynberg, O.; Mattyus, G.; Reinartz, P. Performance of a Real-Time Sensor and Processing System on a Helicopter. In *ISPRS—International Archives of the Photogrammetry, Remote Sensing and Spatial Information Sciences*; ISPRS: Denver, CO, USA, 2014; Volume XL-1, pp. 189–193. [CrossRef]
61. Hu, F.; Gao, X.; Li, G.; Li, M. DEM Extraction from WorldView-3 Stereo-images and Accuracy Evaluation. In Proceedings of the International Archives of the Photogrammetry, Remote Sensing and Spatial Information Sciences, Prague, Czech Republic, 12–19 July 2016; Volume 41.
62. Bosch, M.; Foster, K.; Christie, G.A.; Wang, S.; Hager, G.D.; Brown, M.Z. Semantic Stereo for Incidental Satellite Images. In Proceedings of the 2019 IEEE Winter Conference on Applications of Computer Vision (WACV), Waikoloa Village, HI, USA, 7–11 January 2019; pp. 1524–1532.
63. Michel, J.; Sarrazin, E.; Youssefi, D.; Cournet, M.; Buffe, F.; Delvit, J.M.; Emilien, A.; Bosman, J.; Melet, O.; L'Helguen, C. A New Satellite Imagery Stereo Pipeline Designed for Scalability, Robustness and Performance. In *ISPRS Annals of the Photogrammetry, Remote Sensing and Spatial Information Sciences*; ISPRS: Nice, France, 2020; Volume V-2-2020, pp. 171–178. [CrossRef]
64. Aguilar, M.A.; Bianconi, F.; Aguilar, F.J.; Fernández, I. Object-based greenhouse classification from GeoEye-1 and WorldView-2 stereo imagery. *Remote Sens.* **2014**, *6*, 3554–3582. [CrossRef]

remote sensing

Article

City3D: Large-Scale Building Reconstruction from Airborne LiDAR Point Clouds

Jin Huang, Jantien Stoter, Ravi Peters and Liangliang Nan *

3D Geoinformation Research Group, Faculty of Architecture and the Built Environment, Delft University of Technology, 2628 BL Delft, The Netherlands; j.huang-1@tudelft.nl (J.H.); j.e.stoter@tudelft.nl (J.S.); r.y.peters@tudelft.nl (R.P.)
* Correspondence: liangliang.nan@tudelft.nl

Abstract: We present a fully automatic approach for reconstructing compact 3D building models from large-scale airborne point clouds. A major challenge of urban reconstruction from airborne LiDAR point clouds lies in that the vertical walls are typically missing. Based on the observation that urban buildings typically consist of planar roofs connected with vertical walls to the ground, we propose an approach to infer the vertical walls directly from the data. With the planar segments of both roofs and walls, we hypothesize the faces of the building surface, and the final model is obtained by using an extended hypothesis-and-selection-based polygonal surface reconstruction framework. Specifically, we introduce a new energy term to encourage roof preferences and two additional hard constraints into the optimization step to ensure correct topology and enhance detail recovery. Experiments on various large-scale airborne LiDAR point clouds have demonstrated that the method is superior to the state-of-the-art methods in terms of reconstruction accuracy and robustness. In addition, we have generated a new dataset with our method consisting of the point clouds and 3D models of 20k real-world buildings. We believe this dataset can stimulate research in urban reconstruction from airborne LiDAR point clouds and the use of 3D city models in urban applications.

Keywords: building reconstruction; LiDAR; point clouds; integer programming

Citation: Huang, J.; Stoter, J.; Peters, R.; Nan, L. City3D: Large-Scale Building Reconstruction from Airborne LiDAR Point Clouds. *Remote Sens.* **2022**, *14*, 2254. https://doi.org/10.3390/rs14092254

Academic Editors: Mohammad Awrangjeb, Jiaojiao Tian, Qin Yan, Beril Sirmacek and Nusret Demir

Received: 24 March 2022
Accepted: 2 May 2022
Published: 7 May 2022

1. Introduction

Digitizing urban scenes is an important research problem in computer vision, computer graphics, and photogrammetry communities. Three-dimensional models of urban buildings have become the infrastructure for a variety of real-world applications such as visualization [1], simulation [2–4], navigation [5], and entertainment [6]. These applications typically require high-accuracy and compact 3D building models of large-scale urban environments.

Existing urban building reconstruction methods strive to bring in a great level of detail and automate the process for large-scale urban environments. Interactive reconstruction techniques are successful in reconstructing accurate 3D building models with great detail [7,8], but they require either high-quality laser scans as input or considerable amounts of user interaction. These methods can thus hardly be applied to large-scale urban scenes. To facilitate practical applications that require large-scale 3D building models, researchers have attempted to address the reconstruction challenge using various data sources [9–16]. Existing methods based on aerial images [10,12,13] and dense triangle meshes [11] typically require good coverage of the buildings, which imposes challenges in data acquisition [17]. Approaches based on airborne LiDAR point clouds alleviate data acquisition issues. However, the accuracy and geometric details are usually compromised [9,14–16]. Following previous works using widely available airborne LiDAR point clouds, we strive to recover desired geometric details of real-world buildings while ensuring topological correctness, reconstruction accuracy, and good efficiency.

The challenges for large-scale urban reconstruction from airborne LiDAR point clouds include:

- Building instance segmentation. Urban scenes are populated with diverse objects, such as buildings, trees, city furniture, and dynamic objects (e.g., vehicles and pedestrians). The cluttered nature of urban scenes poses a severe challenge to the identification and separation of individual buildings from the massive point clouds. This has drawn considerable attention in recent years [18,19].
- Incomplete data. Some important structures (e.g., vertical walls) of buildings are typically not captured in airborne LiDAR point clouds due to the restricted positioning and moving trajectories of airborne scanners.
- Complex structures. Real-world buildings demonstrate complex structures with varying styles. However, limited cues about structure can be extracted from the sparse and noisy point clouds, which further introduces ambiguities in obtaining topologically correct surface models.

In this work, we address the above challenges with the following strategies. Firstly, we address the building instance segmentation challenge by separating individual buildings using increasingly-available vectorized building footprint data. Secondly, we exploit prior knowledge about the structures of buildings to infer their vertical planes. Based on the fact that vertical planes in airborne LiDAR point clouds are typically walls connecting the piecewise planar roofs to the ground, we propose an algorithm to infer the vertical planes from incomplete point clouds. Our method has the option to extrude outer walls directly from the given building footprint. Finally, we approach surface reconstruction by introducing the inferred vertical planes as constraints into an existing hypothesis-and-selection-based polygonal surface reconstruction framework [20], which favors good fitting to the input point cloud, encourages compactness, and enforces manifoldness of the final model (see Figure 1 for an example of the reconstruction results). The main contributions of this work include:

- A robust framework for fully automatic reconstruction of large-scale urban buildings from airborne LiDAR point clouds.
- An extension of an existing hypothesis-and-selection-based surface reconstruction method for buildings, which is achieved by introducing a new energy term to encourage roof preferences and two additional hard constraints to ensure correct topology and enhance detail recovery.
- A novel approach for inferring vertical planes of buildings from airborne LiDAR point clouds, for which we introduce an optimal-transport method to extract polylines from 2D bounding contours.
- A new dataset consisting of the point clouds and reconstructed surface models of 20 k real-world buildings.

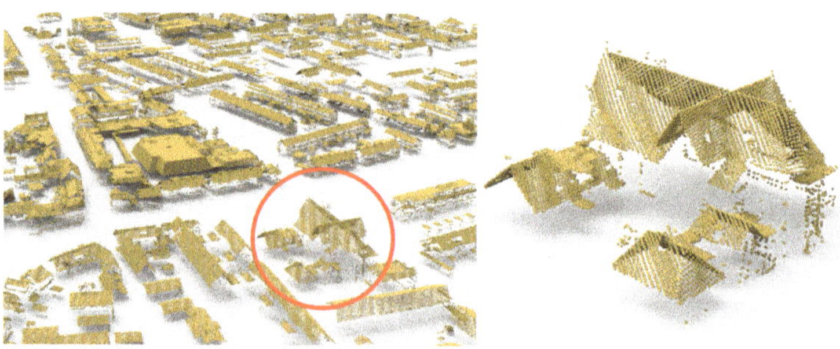

(a) Input airborne LiDAR point cloud.

Figure 1. *Cont.*

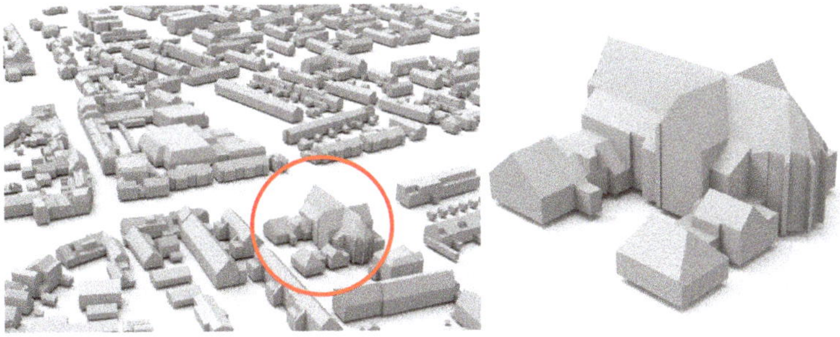

(b) Our reconstruction result

Figure 1. The automatic reconstruction result of all the buildings in a large scene from the AHN3 dataset [21].

2. Related Work

A large volume of methods for urban building reconstruction has been proposed. In this section, we mainly review the techniques relevant to the key components of our method. Since our method relies on footprint data for extracting building instances from the massive point clouds of large scenes, and it can also be used for footprint extraction, we also discuss related techniques in footprint extraction.

Roof primitive extraction. The commonly used method for extracting basic primitives (e.g., planes and cylinders) from point clouds is random sample consensus (RANSAC) [22] and its variants [23,24], which are robust against noise and outliers. Another group of widely used methods is based on region growing [25–27], which assumes roofs are piecewise planar and iteratively propagates planar regions by advancing the boundaries. The main difference between existing region growing methods lies in the generation of seed points and the criteria for region expansion. In this paper, we utilize an existing region growing method to extract roof primitives given its simplicity and robustness, which is detailed in Rabbani et al. [25].

Footprint extraction. Footprints are 2D outlines of buildings, capturing the geometry of outer walls projected onto the ground plane. Methods for footprint extraction commonly project the points to a 2D grid and analyze their distributions [28]. Chen et al. [27] detect rooftop boundaries and cluster them by taking into account topological consistency between the contours. To obtain simplified footprints, polyline simplification methods such as the Douglas-Peucker algorithm [29] are commonly used to reduce the complexity of the extracted contours [12,30,31]. To favor structural regularities, Zhou and Neumann [32] compute the principal directions of a building and regularize the roof boundary polylines along with these directions. Following these works, we infer the vertical planes of a building by detecting its contours from a heightmap generated from a 2D projection of the input points. The contour polylines are then regularized by orientation-based clustering followed by an adjustment step.

Building surface reconstruction. This type of methods aims at obtaining a simplified surface representation of buildings by exploiting geometric cues, e.g., planar primitives and their boundaries [15,32–36]. Zhou and Neumann [37] approached this by simplifying the 2.5D TIN (triangulated irregular network) of buildings, which may result in artifacts in building contours due to its limited capability in capturing complex topology. To address this issue, the authors proposed an extended 2.5D contouring method with improved topology control [38]. To cope with missing walls, Chauve et al. [39] also incorporated additional primitives inferred from the point clouds. Another group of building surface reconstruction methods involves predefined building parts, commonly known as model-driven approaches [40,41]. These methods rely on templates of known roof structures and deform-to-fit the templates to the input points. Therefore, the results are usually limited to the predefined shape templates, regardless of the diverse and complex nature of roof structures or high intraclass variations. Given the fact that buildings demonstrate mainly piecewise planar regions, methods have also been proposed to obtain an arrangement of extracted planar primitives to represent the building geometry [20,42–44]. These methods first detect a set of planar primitives from the input point clouds and then hypothesize a set of polyhedral cells or polygonal faces using the supporting planes of the extracted planar primitives. Finally, a compact polygonal mesh is extracted from the hypothesized cells or faces. These methods focus on the assembly of planar primitives, for which obtaining a complete set of planar primitives from airborne LiDAR point clouds is still a challenge.

In this work, we extend an existing hypothesis-and-selection-based general polygonal surface reconstruction method [20] to reconstruct buildings that consist of piecewise planar roofs connected to the ground by vertical walls. We approach this by introducing a novel energy term and a few hard constraints specially designed for buildings to ensure correct topology and decent details.

3. Methodology

3.1. Overview

The proposed approach takes as input a raw airborne LiDAR point cloud of a large urban scene and the corresponding building footprints, and it outputs 2-manifold and watertight 3D polygonal models of the buildings in the scene. Figure 2 shows the pipeline of the proposed method. It first extracts the point clouds of individual buildings by projecting all points onto the ground plane and collecting the points lying inside the footprint polygon of each building. Then, we reconstruct a compact polygonal model from the point cloud of each building.

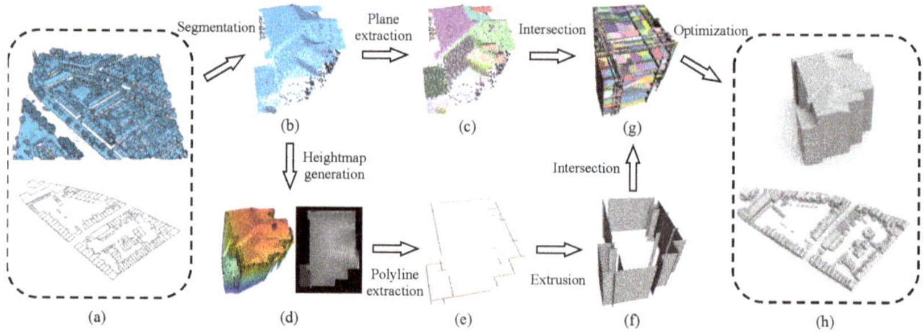

Figure 2. The pipeline of the proposed method (only one building is selected to illustrate the workflow). (**a**) Input point cloud and corresponding footprint data. (**b**) A building extracted from the input point cloud using its footprint polygon. (**c**) Planar segments extracted from the point cloud. (**d**) The heightmap (right) generated from the TIN (left, colored as a height field). (**e**) The polylines extracted from the heightmap. (**f**) The vertical planes obtained by extruding the inferred polylines. (**g**) The hypothesized building faces generated using both the extracted planes and inferred vertical planes. (**h**) The final model obtained through optimization.

Our reconstruction of a single building is based on the hypothesis-and-selection-based framework of PolyFit [20], which is for reconstructing general piecewise-planar objects from a set of planar segments extracted from the point cloud. Our method exploits not only the planar segments directly extracted from the point cloud but also the vertical planes inferred from the point cloud. From these two types of planar primitives, we hypothesize the faces of the building. The final model is then obtained by choosing the optimal subset of the faces through optimization.

The differences between our method and PolyFit are: (1) our method is dedicated to reconstructing urban buildings, and it makes use of vertical planes as hard constraints, for which we propose a novel algorithm for inferring the vertical planes of buildings that are commonly missing in airborne LiDAR point clouds. (2) We introduce a new *roof preference* energy term and two additional hard constraints into the optimization to ensure correct topology and enhance detail recovery. In the following sections, we detail the key steps of our method with an emphasis on the processes that differ from PolyFit [20].

3.2. Inferring Vertical Planes

With airborne LiDAR point clouds, important structures like vertical walls of a building are commonly missed due to the restricted positioning and moving trajectories of the scanner. In contrast, the roof surfaces are usually well captured. This inspired us to infer the missing walls from the available points containing the roof surfaces. We infer the vertical planes representing not only the outer walls but also the vertical walls within the footprint of a building. We achieve this by generating a 2D rasterized height map from its 3D points and looking for the contours that demonstrate considerable variations in the height values. To this end, an optimal-transport method is proposed to extract closed polylines from the contours. The polylines are then extruded to obtain the vertical walls. The process for inferring the vertical planes is outlined in Figure 2d–f.

Specifically, after obtaining the point cloud of a building, we project the points onto the ground plane, from which we create a height map. To cope with the non-uniform distribution of the points (e.g., some regions have holes while others may have repeating points), we construct a Triangulated Irregular Network (TIN) model using 2D Delaunay triangulation. The TIN model is a continuous surface and naturally completes the missing regions. Then, a height map is generated by rasterizing the TIN model with a specified resolution r. The issue of small holes in the height maps (due to uneven distribution of roof points) is further alleviated by image morphological operators while preserving the shape and size of the building [45]. After that, a set of contours are extracted from the height map using the Canny detector [46], which serves as the initial estimation of the vertical planes. We propose an optimal-transport method to extract polylines from the initial set of contours.

Optimal-transport method for polyline extraction. The initial set of contours are discrete pixels, denoted as S, from which we would like to extract simplified polylines that best describe the 2D geometry of S. Our optimal-transport method for extracting polylines from S works as follows. First, a 2D Delaunay triangulation T_0 is constructed from the discrete points in S. Then, the initial triangulation T_0 is simplified through iterative edge collapse and vertex removal operations. In each iteration, the most suitable vertex to be removed is determined in a way such that the following conditions are met:

- The maximum Hausdorff distance from the simplified mesh T_0 to S is less than a distance threshold ϵ_d.
- The increase of the total transport cost [47] between S and T_0 is kept at a minimum.

In each iteration, a vertex satisfying the above conditions is removed from T_0 by edge collapse, and the overall transportation cost is updated.

As the iterative simplification process continues, the overall transportation cost will increase. The edge collapse operation stops until no vertex can be further removed, or the overall transportation cost has increased above a user-specified tolerance ϵ_c. After that, we apply an edge filtering step [47] to eliminate small groups of undesirable edges caused by noise and outliers. Finally, the polylines are derived from the remaining vertices and edges of the simplified triangulation using the procedure described in [47]. Compared to [47], our method not only minimizes the total transport cost but also provides control over local geometry, ensuring that the distance between every vertex in the final polylines and the initial contours is smaller than the specified distance threshold ϵ_d.

Regularity enhancement. Due to noise and uneven point density in the point cloud, the polylines generated by the optimal-transport algorithm are unavoidably inaccurate and irregular (see Figure 3a), which often leads to artifacts in the final reconstruction. We alleviate these artifacts by enforcing structure regularities that commonly dominate urban buildings. We consider the structure regularities, namely parallelism, collinearity, and orthogonality, defined by [48]. Please note that since all the lines will be extruded vertically to obtain the vertical planes, the verticality regularity will inherently be satisfied. We propose a clustering-based method to identify the groups of line segments that potentially satisfy these regularities. Our method achieves structure regularization in two steps: clustering and adjustment.

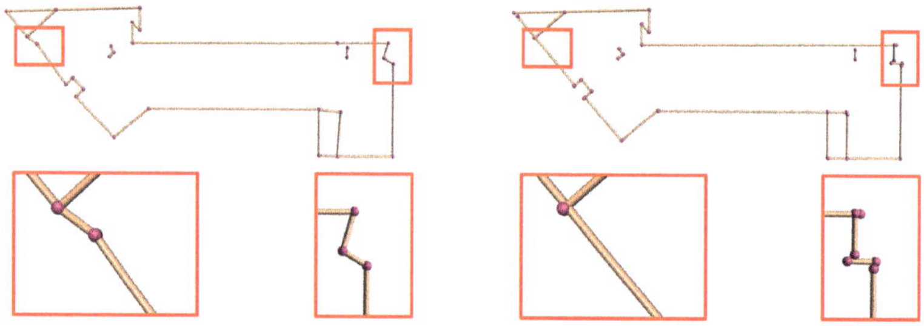

(**a**) Before (28 segments) (**b**) After (22 segments)

Figure 3. The effect of the clustering-based regularity enhancement on the polylines inferring the vertical walls. (**a**) Before regularity enhancement. (**b**) After regularity enhancement.

Clustering. In this work, we cluster the line segments of the polylines generated by the optimal-transport algorithm based on their orientation and pairwise Euclidean distance [49]. The pairwise Euclidean distance is measured by the minimum distance between a line segment and the supporting line of the other line segment.

Adjustment. For each cluster that contains multiple line segments, we compute its average direction. Then each line segment in the cluster is adjusted to align with the average direction. In case the building footprint is provided, the structure regularity can be further improved by aligning the segments with the edges in the footprint. After average adjustment, the near-collinear and near-orthogonal line segments are adjusted to be perfectly collinear and orthogonal, respectively (we use an angle threshold of 20°).

After regularity enhancement, the vertical planes of the building can be obtained by vertical extrusion of the regularized polylines. The effect of the regularity enhancement is demonstrated in Figure 3, from which we can see that it significantly improves structure regularity and reduces the complexity of the building outlines.

3.3. Reconstruction

Our surface reconstruction involves two types of planar primitives, i.e., vertical planes inferred in the previous step (see Section 3.2) and roof planes directly extracted from the point cloud. Unlike PolyFit [20] that hypothesizes faces by computing pairwise intersections using all planar primitives, we compute pairwise intersections using only the roof planes, and then the resulted faces are cropped with the outer vertical planes (see Figure 2g). This process ensures that the roof boundaries of the reconstructed building can be precisely connected with the inferred vertical walls. Additionally, since the object to be reconstructed is a real-world building, we introduce a *roof preference* energy term and a set of new hard constraints specially designed for buildings into the original formulation. Specifically, our objective for obtaining the model faces F^* can be written as

$$F^* = \arg\min_{X} \lambda_d E_d + \lambda_c E_c + \lambda_r E_r, \tag{1}$$

where $X = \{x_i | x_i \in \{0,1\}\}$ denotes the binary variables for the faces (1 for *selected* and 0 otherwise). E_d is the data fitting term that encourages selecting faces supported by more points, and E_c is the model complexity term that favors simple planar structures. For more details about the data fitting term and the model complexity term, please refer to the original paper of PolyFit [20]. In the following part, we elaborate on the new energy term and hard constraints.

New energy term: *roof preference*. We have observed in rare cases that a building in aerial point clouds may demonstrate more than one layer of roofs, e.g., semi-transparent or overhung roofs. In such a case, we assume a higher roof face is always preferable to the ones underneath. We formulate this preference as an additional energy term called *roof preference*, which is defined as

$$E_r = \frac{1}{|F|} \sum_{i=1}^{|F|} x_i \cdot \frac{z_{max} - z_i}{z_{max} - z_{min}} \tag{2}$$

where z_i denotes the Z coordinate of the centroid of a hypothsized face f_i. z_{max} and z_{min} are, respectively, the highest and lowest Z coordinates of the building points. $|F|$ denotes the total number of hypothesized faces.

New hard constraints. We impose two hard constraints to enhance the topological correctness of the final reconstruction.

- *Single-layer roof*. This constraint ensures that the reconstructed 3D model of a real-world building has a single layer of roofs, which can be written as,

$$\sum_{k \in V(f_i)} x_k = 1, (1 \leq i \leq |F|)$$

where $V(f_i)$ denotes the set of hypothesized faces that have overlap with face $f_i \in F$ in the vertical direction.

- *Face prior*. This constraint enforces that for all the derived faces from the same planar segment, the one with the highest confidence value is always selected as a prior. Here, the confidence of a face is measured by the number of its supporting points. This constraint can be simply written as

$$x_l = 1,$$

where x_l is the variable whose value denotes the status of the most confident face f_l of a planar segment. This constraint resolves ambiguities if two hypothesized faces are near coplanar and close to each other, which preserves finer geometric details. The effect of this constraint is demonstrated in Figure 4.

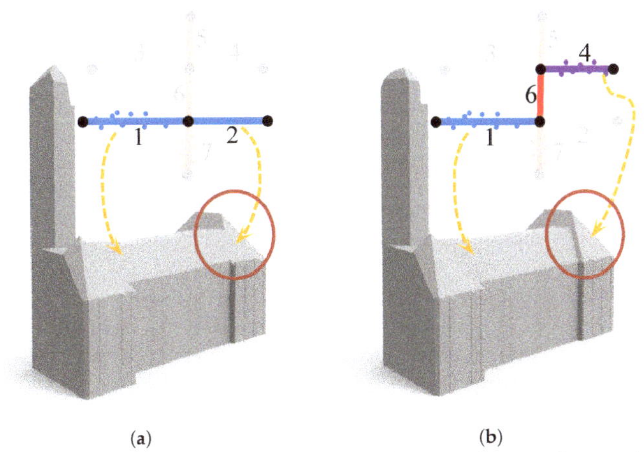

Figure 4. The effect of the *face prior* constraint. The insets illustrate the assembly of the hypothesized faces in the corresponding marked regions (each line segment denotes a hypothesized face, and line segments of the same color represent faces derived from the same planar primitive). (**a**) Reconstruction without the *face prior* constraint. (**b**) Reconstruction with the *face prior* constraint, for which faces 1 and 4 both satisfy the *face prior* constraint. The numbers 1–7 denote the 7 candidate faces.

The final surface model of the building can be obtained by solving the optimization problem given in Equation (A4), subject to the *single-layer roof* and *face prior* hard constraints.

4. Results and Evaluation

Our method is implemented in C++ using CGAL [50]. All experiments were conducted on a desktop PC with a 3.5 GHz AMD Ryzen Threadripper 1920X and 64 GB RAM.

4.1. Test Datasets

We have tested our method on three datasets of large-scale urban point clouds including more than 20 k buildings.

- AHN3 [21]. An openly available country-wide airborne LiDAR point cloud dataset covering the entire Netherlands, with an average point density of 8 points/m^2. The corresponding footprints of the buildings are obtained from the Register of Buildings and Addresses (BAG) [51]. The geometry of footprint is acquired from aerial photos and terrestrial measurements with an accuracy of 0.3 m. The polygons in the BAG represent the outlines of buildings as their outer walls seen from above, which are slightly different from footprints. We still use 'footprint' in this paper.
- DALES [52]. A large-scale aerial point cloud dataset consisting of forty scenes spanning an area of 10 km^2, with instance labels of 6 k buildings. The data was collected using a Riegl Q1560 dual-channel system with a flight altitude of 1300 m above ground and a speed of 72 m/s. Each area was collected by a minimum of 5 laser pulses per meter in four directions. The LiDAR swaths were calibrated using the BayesStripAlign 2.0 software and registered, taking both relative and absolute errors into account and correcting for altitude and positional errors. The average point density is 50 points/m^2. No footprint data is available in this dataset.
- Vaihingen [53]. An airborne LiDAR point cloud dataset published by ISPRS, which has been widely used in semantic segmentation and reconstruction of urban scenes. The data were obtained using a Leica ALS50 system with 45° field of view and a mean flying height above ground of 500 m. The average strip overlap is 30% and multiple pulses were recorded. The point cloud was pre-processed to compensate for systematic offsets between the strips. We use in our experiments a training set that contains footprint information and covers an area of 399 m $\times$ 421 m with 753 k points. The average point density is 4 points/m^2.

4.2. Reconstruction Results

Visual results. We have used our method to reconstruct more than 20 k buildings from the aforementioned three datasets. For the AHN3 [21] and Vaihingen [53] datasets, the provided footprints were used for both building instance segmentation and extrusion of the outer walls. Our inferred vertical planes were used to complete the missed inner walls. For the DALES [52] dataset, we used the provided instance labels to extract building instances, and we used our inferred vertical walls for the reconstruction.

Figures 1 and 5 show the 3D reconstruction of all buildings in two large scenes from the AHN3 dataset [21]. For the buildings reconstructed in Figure 1, their models are simplified polygonal meshes with an average face count of 34. To better reveal the quality of our reconstructed building models, we demonstrate in Figure 6 a set of individual buildings reconstructed from the three test datasets. From these visual results, we can see that although the buildings have diverse structures of different styles, and the input point clouds have varying densities and different levels of noise, outliers, and missing data, our method succeeded in obtaining visually plausible reconstruction results. These experiments also indicate that our approach is successful in inferring the vertical planes of buildings from airborne LiDAR point clouds and it is effective to include these planes in the 3D reconstruction of urban buildings.

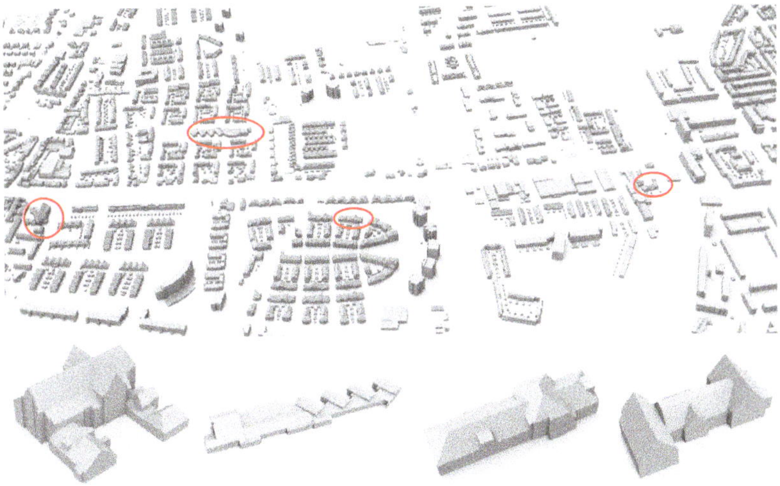

Figure 5. Reconstruction of a large scene from the AHN3 dataset [21].

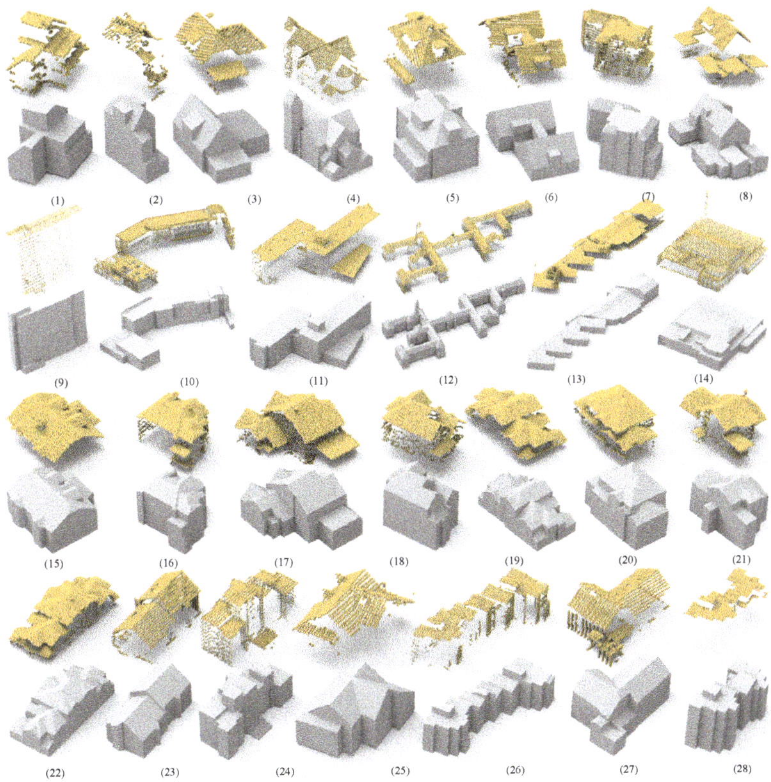

Figure 6. The reconstruction results of a set of buildings from various dataset. (**1–14**) are from the AHN3 dataset [21], (**15–22**) are from the DALES dataset [52], (**23–28**) are from the Vaihingen dataset [53].

Quantitative results. We have also evaluated the reconstruction results quantitatively. Since ground-truth reconstruction is not available for all buildings in the three datasets, we chose to use the commonly used accuracy measure, Root Mean Square Error (RMSE), to quantify the quality of each reconstructed model. In the context of surface reconstruction, RMSE is defined as the square root of the average of squared Euclidean distances from the points to the reconstructed model. In Table 1, we report the statistics of our quantitative results on the buildings shown in Figure 6. We can see that our method has obtained good reconstruction accuracy, i.e., the RMSE for all buildings is between 0.04 m to 0.26 m, which is quite promising for 3D reconstruction of real-world buildings from noisy and sparse airborne LiDAR point clouds. As observed from the number of faces column of Table 1, our results are simplified polygonal models and are more compact than those obtained from commonly used approaches such as the Poisson surface reconstruction method [54] (that produces dense triangles). Table 1 also shows that the running times for most buildings are less than 30 s. The reconstruction of the large complex building shown in Figure 6 (12) took 42 min. This long reconstruction time is due to that our method computes the pairwise intersection of the detected planar primitives and inferred vertical planes, and it generates a large number of candidate faces and results in a large optimization problem [20] (see also Section 4.7). The running time with respect to the number of detected planar segments for the reconstruction of more buildings is reported in Figure 7.

Table 1. Statistics on the reconstructed buildings shown in Figure 6. For each building, the number of points in the input, number of faces in the reconstructed model, fitting error (i.e., RMSE in meters), and running time (in seconds) are reported.

Dataset	Model	#Points	#Faces	RMSE (m)	Time (s)
AHN3	(1)	732	23	0.07	3
	(2)	532	42	0.12	4
	(3)	1165	31	0.04	3
	(4)	20,365	127	0.15	62
	(5)	1371	48	0.04	5
	(6)	1611	45	0.06	4
	(7)	3636	68	0.21	18
	(8)	2545	52	0.04	8
	(9)	15,022	63	0.11	28
	(10)	23,654	262	0.26	115
	(11)	13,269	102	0.11	34
	(12)	155,360	1520	0.09	2520
	(13)	24,027	176	0.24	141
	(14)	28,522	227	0.15	78
DALES	(15)	8662	39	0.04	11
	(16)	11,830	73	0.1	8
	(17)	10,673	47	0.07	7
	(18)	7594	33	0.07	14
	(19)	13,060	278	0.05	145
	(20)	11,114	55	0.06	24
	(21)	8589	51	0.06	15
	(22)	18,909	282	0.08	86

Table 1. *Cont.*

Dataset	Model	#Points	#Faces	RMSE (m)	Time (s)
Vaihingen	(23)	7701	51	0.24	25
	(24)	6845	99	0.12	8
	(25)	1007	24	0.11	2
	(26)	11,591	206	0.17	10
	(27)	4026	42	0.26	6
	(28)	5059	61	0.22	9

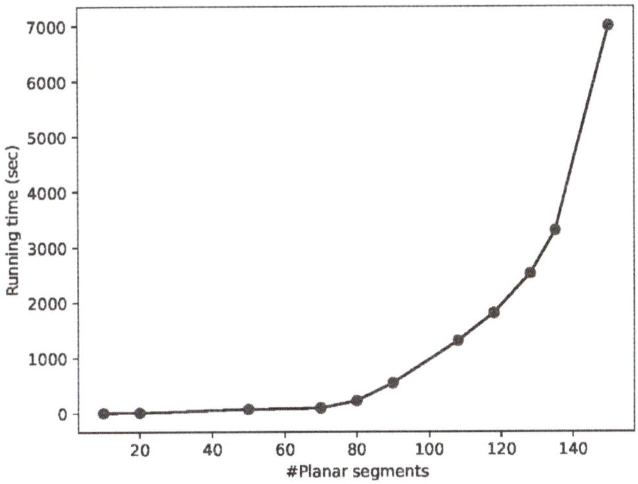

Figure 7. The running time of our method with respect to the number of the detected planar segments. These statistics are obtained by testing on the AHN3 dataset.

New dataset. Our method has been applied to city-scale building reconstruction. The results are released as a new dataset consisting of 20 k buildings (including the reconstructed 3D models and the corresponding airborne LiDAR point clouds). We believe this dataset can stimulate research in urban reconstruction from airborne LiDAR point clouds and the use of 3D city models in urban applications.

4.3. Parameters

Our method involves a few parameters that are empirically set to fixed values for all experiments, i.e., the distance threshold $\epsilon_d = 0.25$ and the tolerance for overall transportation cost $\epsilon_c = 2.0$. The resolution r for the rasterization of the TIN model to generate heightmaps is dataset dependent due to the difference in point density. It is set to 0.20 m from AHN3, 0.15 m for DALES, and 0.25 m for Vaihingen. The weight of the *roof preference* energy term $\lambda_r = 0.04$ (while the weights for the data fitting and model complexity terms are set to $\lambda_d = 0.34$ and $\lambda_c = 0.62$, respectively).

4.4. Comparisons

We have compared our method with two successful open-source methods, i.e., 2.5D Dual Contouring (dedicated for urban buildings) [37] and PolyFit (for general piecewise-planar objects) [20], on the AHN3 [21], DALES [52], and Vaihingen [53] datasets. The city block from the AHN3 dataset [21] is sparse and contains only 80,447 points for 160 buildings (i.e., on average 503 points per building). The city region from DALES is denser and contains

214,601 points for 41 buildings (i.e., on average 5234 points per building). The city area from the Vaihingen dataset contains 69,254 points for 57 buildings (i.e., on average 1215 points per building). The walls of all the point clouds are severely occluded. Figure 8 shows the visual comparison of one of the buildings. PolyFit assumes a complete set of input planar primitives, which is not the case for airborne LiDAR point clouds because the vertical walls are often missing. For PolyFit to be effective, we added our inferred vertical planes to its initial set of planar primitives. From the result, we can observe that both PolyFit and our method can generate compact building models, and the number of faces in the result is an order of magnitude less than that of the 2.5D Dual Contouring method. It is worth noting that even with the additional planes, PolyFit still failed to reconstruct some walls and performed poorly in recovering geometric details. In contrast, our method produces the most plausible 3D models. By inferring missing vertical planes, our method can recover inner walls, which further split the roof planes and bring in more geometric details into the final reconstruction. Table 2 reports the statistics of the comparison, from which we can see that the reconstructed building models from our method have the highest accuracy. In terms of running time, our method is slower than the other two, but it is still acceptable in practical applications (on average 4.9 s per building).

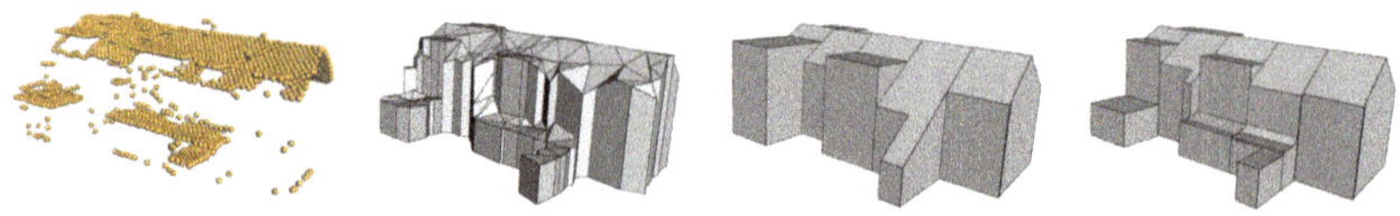

(**a**) Input point cloud (**b**) 2.5DC (296 faces) (**c**) PolyFit (58 faces) (**d**) Ours (86 faces)

Figure 8. Comparison with 2.5D Dual Contouring (2.5DC) [37] and PolyFit [20] on a single building from the AHN3 dataset [21].

Table 2. Statistics on the comparison of 2.5D Dual Contouring [37], PolyFit [20], and our method on the reconstruction from the AHN3 [21], DALES [52], and Vaihingen [53] datasets. Total face numbers, running times, and average errors are reported.

Dataset	Method	#Faces	RMSE (m)	Time (s)
AHN3	2.5D DC [37]	12,781	0.213	13
	PolyFit [20]	1848	0.242	160
	Ours	2453	0.128	380
DALES	2.5D DC [37]	2297	0.204	10
	PolyFit [20]	444	0.287	230
	Ours	583	0.184	670
Vaihingen	2.5D DC [37]	2695	0.168	6
	PolyFit [20]	647	0.275	102
	Ours	798	0.157	212

We also performed an extensive quantitative comparison with the 3D building models from the BAG3D [55], which is a public 3D city platform that provides 3D models of urban buildings at the LoD2 level. For this comparison, we picked four different regions consisting of 1113 buildings in total from the BAG3D. In Figure 9, we demonstrate a visual comparison, from which we can see that our models demonstrate more regularity. The quantitive result is reported in Table 3, from which we can see that our results have higher accuracy.

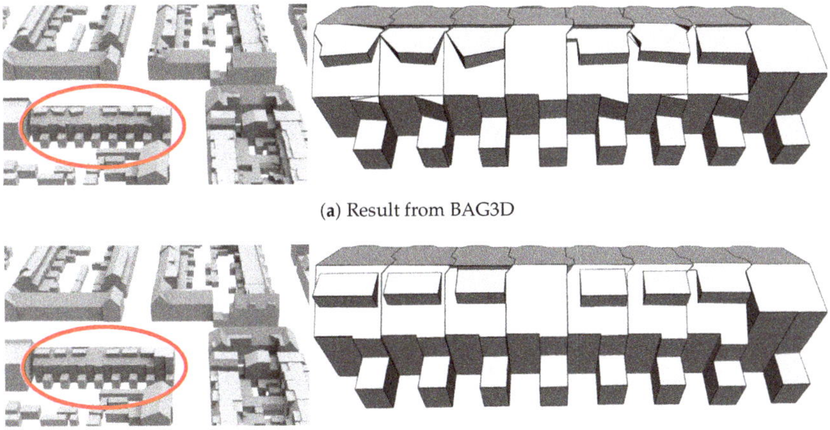

(a) Result from BAG3D

(b) Ours result

Figure 9. A visual comparison with BAG3D [55]. A building from Table 3 (b) is shown.

Table 3. Quantitative comparison with the BAG3D [55] on four urban scenes (a)–(d). Both BAG3D and our method used the point clouds from the AHN3 dataset [21] as input. The bold font indicates smaller RMSE values.

Region	#Points	#Building	RMSE (m) BAG3D	RMSE (m) Ours
(a)	1,694,247	198	0.088	**0.079**
(b)	329,593	387	0.139	**0.138**
(c)	224,970	368	0.140	**0.132**
(d)	80,447	160	0.146	**0.128**

4.5. With vs. Without Footprint

Our method can infer the vertical planes of a building from its roof points, and then the outer walls are completed using the vertical planes. It also has the option to directly use given footprint data for reconstruction. With a given footprint, vertically planes are firstly obtained by extruding the footprint polygons. Then these planes and those extracted from the point clouds are intersected to hypothesize the model faces, followed by the optimization step to obtain the final reconstruction. Figure 10 shows such a comparison on two buildings.

4.6. Reconstruction Using Point Clouds with Vertical Planes

The methodology presented in our paper only focuses on airborne LiDAR point clouds, in which vertical walls of buildings are typically missing. In practice, our method can be easily adapted to work with other types of point clouds that contain points of vertical walls, e.g., point clouds reconstructed from drone images. For such point clouds, our method can still be effective by replacing the inferred vertical planes with those directly detected from the point clouds. Figure 11 shows two such examples.

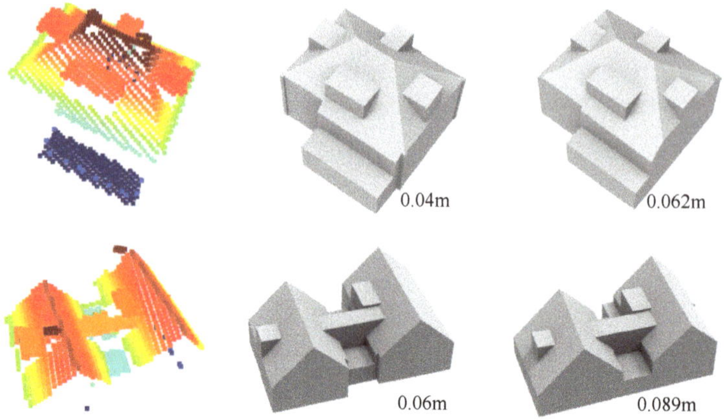

(a) Input (b) With footprint (c) Without footprint

Figure 10. Comparison between the reconstruction *with* (**b**) and *without* (**c**) footprint data on two buildings (**a**) from the AHN3 dataset [21]. The number below each model denotes the root mean square error (RMSE). Using the inferred vertical planes slightly increases reconstruction errors.

Figure 11. Reconstruction from aerial point clouds. In these point clouds, the vertical walls can be extracted from the point clouds and directly used in reconstruction, and thus the vertical plane inference step was skipped. The dataset is obtained from Can et al. [56].

4.7. Limitations

Our method can infer the missing vertical planes of buildings, from which the outer vertical planes serve as outer walls in the reconstruction. Since the vertical planes are inferred from the 3D points of rooftops, the walls in the final models may not perfectly align with the ground-truth footprints (see the figure below). Thus, we recommend the use of high-quality footprint data whenever it is available. Besides, our method extends the hypothesis-and-selection-based surface reconstruction framework of PolyFit [20] by introducing new energy terms and hard constraints. It naturally inherits the limitation of PolyFit, i.e., it may encounter computation bottlenecks for buildings with complex structures (e.g., buildings with more than 100 planar regions). An example has already been shown in Figure 6 (12).

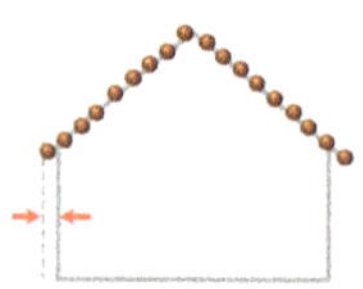

5. Conclusions and Future Work

We have presented a fully automatic approach for large-scale 3D reconstruction of urban buildings from airborne LiDAR point clouds. We propose to infer the vertical planes of buildings that are commonly missing from airborne LiDAR point clouds. The inferred vertical planes play two different roles during the reconstruction. The outer vertical planes directly become part of the exterior walls of the building, and the inner vertical planes enrich building details by splitting the roof planes at proper locations and forming the necessary inner walls in final models. Our method can also incorporate given building footprints for reconstruction. In case footprints are used, they are extruded to serve the exterior walls of the models, and the inferred inner planes enrich building details. Extensive experiments on different datasets have demonstrated that inferring vertical planes is an effective strategy for building reconstruction from airborne LiDAR point clouds, and the proposed *roof preference* energy term and the novel hard constraints ensure topologically correct and accurate reconstruction.

Our current framework uses only planar primitives and it is sufficient for reconstructing most urban buildings. In the real world, there still exist buildings with curved surfaces, which our current implementation could not handle. However, our hypothesize-and-selection strategy is general and can be extended to process different types of primitives. As a future work direction, our method can be extended to incorporate other geometric primitives, such as spheres, cylinders, or even parametric surfaces. With such an extension, buildings with curved surfaces can also be reconstructed.

Author Contributions: J.H. performed the study and implemented the algorithms. R.P. and J.S. provided constructive comments and suggestions. L.N. proposed this topic, provided daily supervision, and wrote the paper together with J.H. All authors have read and agreed to the published version of the manuscript.

Funding: Jin Huang is financially supported by the China Scholarship Council.

Data Availability Statement: Our code and data are available at https://github.com/yidahuang/City3D, accessed on 23 March 2022.

Acknowledgments: We thank Zexin Yang, Zhaiyu Chen, and Noortje van der Horst for proofreading the paper.

Conflicts of Interest: The authors declare no conflict of interest.

Abbreviations

The following abbreviations are used in this manuscript:

LiDAR	Light Detection and Ranging
TIN	Triangular Irregular Network
RMSE	Root Mean Square Error

Appendix A. The Complete Formulation

Our reconstruction is obtained by finding the optimal subset of the hypothesized faces. We formulate this as an optimization problem, with an objective function consisting of three energy terms: *data fitting*, *model complexity*, and *roof preference*. The first two terms are the same as in [20]. In the following, we briefly introduce all these terms and provide the final complete formulation.

- **Data fitting.** It is defined to measure how well the final model (i.e., the assembly of the chosen faces) fits to the input point cloud,

$$E_d = 1 - \frac{1}{|P|} \sum_{i=1}^{|F|} x_i \cdot support(f_i), \tag{A1}$$

where $|P|$ is the number of points in the point cloud. $support(f_i)$ measures the number of points that are ϵ-close to a face $f_i \in F$, and $x_i \in \{0,1\}$ denotes the binary status of the face f_i (1 for *selected* and 0 otherwise). $|F|$ denotes the total number of hypothesized faces.

- **Model complexity**. To avoid defects introduced by noise and outliers, this term is introduced to encourage large planar structures,

$$E_c = \frac{1}{|E|} \sum_{i=1}^{|E|} corner(e_i), \tag{A2}$$

where $|E|$ denotes the total number of pairwise intersections in the hypothesized face set. $corner(e_i)$ is an indicator function denoting if choosing two faces connected by an edge e_i results in a sharp edge in the final model (1 for *sharp* and 0 otherwise).

- **Roof preference**. We have observed in rare cases that a building in aerial point clouds may demonstrate more than one layer of roofs, e.g., semi-transparent or overhung roofs. In such a case, we assume a higher roof face is preferable to the ones underneath. We formulate this preference as an additional *roof preference* energy term,

$$E_r = \frac{1}{|F|} \sum_{i=1}^{|F|} x_i \cdot \frac{z_{max} - z_i}{z_{max} - z_{min}} \tag{A3}$$

where z_i denotes the Z coordinate of the centroid of a face f_i. z_{max} and z_{min} are, respectively, the highest and lowest Z coordinates of the building points.

With all the constraints, the complete optimization problem is written as

$$\begin{aligned} &\min_X \lambda_d E_d + \lambda_c E_c + \lambda_r E_r \\ &\text{s.t.} \begin{cases} \sum\limits_{k \in V(f_i)} x_k = 1, (1 \le i \le |F|) \\ \sum\limits_{j \in N(e_i)} x_j = 0 \quad or \quad 2, \quad (1 \le j \le |E|) \\ x_l = 1, \\ x_i \in \{0,1\}, \quad \forall i \in N \end{cases} \end{aligned} \tag{A4}$$

where the first constraint is call *single roof*, which ensures that the reconstructed building model has a single layer of roofs. The second constraint enforces that in the final model an edge is associated with two adjacent faces, ensuring the final model to be watertight and manifold. The third constraint is call *face prior*, which ensures that, for the faces derived from the same planar segment, the one with the highest confidence value is selected as a prior.

By solving the above optimization problem, the set of selected faces $\{f_i | x_i = 1\}$ forms the final surface model of a building.

References

1. Yao, Z.; Nagel, C.; Kunde, F.; Hudra, G.; Willkomm, P.; Donaubauer, A.; Adolphi, T.; Kolbe, T.H. 3DCityDB—A 3D geodatabase solution for the management, analysis, and visualization of semantic 3D city models based on CityGML. *Open Geospat. Data Softw. Stand.* **2018**, *3*, 1–26. [CrossRef]
2. Zhivov, A.M.; Case, M.P.; Jank, R.; Eicker, U.; Booth, S. Planning tools to simulate and optimize neighborhood energy systems. In *Green Defense Technology*; Springer: Dordrecht, The Netherlands, 2017; pp. 137–163.
3. Stoter, J.; Peters, R.; Commandeur, T.; Dukai, B.; Kumar, K.; Ledoux, H. Automated reconstruction of 3D input data for noise simulation. *Comput. Environ. Urban Syst.* **2020**, *80*, 101424. [CrossRef]
4. Widl, E.; Agugiaro, G.; Peters-Anders, J. Linking Semantic 3D City Models with Domain-Specific Simulation Tools for the Planning and Validation of Energy Applications at District Level. *Sustainability* **2021**, *13*, 8782. [CrossRef]
5. Cappelle, C.; El Najjar, M.E.; Charpillet, F.; Pomorski, D. Virtual 3D city model for navigation in urban areas. *J. Intell. Robot. Syst.* **2012**, *66*, 377–399. [CrossRef]

6. Kargas, A.; Loumos, G.; Varoutas, D. Using different ways of 3D reconstruction of historical cities for gaming purposes: The case study of Nafplio. *Heritage* **2019**, *2*, 1799–1811. [CrossRef]
7. Nan, L.; Sharf, A.; Zhang, H.; Cohen-Or, D.; Chen, B. Smartboxes for interactive urban reconstruction. In *ACM Siggraph 2010 Papers*; ACM: New Yrok, NY, USA, 2010; pp. 1–10.
8. Nan, L.; Jiang, C.; Ghanem, B.; Wonka, P. Template assembly for detailed urban reconstruction. In *Computer Graphics Forum*; Wiley Online Library: Zurich, Switzerland, 2015; Volume 34, pp. 217–228.
9. Zhou, Q.Y. *3D Urban Modeling from City-Scale Aerial LiDAR Data*; University of Southern California: Los Angeles, CA, USA, 2012.
10. Haala, N.; Rothermel, M.; Cavegn, S. Extracting 3D urban models from oblique aerial images. In Proceedings of the 2015 Joint Urban Remote Sensing Event (JURSE), Lausanne, Switzerland, 30 March–1 April 2015; pp. 1–4.
11. Verdie, Y.; Lafarge, F.; Alliez, P. LOD generation for urban scenes. *ACM Trans. Graph.* **2015**, *34*, 30. [CrossRef]
12. Li, M.; Nan, L.; Smith, N.; Wonka, P. Reconstructing building mass models from UAV images. *Comput. Graph.* **2016**, *54*, 84–93. [CrossRef]
13. Buyukdemircioglu, M.; Kocaman, S.; Isikdag, U. Semi-automatic 3D city model generation from large-format aerial images. *ISPRS Int. J.-Geo-Inf.* **2018**, *7*, 339. [CrossRef]
14. Bauchet, J.P.; Lafarge, F. City Reconstruction from Airborne Lidar: A Computational Geometry Approach. In Proceedings of the 3D GeoInfo 2019—14thConference 3D GeoInfo, Singapore, 26–27 September 2019.
15. Li, M.; Rottensteiner, F.; Heipke, C. Modelling of buildings from aerial LiDAR point clouds using TINs and label maps. *ISPRS J. Photogramm. Remote Sens.* **2019**, *154*, 127–138. [CrossRef]
16. Ledoux, H.; Biljecki, F.; Dukai, B.; Kumar, K.; Peters, R.; Stoter, J.; Commandeur, T. 3dfier: Automatic reconstruction of 3D city models. *J. Open Source Softw.* **2021**, *6*, 2866. [CrossRef]
17. Zhou, X.; Yi, Z.; Liu, Y.; Huang, K.; Huang, H. Survey on path and view planning for UAVs. *Virtual Real. Intell. Hardw.* **2020**, *2*, 56–69. [CrossRef]
18. Qi, C.R.; Su, H.; Mo, K.; Guibas, L.J. Pointnet: Deep learning on point sets for 3d classification and segmentation. In Proceedings of the IEEE Conference on Computer Vision and Pattern Recognition, Honolulu, HI, USA, 21–26 July 2017; pp. 652–660.
19. Thomas, H.; Qi, C.R.; Deschaud, J.E.; Marcotegui, B.; Goulette, F.; Guibas, L.J. Kpconv: Flexible and deformable convolution for point clouds. In Proceedings of the IEEE/CVF International Conference on Computer Vision, Seoul, Korea, 27 October–2 November 2019; pp. 6411–6420.
20. Nan, L.; Wonka, P. PolyFit: Polygonal Surface Reconstruction from Point Clouds. In Proceedings of the IEEE International Conference on Computer Vision, Venice, Italy, 22–29 October, 2017.
21. AHN3. Actueel Hoogtebestand Nederland (AHN). 2018. Available online: https://www.pdok.nl/nl/ahn3-downloads (accessed on 13 November 2021).
22. Fischler, M.A.; Bolles, R.C. Random sample consensus: A paradigm for model fitting with applications to image analysis and automated cartography. *Commun. ACM* **1981**, *24*, 381–395. [CrossRef]
23. Schnabel, R.; Wahl, R.; Klein, R. Efficient RANSAC for point-cloud shape detection. In *Computer Graphics Forum*; Wiley Online Library: Oxford, UK, 2007; Volume 26, pp. 214–226.
24. Zuliani, M.; Kenney, C.S.; Manjunath, B. The multiransac algorithm and its application to detect planar homographies. In Proceedings of the IEEE International Conference on Image Processing 2005, Genova, Italy, 14 September 2005; Volume 3, p. III-153.
25. Rabbani, T.; Van Den Heuvel, F.; Vosselmann, G. Segmentation of point clouds using smoothness constraint. *Int. Arch. Photogramm. Remote Sens. Spat. Inf. Sci.* **2006**, *36*, 248–253.
26. Sun, S.; Salvaggio, C. Aerial 3D building detection and modeling from airborne LiDAR point clouds. *IEEE J. Sel. Top. Appl. Earth Obs. Remote Sens.* **2013**, *6*, 1440–1449. [CrossRef]
27. Chen, D.; Wang, R.; Peethambaran, J. Topologically aware building rooftop reconstruction from airborne laser scanning point clouds. *IEEE Trans. Geosci. Remote Sens.* **2017**, *55*, 7032–7052. [CrossRef]
28. Meng, X.; Wang, L.; Currit, N. Morphology-based building detection from airborne LIDAR data. *Photogramm. Eng. Remote Sens.* **2009**, *75*, 437–442. [CrossRef]
29. Douglas, D.H.; Peucker, T.K. Algorithms for the reduction of the number of points required to represent a digitized line or its caricature. *Cartogr. Int. J. Geogr. Inf. Geovis.* **1973**, *10*, 112–122. [CrossRef]
30. Zhang, K.; Yan, J.; Chen, S.C. Automatic construction of building footprints from airborne LIDAR data. *IEEE Trans. Geosci. Remote Sens.* **2006**, *44*, 2523–2533. [CrossRef]
31. Xiong, B.; Elberink, S.O.; Vosselman, G. Footprint map partitioning using airborne laser scanning data. *ISPRS Ann. Photogramm. Remote Sens. Spat. Inf. Sci.* **2016**, *3*, 241–247. [CrossRef]
32. Zhou, Q.Y.; Neumann, U. Fast and extensible building modeling from airborne LiDAR data. In Proceedings of the 16th ACM SIGSPATIAL International Conference on Advances in Geographic Information Systems, Irvine, CA, USA, 5–7 November 2008; pp. 1–8.
33. Dorninger, P.; Pfeifer, N. A comprehensive automated 3D approach for building extraction, reconstruction, and regularization from airborne laser scanning point clouds. *Sensors* **2008**, *8*, 7323–7343. [CrossRef]
34. Lafarge, F.; Mallet, C. Creating large-scale city models from 3D-point clouds: A robust approach with hybrid representation. *Int. J. Comput. Vis.* **2012**, *99*, 69–85. [CrossRef]

35. Xiao, Y.; Wang, C.; Li, J.; Zhang, W.; Xi, X.; Wang, C.; Dong, P. Building segmentation and modeling from airborne LiDAR data. *Int. J. Digit. Earth* **2015**, *8*, 694–709. [CrossRef]
36. Yi, C.; Zhang, Y.; Wu, Q.; Xu, Y.; Remil, O.; Wei, M.; Wang, J. Urban building reconstruction from raw LiDAR point data. *Comput.-Aided Des.* **2017**, *93*, 1–14. [CrossRef]
37. Zhou, Q.Y.; Neumann, U. 2.5 d dual contouring: A robust approach to creating building models from aerial lidar point clouds. In *European Conference on Computer Vision*; Springer: Cham, Switzerland, 2010; pp. 115–128.
38. Zhou, Q.Y.; Neumann, U. 2.5 D building modeling with topology control. In Proceedings of the CVPR 2011, Colorado Springs, CO, USA, 20–25 June 2011; pp. 2489–2496.
39. Chauve, A.L.; Labatut, P.; Pons, J.P. Robust piecewise-planar 3D reconstruction and completion from large-scale unstructured point data. In Proceedings of the 2010 IEEE Computer Society Conference on Computer Vision and Pattern Recognition, San Francisco, CA, USA, 13–18 June 2010; pp. 1261–1268.
40. Lafarge, F.; Descombes, X.; Zerubia, J.; Pierrot-Deseilligny, M. Structural approach for building reconstruction from a single DSM. *IEEE Trans. Pattern Anal. Mach. Intell.* **2008**, *32*, 135–147. [CrossRef]
41. Xiong, B.; Elberink, S.O.; Vosselman, G. A graph edit dictionary for correcting errors in roof topology graphs reconstructed from point clouds. *ISPRS J. Photogramm. Remote Sens.* **2014**, *93*, 227–242. [CrossRef]
42. Li, M.; Wonka, P.; Nan, L. Manhattan-world Urban Reconstruction from Point Clouds. In *European Conference on Computer Vision*; Springer: Cham, Switzerland, 2016.
43. Bauchet, J.P.; Lafarge, F. Kinetic shape reconstruction. *ACM Trans. Graph. (TOG)* **2020**, *39*, 1–14. [CrossRef]
44. Fang, H.; Lafarge, F. Connect-and-Slice: An hybrid approach for reconstructing 3D objects. In Proceedings of the IEEE/CVF Conference on Computer Vision and Pattern Recognition, Seattle, WA, USA, 14–19 June 2020; pp. 13490–13498.
45. Huang, H.; Brenner, C.; Sester, M. A generative statistical approach to automatic 3D building roof reconstruction from laser scanning data. *ISPRS J. Photogramm. Remote Sens.* **2013**, *79*, 29–43. [CrossRef]
46. Canny, J. A computational approach to edge detection. *IEEE Trans. Pattern Anal. Mach. Intell.* **1986**, *PAMI-8*, 679–698. [CrossRef]
47. De Goes, F.; Cohen-Steiner, D.; Alliez, P.; Desbrun, M. An optimal transport approach to robust reconstruction and simplification of 2D shapes. In *Computer Graphics Forum*; Wiley Online Library: Oxford, UK, 2011; Volume 30, pp. 1593–1602.
48. Li, Y.; Wu, B. Relation-Constrained 3D Reconstruction of Buildings in Metropolitan Areas from Photogrammetric Point Clouds. *Remote Sens.* **2021**, *13*, 129. [CrossRef]
49. Schubert, E.; Sander, J.; Ester, M.; Kriegel, H.P.; Xu, X. DBSCAN revisited, revisited: Why and how you should (still) use DBSCAN. *ACM Trans. Database Syst. (TODS)* **2017**, *42*, 1–21. [CrossRef]
50. CGAL Library. *CGAL User and Reference Manual*, 5.0.2 ed.; CGAL Editorial Board: Valbonne, French, 2020.
51. BAG. Basisregistratie Adressen en Gebouwen (BAG). 2019. Available online: https://bag.basisregistraties.overheid.nl/datamodel (accessed on 13 November 2021).
52. Varney, N.; Asari, V.K.; Graehling, Q. DALES: A large-scale aerial LiDAR data set for semantic segmentation. In Proceedings of the IEEE/CVF Conference on Computer Vision and Pattern Recognition Workshops, Seattle, WA, USA, 14–19 June 2020; pp. 186–187.
53. Rottensteiner, F.; Sohn, G.; Jung, J.; Gerke, M.; Baillard, C.; Benitez, S.; Breitkopf, U. The ISPRS benchmark on urban object classification and 3D building reconstruction. *ISPRS Ann. Photogramm. Remote Sens. Spat. Inf. Sci. I-3* **2012**, *1*, 293–298. [CrossRef]
54. Kazhdan, M.; Bolitho, M.; Hoppe, H. Poisson surface reconstruction. In Proceedings of the Fourth Eurographics Symposium on Geometry Processing, Cagliari, Italy, 26–28 June 2006; Volume 7.
55. 3D BAG (v21.09.8). 2021. Available online: https://3dbag.nl/en/viewer (accessed on 13 November 2021).
56. Can, G.; Mantegazza, D.; Abbate, G.; Chappuis, S.; Giusti, A. Semantic segmentation on Swiss3DCities: A benchmark study on aerial photogrammetric 3D pointcloud dataset. *Pattern Recognit. Lett.* **2021**, *150*, 108–114. [CrossRef]

remote sensing

Article

Combining Deep Semantic Edge and Object Segmentation for Large-Scale Roof-Part Polygon Extraction from Ultrahigh-Resolution Aerial Imagery

Wouter A. J. Van den Broeck * and Toon Goedemé

ESAT-PSI-EAVISE, KU Leuven, Jan Pieter De Nayerlaan 5, 2860 Sint-Katelijne-Waver, Belgium; toon.goedeme@kuleuven.be
* Correspondence: wouter.vandenbroeck@kuleuven.be; Tel.: +32-476-72-14-73

Abstract: The roofscape plays a vital role in the support of sustainable urban planning and development. However, availability of detailed and up-to-date information on the level of individual roof-part topology remains a bottleneck for reliable assessment of its present status and future potential. Motivated by the need for automation, the current state-of-the-art focuses on applying deep learning techniques for roof-plane segmentation from light-detection-and-ranging (LiDAR) point clouds, but fails to deliver on criteria such as scalability, spatial predictive continuity, and vectorization for use in geographic information systems (GISs). Therefore, this paper proposes a fully automated end-to-end workflow capable of extracting large-scale continuous polygon maps of roof-part instances from ultra-high-resolution (UHR) aerial imagery. In summary, the workflow consists of three main steps: (1) use a multitask fully convolutional network (FCN) to infer semantic roof-part edges and objects, (2) extract distinct closed shapes given the edges and objects, and (3) vectorize to obtain roof-part polygons. The methodology is trained and tested on a challenging dataset comprising of UHR aerial RGB orthoimagery (0.03 m GSD) and LiDAR-derived digital elevation models (DEMs) (0.25 m GSD) of three Belgian urban areas (including the famous touristic city of Bruges). We argue that UHR optical imagery may provide a competing alternative for this task over classically used LiDAR data, and investigate the added value of combining these two data sources. Further, we conduct an ablation study to optimize various components of the workflow, reaching a final panoptic quality of 54.8% (segmentation quality = 87.7%, recognition quality = 62.6%). In combination with human validation, our methodology can provide automated support for the efficient and detailed mapping of roofscapes.

Keywords: airborne Earth observation; ultrahigh spatial resolution; semantic segmentation; instance segmentation; fully convolutional neural networks; roofscape

Citation: Van den Broeck, W.A.J.; Goedemé, T. Combining Deep Semantic Edge and Object Segmentation for Large-Scale Roof-Part Polygon Extraction from Ultrahigh-Resolution Aerial Imagery. *Remote Sens.* **2022**, *14*, 4722. https://doi.org/10.3390/rs14194722

Academic Editors: Mohammad Awrangjeb, Jiaojiao Tian, Qin Yan, Beril Sirmacek and Nusret Demir

Received: 30 June 2022
Accepted: 16 September 2022
Published: 21 September 2022

1. Introduction

As buildings cover a considerable area of the urban environment, the roofscape plays a vital role in supporting sustainable urban development [1]. Roofs hold a myriad of potential uses, including civic spaces, urban farms, rainwater harvesting, and solar power. Strategic design choices of the roofscape, such as material type and coverage, are therefore instrumental in regulating the urban energy budget, storm water management, and local food systems. For instance, popular incentives include installing solar panels to aid in achieving a net-zero energy consumption, thermal insulation to reduce building energy use and thus their carbon footprint, and green roofs to contribute to reducing traffic noise, improving air quality and urban biodiversity, and mitigating the urban heat island effect.

To assess the current status and future potential of the sustainable roofscape, accurate and up-to-date information on the spatial distribution and topology of rooftops is needed. Simple building contour maps can often be found from open-source databases (e.g., OpenStreetMap or local governments). However, more detailed maps of individual *roofparts* are lacking. We define a roof part here as a semantic instance within a larger roof complex, distinguishable from other roof parts by a difference in inclination, height, roof material, or any other kind of roof-part edge. Following this definition, we consider a roof part to be different from a *roofplane*, as a roofplane may, for instance, be constituted of different roof parts, e.g., neighboring houses in ribbon development may share the same roofplane that comprises different roof parts separated by gutters. Such more refined maps are needed to facilitate the construction of geographic information systems (GISs) linking attributes to the roof parts, including the surface area, inclination, position, roof material, thermal insulation, degradation state, and mounted objects (e.g., solar panels, air conditioners, satellite dishes), and would allow for better estimates of the solar photovoltaic and green potential of a roofscape. Since mapping roof parts for large areas is a tedious and time-consuming task, automated methodologies are necessary. However, to the best of our knowledge, no end-to-end methodology has yet been proposed for the efficient automated extraction of roof-part maps on a large scale.

Undeniably, deep learning has become the tool of choice for the automated supervised image analysis of airborne and spaceborne Earth observation (EO) imagery, be it for object detection, semantic segmentation, or instance segmentation [2]. Largely driven by data availability, the majority of research concerned with applying deep learning for rooftop mapping focuses on either (1) building footprint extraction from RGB (ortho-)images [3], sometimes combined with LiDAR-derived 2D height maps [4,5], or (2) roofplane segmentation from LiDAR point clouds [6–10]. Research on the first application is predominantly propelled by well-known open-source datasets such as Vaihingen [11], the Inria Aerial Image Labelling Dataset [12] or the more recent SemCity Toulouse [13]. The scope of these studies is solely to segment complete rooftop instances and not individual roof parts that constitute the rooftops. The smaller body of research focusing on the second application uses smaller datasets, as LiDAR is more costly and complex to acquire than optical imagery. Arguably, this render LiDAR a suboptimal choice for large-scale high-resolution problem settings. Therefore, in this paper we compare using only LiDAR-derived height data (i.e., a digital elevation model (DEM)) with using only RGB orthoimagery, and show that, for our case, the latter outperformed the former. Additionally, we investigate the added value of combining these two data sources (RGB + DEM).

Furthermore, an increasing number of rooftop or roofplane extraction methodologies are recognizing the importance of having polygon maps as the workflow output as opposed to only providing the identified rooftops as pixel-based areas (i.e., a raster map) [14,15]. The difficulty with raster maps is that they require considerable storage memory, and are inconvenient for further processing and usage in a GIS. Hence, this paper also advocates that automated mapping approaches should target polygon maps as the workflow output rather than raster maps. This goes in hand with two crucial considerations. First, although segmentation algorithms are becoming increasingly accurate, they often cannot ensure closed object shapes, hence hampering the vectorization of the raster object [16]. Second, due to the patchwise processing of large EO imagery because of computational constraints, predictive spatial continuity is not always ensured. More specifically, it is obvious that methodologies that assume either a single object instance per image patch (e.g., a single building) or only a limited number of scattered instances are not applicable for large-scale roof-part extraction. As such, there is a need for paradigms that take these consideration into account.

Having identified some key requirements for large-scale roof-part polygon extraction, this paper proposes a workflow based on three steps: (1) use a deep neural network to predict roof-part objects and edges, (2) use a bottom–up clustering algorithm given the predicted edges to derive distinct closed shapes, and (3) vectorize and simplify the roof-part shapes. For Step 1, we opted for a semantic segmentation approach using a fully convolutional network (FCN). We did not use a detection-based instance segmentation method (i.e., using bounding boxes), such as the famous mask R-CNN [17], because they have difficulty with highly clustered instances, which is evidently the case for urban roofscapes. Therefore, the emphasis of the workflow lies on finding roof-part edges rather than roof-part objects. To this end, we briefly review the related work focusing on using FCNs for rooftop edge segmentation. The overall objective of these studies is to optimize semantic object and edge predictions by designing specific FCN architectures or training loss functions that explicitly account for both targets. For example, Marmanis et al. (2018) integrated a separate edge detector (holistically nested edge detection (HED)) into their semantic segmentation model to build in semantic boundary awareness [18]. Wu et al. (2018) proposed a boundary regulated network (BR-Net) to simultaneously perform building segmentation and outline extraction on the basis of a shared feature representation and a multilabel optimization approach [19]. Diakogiannis et al. (2020) also built on the idea of multitask inference and created ResUnet-a, a FCN that simultaneously outputs a semantic object mask, edge mask, object center distance map, and an HSV colored reconstruction of the input [20]. To promote edge connectivity and clear object boundaries, Xia et al. (2021) designed specific edge guidance modules and applied multiscale supervision for training their network (DDLNet) [16]. Additionally, they used a multilabel target approach concatenating the edge prediction as a feature map for predicting the semantic objects. Next, a number of studies proposed custom loss functions to improve edge and object-boundary predictions, of which most are weighted flavors of the Dice loss, cross-entropy loss, or combinations of the latter [14,21]. Here, the chief consideration is to cope with class imbalance, as the number of edge pixels is typically a number of magnitudes lower than the number of non-edge pixels. Further, some works address the problem of building instance segmentation from a different angle by combining FCNs with other deep-learning paradigms to predict building corners as key points, which can more easily be processed into regular polygons [22,23]. Lastly, an alternative direction is to train directly on extracting polygon coordinates. For example, Chen et al. (2020) proposed a modeling framework for vectorized building outline extraction using a combination of FCN-based segmentation and a modified PointNet to learn shape priors and predict polygon vertex deformation [24].

However, the aforementioned studies focused on complete *roof* extraction and not on individual *roof-part* extraction. Moreover, no existing framework could be readily applied to our application as they failed to fulfil at least one of our identified criteria, i.e., (i) scalable to large areas, (ii) predictive spatial continuity, and (iii) polygon-oriented. Therefore, the novelty of this paper is manifold: (i) we propose a fully automated workflow for the extraction of individual roof parts on a large scale; (ii) we suggest a multiclass FCN approach for combined semantic edge and object prediction, as opposed to the more common multilabel approach; (iii) spatial predictive continuity is taken into account by saving intermediate output mosaics; and (iv) the methodology is polygon-oriented, i.e., the FCN ground truth is generated on the fly on the basis of polygon annotations, the workflow output is a polygon map, and the quality assessment is polygon-based. The methodology was trained and tested on a new challenging dataset comprising of UHR aerial RGB orthoimagery (0.03 m GSD) and LiDAR-derived DEMs (0.25 m GSD) of three Belgian urban areas (including the famous touristic city of Bruges). We conducted an ablation study to optimize various components of the workflow, reaching a final panoptic quality of 54.8% (segmentation quality of 87.7% and recognition quality of 62.6%). Lastly, we highlight some current shortcomings, potential improvements, and future opportunities.

2. Materials and Methods

2.1. Study Area and Dataset

The study area under consideration is the region of Flanders, i.e., the northern part of Belgium (Figure 1). Geographically, Flanders is an agriculturally fertile and densely populated region with little to no relief. The landscape is predominantly characterized by cropland (ca. 31%), agricultural grassland (ca. 20%), residential areas (ca. 13%), and tree cover or forests (ca. 10%). Another ca. 15% is covered by human infrastructure for transport, services, built-op areas, industry, agriculture, and airports [25]. The cities and villages are mainly organized as dense smaller city centers with little to no high-rise buildings and surrounding ribbon development. The building architecture and structure, and hence the rooftop appearance, are highly diverse.

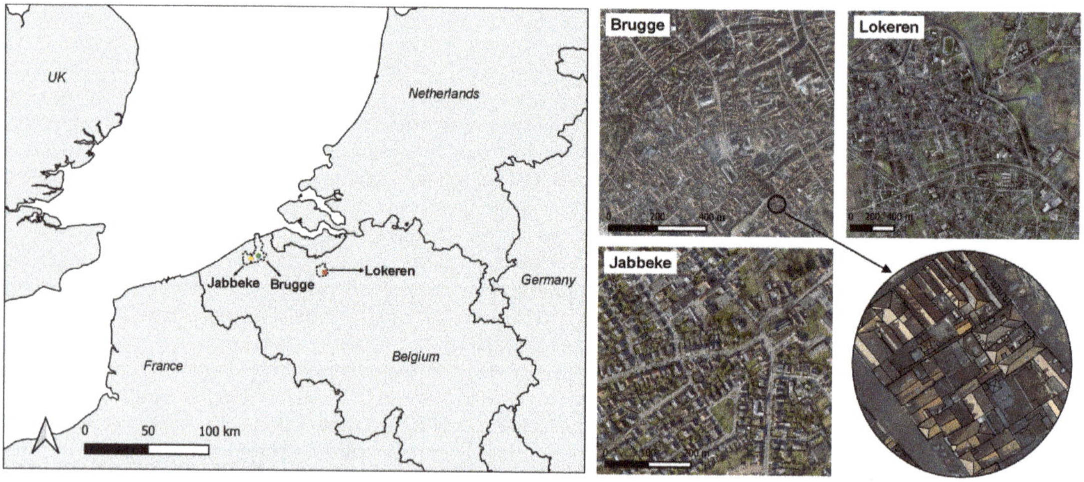

Figure 1. Geographic location and visual overview of the training (Brugge), validation (Jabbeke), and test (Lokeren) sets. Colored dots delineate the considered regions within the municipality borders (dotted lines). The magnification at the bottom right shows an example of the complexity of the rooftop structure and its overlying polygon labels.

To develop and test our methodology, we selected three municipalities within Flanders to serve as training, validation, and test regions: Brugge, Jabbeke, and Lokeren, respectively (Figure 1). For each of the three towns, the following data were available: (1) an aerial RGB orthophoto with ultrahigh resolution of 0.03 m GSD stored as an ERDAS JPEG2000 uint8 compressed file. To give a notion of the magnitude, the Lokeren orthophoto, covering 3.65 km^2, has an image resolution of 320,716 $\times$ 297,190 pixels. The used coordinate reference system (CRS) was the Belgian Lambert 72 (EPSG:31370). (2) A digital elevation model (DEM) of 0.25 m GSD stored as float32 GeoTIFF. (3) Polygon labels delineating all distinct roof parts within the three considered regions, stored as GeoJSON files. The labels were generated by (nonexpert) human annotators on the basis of the visual interpretation of the RGB orthophotos. Inevitably, the latter gives rise to some degree of interpretive variability and erroneous labels. Nonetheless, visual inspection confirmed that the annotations were of sufficient quality for training and evaluating our methodology.

Table 1 provides a quantitative overview of the dataset. Figures 2 and 3 further show the distribution of roof-part types and roof-part sizes for the three partitions, respectively. The latter reflect that Brugge (training set) has a highly compact, complex, and densely clustered rooftop structure, i.e., approximately half of the total area (0.72 km^2) is covered by roofs (0.34 km^2), with strongly right-skewed roof-part size distribution and a median roof size of only 7.1 m^2. This feature of many different examples on a small area is advantageous, as it allows for faster model training (see Section 2.3.3). In contrast, Lokeren (test set) is more characterized by a combination of densely clustered roofs, isolated houses, and some very large roofs. Further, Lokeren has approximately the same number of roof parts as Brugge (ca. 27 $\times$ 10^3), but spread out over a region almost five times as large. As such, this allows for evaluating the model on unseen scenery and areas without roofs, rendering our test set truly honest and challenging. Lastly, Jabbeke (validation set) is a smaller town with suburban appearance. It was used for the validation of the performance and generalizability of the models during training.

Table 1. Dataset overview.

Partition	Location	Total Area [km^2]	Roof Parts	Roof-Part Area [km^2]
Training	Brugge	0.72	26,984	0.34
Validation	Jabbeke	0.18	1629	0.04
Test	Lokeren	3.65	27,303	0.68

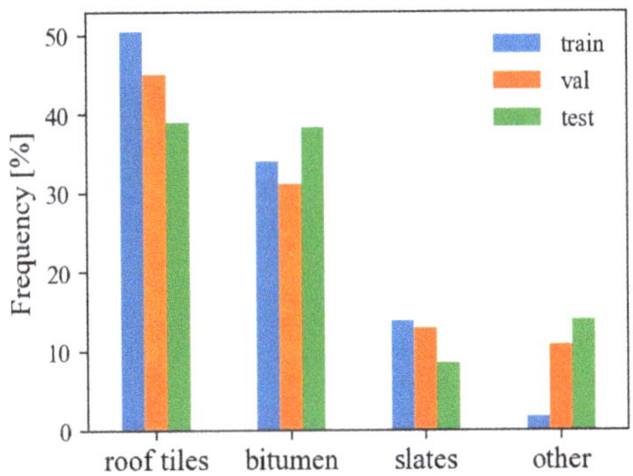

Figure 2. Roof type class frequencies.

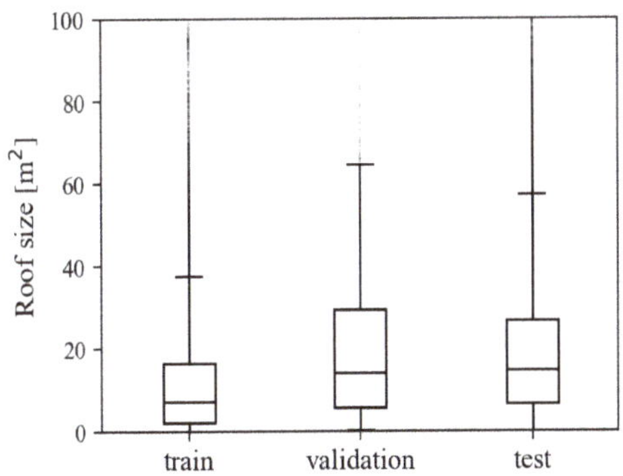

Figure 3. Roof size distribution.

2.2. Workflow Overview

Here, we elaborate on our proposed workflow for large-scale roof-part polygon generation given UHR orthoimagery (Figure 4). In general, the workflow consisted of three main steps: (1) A trained FCN was used to infer semantic roof-part edges and objects. (2) Distinct closed shapes were extracted given the edges and objects. (3) The closed shapes were vectorized to obtain roof-part polygons. As depicted in Figure 4, each step was computationally constrained by a certain patch size (PS), i.e., the maximal image resolution that could be algorithmically processed in terms of time and/or memory. However, in contrast to natural imagery, EO imagery is spatially continuous. Therefore, it is important to ensure the continuity of the overall output despite patchwise processing. As a larger PS corresponds to a smoother overall output, saving intermediate results allows for setting the PSs to a different optimal value for each step. Usually, $PS_{FCN} < PS_{CSE} < PS_{VEC}$, where PS_{FCN} is the input patch size of the FCN model, PS_{CSE} of the closed shape extraction step, and PS_{VEC} of the vectorizing step. Each of the three steps is described in more detail in the subsequent sections.

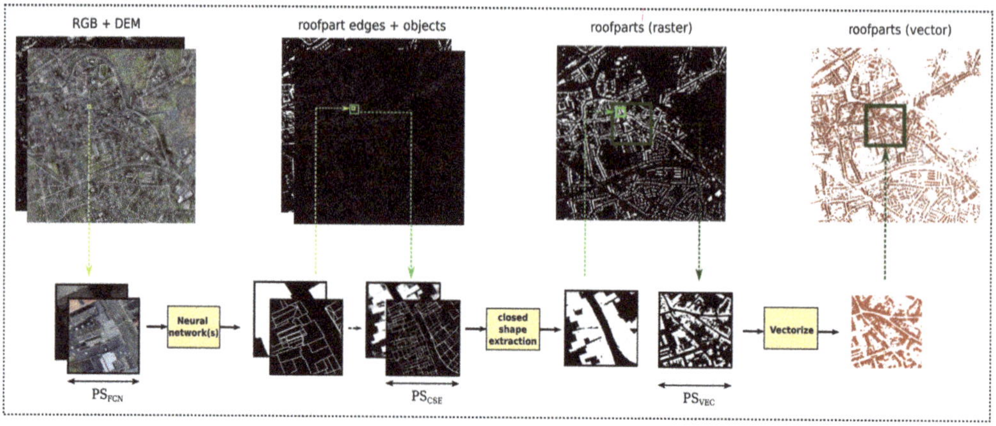

Figure 4. Schematic workflow overview for large scale roof-part polygon extraction from optical (RGB) and height/depth (D) orthoimagery. The workflow consists of three main steps: (**1**) a fully convolutional neural network (FCN) model is used to predict semantic roof edges and objects. (**2**) Distinct closed shapes are extracted given the roof edges and objects. (**3**) The closed shapes are vectorized to obtain roof-part polygons. Each step is computationally constrained by a certain patch size (PS), where usually $PS_{FCN} < PS_{CSE} < PS_{VEC}$. As a larger PS corresponds to a smoother global prediction, intermediate results can be saved to allow for the three PSs to be different.

2.3. Semantic Edge and Object Segmentation

As a first step, we used an FCN to learn a function $\widehat{\boldsymbol{Y}} = f(\boldsymbol{X}, \boldsymbol{\theta})$ with parameters $\boldsymbol{\theta}$ to map an input image $\boldsymbol{X} \in \mathbb{R}_{[0,1]}^{C_i \times H \times W}$ to predicted semantic probability maps $\widehat{\boldsymbol{Y}} \in \mathbb{R}_{[0,1]}^{C_o \times H \times W}$, with C_i being the number of input channels, C_o the number of output channels, H the image height, and W the image width. Each pixel in the output $\widehat{\boldsymbol{Y}}$ represents the confidence with which the corresponding pixel in the input $\boldsymbol{X}$ belongs to a certain semantic class, such as, in our case, roof vs. non-roof and/or roof edge vs. non-roof edge. We compared three paradigms for deriving roof-object and roof-edge probability maps (Figure 5): (i) two independent FCNs with binary output; (ii) a single FCN with multilabel output, i.e., each pixel could belong to multiple classes; and (iii) a single FCN with multiclass output, i.e., each pixel could belong to only one class. Below, we briefly describe the FCN architecture, and the training and validation procedure.

2.3.1. FCN Architectures

The used FCN is the well-known UNet architecture [26]. To date, it remains the standard for semantic segmentation of EO imagery [2]. More specifically, we use a UNet with an EfficientNet-B5 [27] 5-stage encoder ($H = W = 2^x \rightarrow 2^{x/5}$) pretrained on Imagenet [28], and a decoder with batch normalization and spatial and channel squeeze & excitation (scSE) attention modules after every decoder stage [29], as implemented in the *pytorch-segmentation-models* Python package [30]. The input of the FCN is a 1-channel DEM ($C_i = 1$), 3-channel RGB ($C_i = 3$) or 4-channel RGB + DEM ($C_i = 4$) image patch. The used PS_{FCN} is $H = W = 1024$ pix, corresponding to a ground resolution of $\sim 30 \times 30$ m^2. This is the largest PS that could fit into GPU memory during FCN training. The output of the FCN is different for each of the three approaches (Figure 5). For the double binary case, the output is twice a single channel ($C_o = 1$: roof object; $C_o = 1$: roof edge) with pixelwise sigmoid activation to obtain the probability maps. For the single multilabel case, the output is two-channeled ($C_o = 2$: roof object, roof edge) with sigmoid activation. For the multiclass case, the output is three-channeled ($C_o = 3$: roof object, roof edge, and neither) with softmax activation over the channels to ensure mutual exclusiveness. To save memory, the predicted probability patches can simply be saved into the complete output mosaic as integer maps: $X_{uint8} = \lfloor X_{float} \times 255 \rceil$.

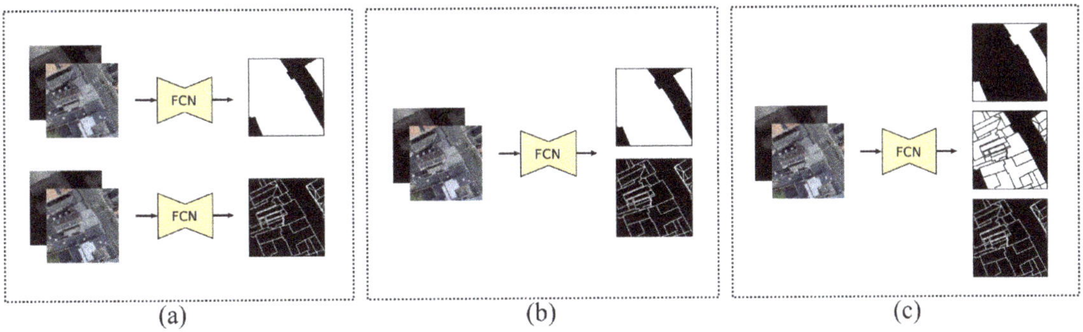

Figure 5. Comparison of modelling approaches using a fully convolutional network (FCN). All approaches take an RGB and/or DEM image patch as input and return a roof-part object and edge probability map. (**a**) Separate models for roof and roof-part edge segmentation. (**b**) Single multilabel model. (**c**) Single multiclass model.

2.3.2. Data Preprocessing

The models are trained by feeding them batches of image patches and the corresponding ground truth. The ground truth is generated on the fly by rasterizing the polygons occurring within the input patches to binary maps $\mathbf{Y} \in \mathbb{B}^{C_o \times H \times W}$. To obtain a roof-object target, the full polygon surface is rasterized, while to obtain a roof-edge target, the polygons are first converted into their line-string format and subsequently dilated by a certain number of pixels (in the vector space), again resulting in a polygon that is then rasterized. For the multiclass model, the roof-edge channel is subtracted from the roof-object channel, and the background channel is obtained as the logical negation of the roof object and edge channels.

To match input dimensions, the DEM is bilinearly upsampled to the resolution of the RGB. Further, DEM values are transformed by clipping all values to the range of 0–30 m and subsequently rescaling them to the 0–1 domain: $X_{DEM,[0,1]} = \max(0, \min(30, X_{DEM}))/30$. This range was arbitrarily chosen, as most buildings in the considered study area are smaller than 30 m, and height values cannot be negative.

Image augmentation techniques were used for the training set, including random flipping, minor (HSV) color variations, jitter (random small translation of the considered patch), and an overlap of 128 pix between adjacent patches. Moreover, to balance and limit the training and validation sets, only patches with occurrence of roofs were included. In the test set, patches without roofs were also included. This resulted in 982 training patches (759 in the case of no overlap), 187 validation patches, and 3865 test patches.

2.3.3. FCN Training

All FCN models were trained for 30 epochs using the Adam optimizer, a fixed learning rate of 2×10^{-4}, and a batch size of 3. The used hardware was an NVIDIA Tesla V100 GPU with 32 GB memory (CUDA 11.0). The model architectures, and training and validation loops were implemented in Python using the Pytorch framework [31]. The FCN parameters were optimized by minimizing the weighted categorical cross-entropy (CCE) loss between the prediction and ground truth:

$$\mathcal{L}_{CCE} = -\frac{1}{N}\sum_{n=1}^{N}\sum_{c=1}^{C_o} w_c \, y_{c,n} \log \widehat{y}_{c,n} \quad (1)$$

where $y_{c,n} \in \{0,1\}$ is the ground-truth value of the n-th pixel belonging to output channel c; $\widehat{y}_{c,n} \in [0,1]$ is the predicted value for that pixel; w_c is the loss weight for output channel c; and $N = H \times W \times BS$ is the total number of pixels for each channel, with BS being the batch size. For a binary model output ($C_o = 1$), the CCE loss reduces to the binary cross-entropy (BCE) loss:

$$\mathcal{L}_{BCE} = -\frac{1}{N}\sum_{n=1}^{N} w_+ \, y_n \log \widehat{y_n} + w_- \, (1 - y_n) \log (1 - \widehat{y_n}) \tag{2}$$

with w_+ and w_- being the weights for the target class and background, respectively. The loss for the multilabel (ML) model is calculated as the average of the binary cross-entropy loss for each target:

$$\mathcal{L}_{ML} = \sum_{C_o} \lambda_c \mathcal{L}_{BCE,c} \tag{3}$$

where λ_c are potential scaling factors to give more or less importance to certain channels. We simply set $\lambda_c = \frac{1}{C_o}$.

The class weights in Equations (1) and (2) were calculated inversely proportionally to the probability of class occurrence p_c, i.e., $w_c \sim 1/p_c$. If it is further assumed that the expected probability for all classes is uniform, i.e., $\mathbb{E}[p_c] = 1/C_o$, then w_c is calculated as follows:

$$w_c = \frac{1}{p_c/\mathbb{E}[p_c]} = \frac{1}{p_c \, C_o} \tag{4}$$

In words, classes that occur less frequently than in a class balanced case are upweighted in the loss function ($p_c < 1/C_0 \rightarrow w_c > 1$), while classes that occur more are down-weighted ($p_c > 1/C_0 \rightarrow w_c < 1$). The class probabilities p_c are estimated as the relative class frequencies, i.e., $p_c = f_c / \sum_C f_c$.

2.3.4. Validation

The best model during training was selected on the basis of the intersection over union (IoU) on the validation set:

$$\text{IoU} = \frac{\text{TP}}{\text{TP} + \text{FP} + \text{FN}} \tag{5}$$

where TP = true positive, FP = false positive, and FN = false negative. To calculate the latter, a threshold of 0.5 was used as the decision boundary for the predictions in the case of the double binary model and single multilabel model, and the maximal confidence was used in the case of the multiclass model. In the case of multichannel output ($C_o > 1$), the mean IoU over the channels was used: $\text{mIoU} = \frac{1}{C_o} \sum_{C_o} \text{IoU}_c$. The IoU was calculated every 200 batches during training.

2.4. Closed Shape Extraction

Given the predicted roof-part edge and roof-part object confidence maps, the next step is to use these to extract closed roof-part shapes. The aim is to obtain a single channel patch of the same dimensions where every pixel is assigned to a unique roof-part cluster, i.e., $f_{CSE} : \widehat{\boldsymbol{Y}} \rightarrow \widehat{\boldsymbol{R}} \in \{1, 2, ..., M\}^{1 \times H \times W}$, with M being the number of roof-part clusters. Figure 6 provides a schematic overview of the closed-shape extraction workflow. The rationale is to apply a bottom–up clustering algorithm starting from markers within the areas delineated by the roof-part edges. To find these markers, the roof-part edge probability map was first thresholded and subsequently eroded to find pixels that had a high probability of *not* being a roof-part edge. These pixels could then be grouped together into a number of connected components that could be used as the markers. On the basis of some related studies [16,32], the watershed clustering algorithm was chosen, which owes its name to the fact that it mimics the flooding of basins where, in this case, the height of the basin is the edge probability. The result is distinct clusters with a one-pixel wide line separating the clusters.

Next, the roof-part object prediction was used to filter out the non-roof clusters. Two options were considered: (1) The predicted roof-part area is used as watershed mask (WM), i.e., the watershed clustering was only performed within this mask. The drawback is that the quality of the resulting clusters strongly depends on the quality of the roof-part object prediction. The advantage is that this reduces the computational need and, if an accurate building or rooftop map of the area under consideration is already available, it can be incorporated into the pipeline as a WM. (2) The predicted roof-part area is used as area threshold (AT), i.e., only clusters are retained where a minimal percentage of the cluster is predicted as a roof. This approach is more dependent on the quality of the roof-part edge prediction. Note, The connected components themselves could already be considered to be a prediction of the roof-part clusters when used in combination with filtering out the background clusters. However, because Chen et al. (2021) found that additionally using the watershed algorithm consistently improved the results, we do not report results on the latter [32].

The following configurations are used in this study: a marker threshold of 0.2, erosion with a single pass of a 3 × 3 kernel, connected components detection with 8-connectivity, the watershed algorithm as implemented in the *scikit-image* library (v.0.17.2) [33], and an area threshold of 0.5. Optionally, the cluster output can be saved as a binary patch into the full output mosaic to ensure a spatially continues predicted roof-part cluster map. For most experiments, PS_{CSE} is set to 1024 pix to allow for direct CSE postprocessing after FCN inference.

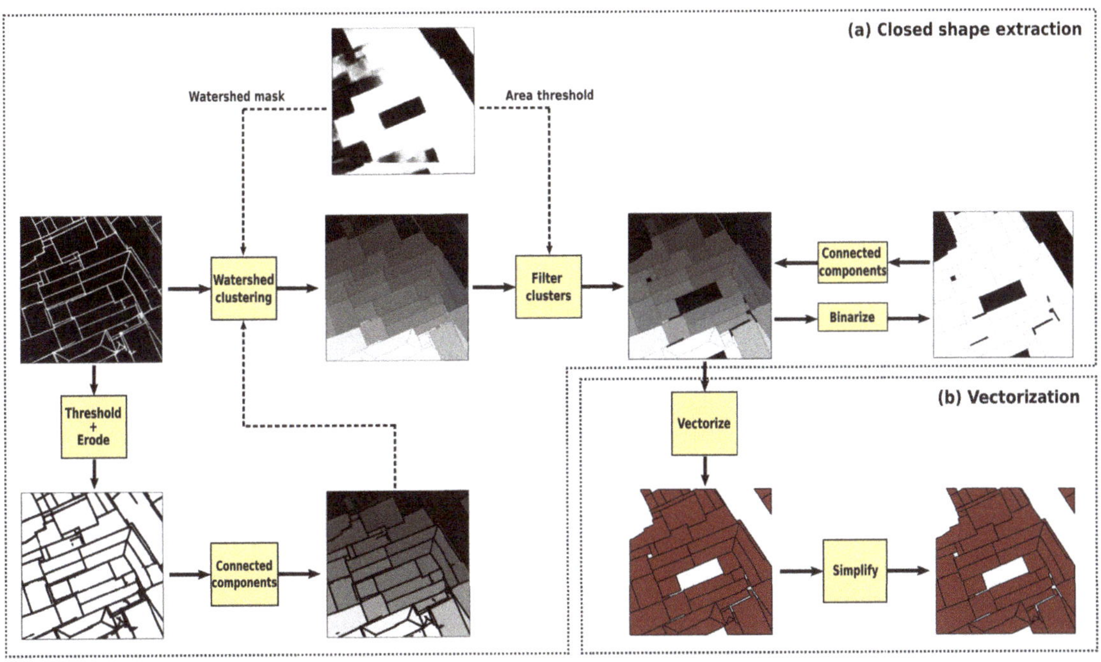

Figure 6. (**a**) Closed shape extraction workflow to convert roof-part object and roof-part edge probability maps to roof-part clusters. (**b**) Vectorization step to convert roof-part clusters into distinct simplified polygons.

2.5. Vectorization

As a final step, the predicted roof-part clusters $\widehat{R}$ are converted into distinct polygons $\widehat{P} : \{(x_1, y_1), (x_2, y_2), ..., (x_n, y_n)\}^M$ with M being the number of polygons. The polygons are derived by simply running vectorization on the connected components of the binary image resulting from CSE (Figure 6b). Subsequently, the polygons are simplified using the Douglas–Peucker algorithm to reduce the number of vertices (x_i, y_i) and hence the required storage memory. The Douglas–Peucker algorithm was selected because of its simplicity and efficiency [34]. The degree of simplification was controlled with a tolerance parameter determining the maximal deviation of the simplified geometry from the original. On the basis of a visual inspection, the tolerance was set to 0.1 m. Further, polygons with an area smaller than 0.8 m^2 were discarded.

The vectorization step can be performed with a much larger PS than that in the two prior steps, and is only constrained by the memory requirement as the overall execution time remains constant. A larger PS leads to less half-vectorized clusters at the patch edges. Therefore we set $\mathrm{PS}_{VEC} = 10{,}240$, i.e., ten times as high as PS_{FCN}. Alternatively, to prevent from these cut edge-clusters, vectorization can also be performed patchwise with overlap. In this case, clusters at the patch edge were not vectorized as they were incomplete, and duplicate polygons had to be removed for clusters that occurred in the overlapping patch regions as they were vectorized twice.

2.6. Evaluation

The final predictions were evaluated by comparing the predicted and ground-truth polygons. The used evaluation metric is the panoptic quality ($\mathrm{PQ} \in [0, 1]$), as introduced by the COCO panoptic segmentation challenge [35]. The PQ is calculated as the product of the recognition quality (RQ) and the segmentation quality (SQ):

$$PQ = \underbrace{\frac{\sum_{(p,\hat{p})\in \mathrm{TP}} \mathrm{IoU}(p,\hat{p})}{|\mathrm{TP}|}}_{\text{segmentation quality (SQ)}} \times \underbrace{\frac{|\mathrm{TP}|}{|\mathrm{TP}| + \frac{1}{2}|\mathrm{FP}| + \frac{1}{2}|\mathrm{FN}|}}_{\text{recognition quality (RQ)}} \quad (6)$$

where p and $\hat{p}$ are the ground-truth and predicted polygons, respectively. The $\mathrm{RQ} \in [0, 1]$ is the widely used F1 score for quality estimation in detection settings. The $\mathrm{SQ} \in [0, 1]$ is simply the average IoU of matching polygons. To calculate this metric, the IoU for each pair of polygons $(p, \hat{p})$ was first computed as $(p \cap \hat{p})/(p \cup \hat{p})$. The set of TP then comprises all uniquely matching pairs for which $\mathrm{IoU}(p, \hat{p}) > 0.5$, the set of FP comprises all *predicted* polygons that did not belong to any pair in TP, and the set of FN comprises all *target* polygons that did not belong to any pair in TP.

In contrast to the training phase, the inference pipeline was run on a laptop with a 12-core Intel i7-8750H (2.20 GHz) processor, 32 GiB RAM, and a 6 GiB NVIDIA GeForce RTX 2060 GPU. When using the settings $PS_{FCN} = 1024$, $PS_{CSE} = 1024$ and $PS_{VEC} = 10{,}240$, running the workflow for the entire test set (9.5×10^{10} pix^2) takes around 1 h, i.e., ~40 min for FCN inference (for a single model), ~10 min for the CSE step, and ~7 min for vectorization.

3. Experiments and Results

To gain insight on the effect of some of the hyperparameters along our workflow, we conducted an ablation study on different levels. First, the influence of the ground-truth generation and loss weighting on the edge prediction was studied. Second, the influence of the patch size and roof-object prediction usage on the CSE step was evaluated in terms of computational effort and prediction quality. Third, the effect of the modelling approach and input sources was investigated. Lastly, multiple common FCN architectures were compared to investigate their suitability for multiclass object and edge segmentation. Visual results are shown for the best found configuration.

3.1. Influence of Ground Truth and Loss Weighting

The generated edge ground truth was varied by choosing different values of edge dilation, i.e., the pixel width of the roof edge, and whether to apply Gaussian smoothing (Table 2). A wider edge corresponds with more pixel examples for the model to train, but may also lead to less well-defined and accurate edge-prediction. In particular, dilations of 5 and 11 pixels were compared corresponding to approximately 0.15 and 0.30 m, respectively. The rationale of smoothing the ground truth is to try to force the model to predict edge probabilities as a Gaussian curve: high in the middle of the edge and gradually lower when moving away from the edge. A Gaussian kernel of size 5 and standard deviation of 1 was used. Further, the idea of varying the weight in the loss function given to the edge is similar to varying the dilation: a higher weight leads to thicker edge predictions. In particular, weights, as calculated in Equation (4) with $\mathbb{E}[p_{\text{roof-edge}}] = \frac{1}{C_o} = 0.5$ (indicated as '++') were compared with weights calculated by setting $\mathbb{E}[p_{\text{roof-edge}}] = 0.2$ (indicated as '+'), which attributed less weight to the edge pixels. The above experiments were conducted using the separate model approach, the roof-object prediction as watershed mask, and the standard settings as previously described. Note, as this 'intermediate' ground truth is variable, we evaluated these influences on the final polygon predictions.

Table 2 shows that the larger dilation of 11 consistently led to a higher IoU on the validation set (calculated on the FCN output). This seems logical, as the fraction of edge pixels on which to train was higher, and it is easier to have a larger relative overlap between prediction and ground truth – and thus a larger IoU – for a wider than for a thinner ground truth. In addition, a smaller loss weight for the edge is associated with a higher IoU_{val}. In contrast, smoothing the ground truth seemed to have no pronounced effect. However, none of the above appeared to largely influence the final PQ. Hence, it seemed preferable to first prioritize optimizing other parts of the workflow. Especially the RQ seemed to be a bottleneck to obtaining a higher PQ. Nonetheless, the proceeding experiments were conducted using the configuration yielding the highest PQ, i.e., a wider dilation, no smoothing, and a tempered loss weight.

Table 2. Influence of edge ground truth (either with (✓) or without (-) Gaussian smoothing) and loss weighting (either strong (++) or limited (+) focus on edge pixels) on panoptic quality (PQ), segmentation quality (SQ), and recognition quality (RQ).

Dilation [pix]	Smoothing	Loss Weight	Edge IoU_{val} [%]	PQ [%]	SQ [%]	RQ [%]
5	-	+	24.87	34.67	83.89	41.33
5	✓	+	24.72	36.19	84.26	42.95
5	-	++	17.88	37.27	84.26	44.23
5	✓	++	18.02	37.20	84.48	44.04
11	-	+	**39.80**	**37.84**	**84.55**	**44.75**
11	✓	+	39.64	37.84	84.55	44.75
11	-	++	32.01	37.84	84.54	44.75
11	✓	++	31.22	36.93	83.98	43.97

3.2. Influence of Patch Size and Object Mask Usage

Keeping the settings from above, we now vary the PS_{CSE} while using the roof-object prediction a first time as watershed mask and a second time as area threshold. Figure 7 plots the PQ and execution time in function of the $\text{PS}_{CSE} \in \{512, 1024, 2048, 4096\}$. Remark, the shown time is the time necessary for the patchwise processing of the complete image and not the processing time for a single patch. Using the roof-object prediction as area threshold consistently outperformed using it as a watershed mask. This means that the roof-edge prediction is more accurate for delineating roofs than the roof-object prediction. On the downside, when opting for using it as the area threshold, the watershed algorithm has to consider the complete patch instead of only the identified rooftops, increasing the computation time. Moreover, the time increases exponentially with an increasing patch size. Similarly, the PQ also increases with a larger PS_{CSE}, as a larger PS leads to fewer poorly defined clusters at the patch-boundary, but rather logarithmically. As such, there is a quality vs. time trade-off for using a larger PS_{CSE}. Using the area threshold option and a

$PS_{CSE} = 4096$ can increase the PQ on the test set to 52.5%. However, for time consideration, in the subsequent experiments we continued using $PS_{CSE} = 1024$.

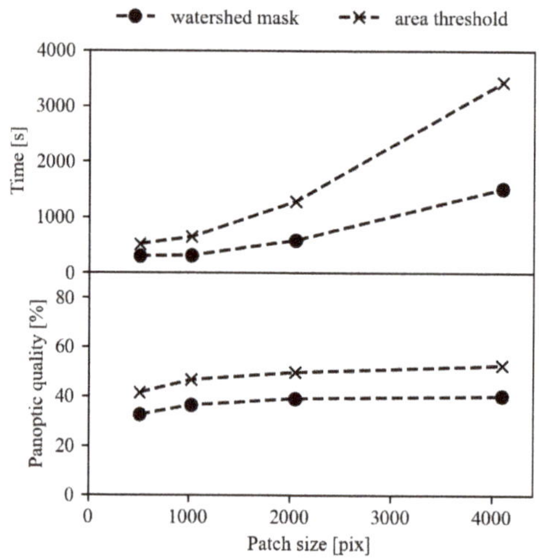

Figure 7. Influence of patch size and roof mask usage in the closed shape extraction step on computation time and final quality.

3.3. *Influence of the Modelling Approach and Input Sources*

Table 3 summarizes the experiments comparing the three modelling approaches as described in Section 2.3. For each approach, the IoU on the FCN output for the test set and the final PQ was computed when using different sources of input information, i.e., only LiDAR-derived height information (DEM = D), only optical spectral information (RGB), or a combination (RGB + D).

Looking at the results of the double binary model, it is clear that the baseline of only using elevation data corresponds with the worst result. Especially the edge prediction seems to suffer from the the lower resolution of the DEM. Using just UHR RGB data led to more than double the PQ (47.1% vs. 22.2%). Interestingly, using RGB + D input did not yield a better PQ than that using just RGB input (46.6% vs. 47.1%) despite an increment in IoU_{test} of around 3% for both roof-object (71.9% vs. 68.3%) and roof-edge (34.4% vs. 31.7%) predictions. This also emphasizes the importance of evaluating the performance at the very end of the workflow, i.e., on the final polygon predictions. Further, as concluded before, using the roof-object prediction as AT was the choice of preference, consistently outperforming the usage as WM.

One advantage of the two-model approach is that the roof-object model can be trained on a much larger dataset. Datasets for rooftop or building segmentation are much more prevalent and available than datasets for individual roof-part segmentation are. To explore this possibility, an FCN for rooftop segmentation was trained on the same type of RGB imagery but now for 17 municipalities, and using the building-class in the large-scale reference database (LRD) of Flanders [36], a GIS serving as a topographical reference for Flanders, as ground truth polygons. A version of the LRD up-to-date with the RGB orthoimages was used. The outcome was a high-performance rooftop segmentation model ($IoU_{test} = 83.5\%$). However, this did not result in a higher final PQ when used in combination with AT (45.8% vs. 47.1%). On the other hand, it significantly raised the PQ when the predictions were used as WM (44.3% vs. 37.8%), even almost to the level of AT.

Focusing on multitask approaches shows that, for RGB input, the multilabel model corresponds with a lower PQ (43.3%) and the multiclass with a very similar PQ (47.0%) compared to the two-model approach. The multilabel model especially seemed to underperform for edge prediction (IoU_{test} = 23.5%). Potential improvement may be achieved by tuning the importance of the channels in the loss function (Equation (3)).

For the multiclass approach, using additional DEM information on top of RGB resulted in an increase in PQ, even reaching the highest PQ (50.0%). Hence, employing a single multiclass model (trained for 30 epochs) could outperform the two single-task models (both trained for 30 epochs). A major advantage is that the former only requires to train and deploy a single model, reducing computation time in half.

As an additional experiment, remark that the roof-object output of the multiclass model (see Figure 5) is already in a format that can be readily vectorized into roof-part polygons. Calculating the PQ on the basis of this direct vectorization (DV) shows that an acceptable quality could be attained (40.3%), but that using the watershed clustering remains advantageous, with it having a quality gain of roughly 6.5%. However, considering that this approach eliminates the need for a clustering postprocessing step, it may be a research direction worth exploring and optimizing.

Table 3. Influence of the model type and input sources on panoptic quality (PQ).

Model Type	Input		IoU_{test} [%]		Roof Mask Usage	PQ [%]
	Object	Edge	Object	Edge		
double	D [1]	D	49.4	16.2	WM [3]	10.6
					AT [4]	22.2
	RGB	RGB	68.3	31.7	WM	37.8
					AT	47.1
	RGB-LRD [2]	RGB	83.5	31.7	WM	44.3
					AT	45.8
	RGB + D	RGB + D	71.9	34.4	WM	35.6
					AT	46.6
Single multilabel	RGB		67.2	23.5	AT	43.3
	RGB + D		74.0	25.4	AT	46.6
Single multiclass	RGB		67.2	31.0	DV [5]	40.3
					AT	47.0
	RGB + D		70.4	31.3	DV	42.4
					AT	**50.0**

[1] D = DEM; [2] LRD = Large-Scale Reference Database Flanders; [3] WM = watershed mask; [4] AT = area threshold; [5] DV = direct vectorization.

3.4. Influence of the FCN Architecture

As a final experiment, we examined the applicability of different common FCN architectures for the case of roof-part object and edge prediction. The optimal settings inferred in the previous section were used, i.e., a multiclass approach with RGB + D input. The considered networks were DeepLabV3+ [37], FPN [38], MAnet [39], PAN [40], Pyramid Scene Parsing Network (PSPNet) [41], and UNet++ [42]. All models had the same imagenet-pretrained EfficientNet-B5 backbone (28.34×10^6 parameters) and the default settings from the *segmentation-models-pytorch* package (v.0.2.0) [30]. For UNet++, scSE attention modules were added. All models were trained using the settings as described in Section 2.3.3 except for UNet++, which was trained with a batch size of 2 due to its larger memory requirement.

Results are reported in Table 4. The best scoring network was UNet, closely followed by MAnet and UNet++. The lowest ranking network was PSPNet, followed by PAN, both which consecutively have the lowest number of parameters. A UNet type FCN model thus appears to be an acceptable choice for the task presented in this study.

Table 4. Influence of various model architectures on panoptic quality (PQ).

Model	Decoder Parameters	PQ [%]
DeepLabV3+	1.2×10^6	46.7
FPN	1.8×10^6	47.9
MAnet	9.9×10^6	49.7
PAN	1.4×10^5	45.2
PSPNet	8.5×10^4	32.4
UNet	3.0×10^6	**50.0**
UNet++	3.7×10^6	49.3

3.5. Visual Examples of the Optimized Workflow

On the basis of the above experiments, the best identified configuration was a UNet with RGB + DEM input, multiclass output, and watershed clustering postprocessing using the predicted roof area as decision threshold. A final roof-part polygon prediction was performed for the test set using the latter approach and setting: $PS_{FCN} = 2048$, $PS_{CSE} = 4096$ and $PS_{VEC} = 30{,}000$. Setting $PS_{FCN} = 2048$ during inference while the model was trained on patches of size 1024 slightly increased the PQ, presumably because it widens the patch context and decreases the number of patch boundary pixels, while the FCN is rather robust to a limited scale increase. This final prediction reached a quality of **PQ = 54.8% (RQ = 62.6%, SQ = 87.7%)**.

Visual results are shown in Figures 8 and 9. Figure 8 displays the predicted roof-part polygons for the entire test region divided into true positives (green), false positives (red), and false negatives (blue). Our proposed approach correctly identified rooftops in the larger landscape. Moreover, it could distinguish individual roof parts within the roofs to a degree in which it is useful for automated support of GIS database construction, albeit in combination with human interaction. Erroneous predictions are also often reasonable. For instance, the red border of false positives in Figure 8a was caused by running patch-based prediction, making the region of roof-part predictions exceed the region of ground-truth polygons. Further, as seen in Figure 8d,e, false positives and false negatives are often related. For example, in the lower left of the patch the predicted polygon covers a roof part that was interpreted as multiple smaller roof parts in the ground truth. However, the pipeline had trouble with scenery on which the FCN was not trained, such as railroads, large industrial sites, and sport and recreational areas. Although this was to be expected, it stresses the need for a diverse training set when upscaling the methodology.

Figure 9 provides some more examples. The example on Row 2 illustrates the potential problem of disconnected line predictions (see red circles), causing the watershed algorithm to flood both adjacent roof parts. The example on Row 3 exemplifies the confusion of the FCN model when confronted with unseen scenery (i.e., not occurring in the training set), such as a tennis court. Moreover, it shows that the FCN did not seem to have fully learned to incorporate the absolute height information to distinguish roof from non-roof. The examples on Rows 1 and 5 are more comparable to the training set (many small clustered rooftops), and as such show more adequate predictions. Lastly, the example on Row 4 demonstrates the diversity in roof-part structure with which the workflow has to cope.

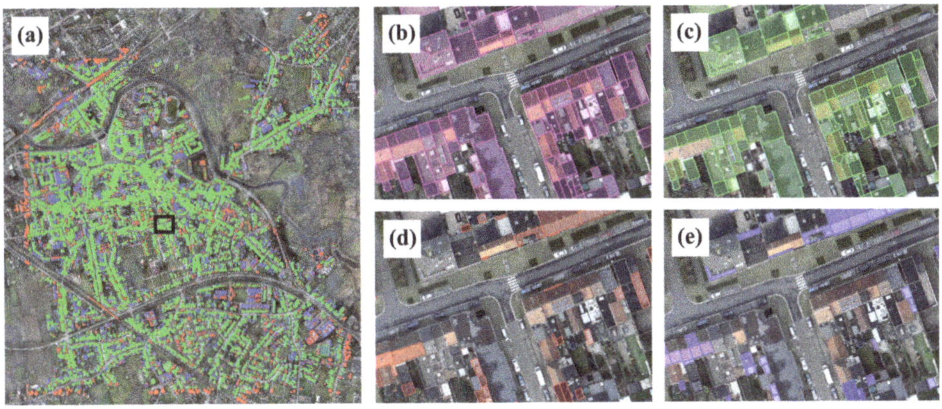

Figure 8. (**a**) Extracted polygon predictions for the whole test region. The black rectangle indicates the zoomed region in (**b–e**). (**b**) Roof-part polygon predictions. (**c**) True positives. (**d**) False positives. (**e**) False negatives.

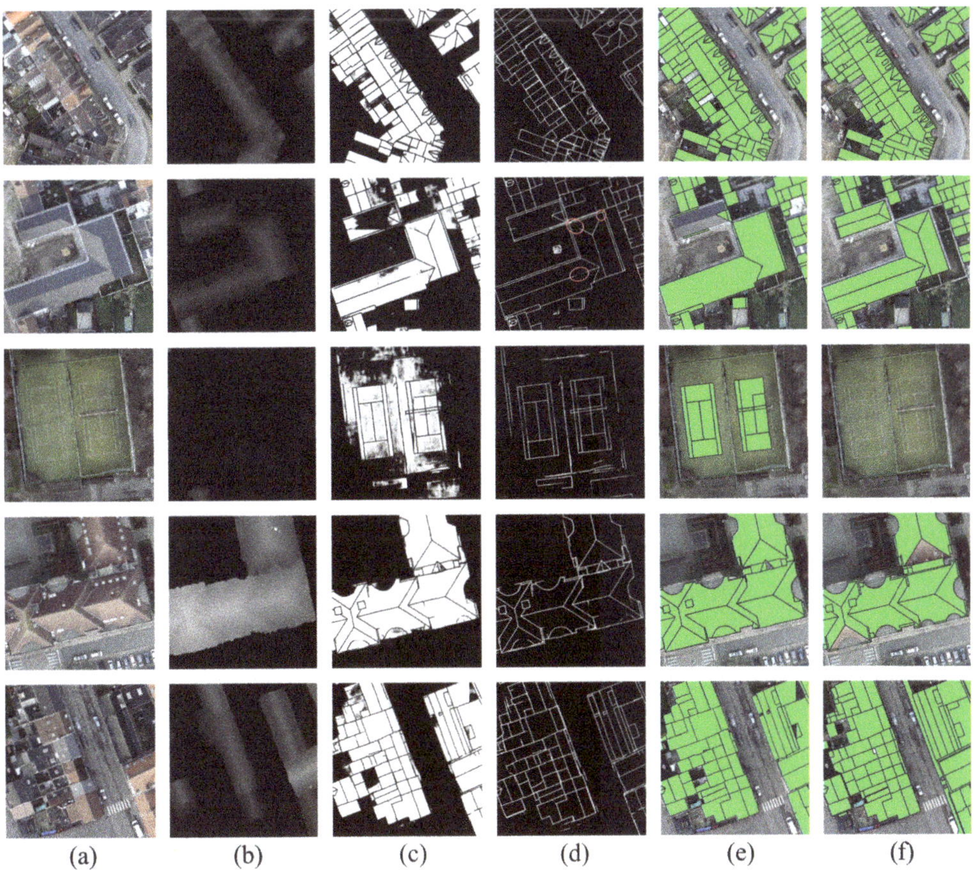

Figure 9. Illustrative results from the test set. Patches were 50 by 50 m (1667 × 1667 pixels). (**a**) RGB image. (**b**) DEM image. (**c**) Predicted roof-part objects. (**d**) Predicted roof-part edges. (**e**) Predicted roof-part polygons. (**f**) Ground truth.

4. Discussion

We here further discuss our results and suggest avenues for future research. First, the FCN model coped well with the complexity of the roof parts as caused by, among others, the diverse rooftop topology, roof structures (chimneys, solar panels, etc.), and shaded surfaces. However, difficulty was experienced with predicting large roof parts, i.e., roof parts with a larger height or width than the PS. One solution may be to sacrifice some order of spatial resolution in favor of spatial range. For example, upsampling patches with PS = 1024 pix from 0.03 to 0.12 m GSD would correspond with a spatial range increase from 31 to 123 m. Investigating this influence of the spatial resolution may be interesting future research. Additionally, the model produced errors on unseen landscape elements such railroads and recreational areas. To combat this, more diverse scenery and a fraction of the area not including the target could be included in the training set.

Furthermore, a pivotal component of the methodology is obtaining connected roof-part edges. If the predicted edges have disconnections, the watershed clustering is not able to distinguish between individual roof parts. Optimizing the hyper-parameters of the watershed algorithm or examining other clustering algorithms may enhance the CSE. Moreover, since this study only considered commonly used FCNs in the context of natural image segmentation, a gain in edge prediction quality may be achieved by employing more specialized FCNs, tailored for promoting edge connectivity. Further improvements may also be found by experimenting with various flavors of the loss function or by choosing a more suited validation metric than the IoU to evaluate edge segmentation quality.

Concerning the workflow input, using only RGB imagery produced significantly better results than using only height information, even though the majority of research on the automated extraction of roof parts is focused on the latter. Moreover, using RGB combined with height information did not necessarily lead to an improved final polygon map quality for all modelling approaches. Reasons may be the lower resolution of the DEM and the difficulty of generalizing absolute elevation. Of course, the DEM was simply concatenated as an additional input channel, while extensive fusion paradigms exist. Also, the FCN was initialized from Imagenet (i.e., natural RGB images)-pretrained weights, which may have led to an initialization bias towards RGB features. Investigating the optimal usage or fusion of the DEM data may, therefore, further improve performance.

Concerning the workflow output, considerable attention was paid to ensuring spatial continuity of the output by allowing for different patch sizes along the workflow and saving the intermediate results. However, additional quality improvement could likely be attained by running patchwise processing in each of the three workflow steps with some patch overlap, such that each pixel is seen from multiple perspectives. Furthermore, the predicted polygon map can be simplified using alternative algorithms. For instance, joint polygon simplification ensuring that adjacent polygons have shared vertices may improve the practicality of the map in a GIS.

Comparing a two single-task FCN model with a single multitask FCN model approach confirmed the benefits of the latter, in line with the existing literature [20]. More specifically, the multiclass model outperformed the multilabel approach, which may be explained by the idea that defining target classes as mutually exclusive constrains the problem complexity. However, the multitask behavior could be further exploited. For example, the object-center distance map could be added as an additional target channel to promote closed object shapes and a smooth conical object's probability surface. Roof corners could also be added as a semantic target, which could then be used in a Delaunay triangulation postprocessing step to extract more connected and regularly shaped polygons. Lastly, the multiclass FCN can be easily extended to a panoptic segmentation problem, i.e., roof materials could be incorporated as additional target classes by considering them as additional output channels and accordingly adapting the weighted loss function.

5. Conclusions

This paper proposed a fully automated workflow for large-scale roof-part polygon extraction from UHR orthoimagery (0.03 m GSD). The workflow comprised three steps: (1) An FCN was utilized for the semantic segmentation of roof-part objects and edges. (2) A bottom–up clustering algorithm was used, given the predicted roof-part edges, to derive individual roof-part clusters, where the predicted roof-part object area distinguish roof from non-roof. (3) The roof-part clusters were vectorized and simplified into polygons. By conducting an ablation study, various components of the workflow were optimized, leading to the conclusion that a single multiclass UNet with RGB + DEM input coupled with a clustering algorithm, and using the predicted roof area as decision threshold corresponded with the best quality among all experiments. The workflow was trained on the touristic medieval city of Brugge and tested on the more distant city of Lokeren (Belgium), which is an honest and challenging setup. Our final best prediction reached a panoptic quality of PQ = 54.8% (RQ = 62.6%, SQ = 87.7%). Notwithstanding the opportunity for further optimization, when trained, the pipeline can produce continuous and up-to-date vector maps of individual roof parts at UHR for entire municipalities within a matter of hours. Roof-part attributes such as inclination, orientation, area, and roof material may relatively be easily linked to polygon instances. Hence, combined with human validation, it can readily serve as a tool for (semi)automated geographic database construction, instrumental for urban monitoring, capacity assessment, and policy making.

Author Contributions: Conceptualization and methodology: W.A.J.V.d.B. and T.G.; software, validation, formal analysis, investigation, data curation, visualization and writing—original draft preparation: W.A.J.V.d.B.; resources, writing—review and editing, supervision, project administration and funding acquisition: T.G. All authors have read and agreed to the published version of the manuscript.

Funding: This research was funded by VLAIO.

Data Availability Statement: All produced code and data used within this study are property of Vansteelandt bv.

Acknowledgments: We thank Vansteelandt bv for preparing and providing the data used in this research. We further acknowledge Tanguy Ophoff for his technical support.

Conflicts of Interest: The authors declare no conflict of interest. The funders had no role in the design of the study; in the collection, analyses, or interpretation of data; in the writing of the manuscript, or in the decision to publish the results.

Abbreviations

The following abbreviations are used in this manuscript:

AT	Area threshold
CCE	Categorical cross-entropy
CRS	Coordinate reference system
CSE	Closed shape extraction
DEM	Digital elevation model
DV	Direct vectorization
EO	Earth observation
FCN	Fully convolutional network
GSD	Ground sampling distance
IoU	Intersection over union
PQ	Panoptic quality
PS	Patch size
RQ	Recognition quality
SQ	Segmentation quality
UHR	Ultrahigh resolution
WM	Watershed mask

References

1. Wu, A.N.; Biljecki, F. Roofpedia: Automatic mapping of green and solar roofs for an open roofscape registry and evaluation of urban sustainability. *Landsc. Urban Plan.* **2021**, *214*, 104167.
2. Hoeser, T.; Kuenzer, C. Object Detection and Image Segmentation with Deep Learning on Earth Observation Data: A Review—Part I: Evolution and Recent Trends. *Remote Sens.* **2020**, *12*, 1667. [CrossRef]
3. Hoeser, T.; Bachofer, F.; Kuenzer, C. Object Detection and Image Segmentation with Deep Learning on Earth Observation Data: A Review—Part II: Applications. *Remote Sens.* **2020**, *12*, 3053. [CrossRef]
4. Huang, J.; Zhang, X.; Xin, Q.; Sun, Y.; Zhang, P. Automatic building extraction from high-resolution aerial images and LiDAR data using gated residual refinement network. *ISPRS J. Photogramm. Remote Sens.* **2019**, *151*, 91–105. [CrossRef]
5. Wierzbicki, D.; Matuk, O.; Bielecka, E. Polish Cadastre Modernization with Remotely Extracted Buildings from High-Resolution Aerial Orthoimagery and Airborne LiDAR. *Remote Sens.* **2021**, *13*, 611. [CrossRef]
6. Chen, H.; Chen, W.; Wu, R.; Huang, Y. Plane segmentation for a building roof combining deep learning and the RANSAC method from a 3D point cloud. *J. Electron. Imaging* **2021**, *30*, 053022. [CrossRef]
7. Jochem, A.; Höfle, B.; Rutzinger, M.; Pfeifer, N. Automatic Roof Plane Detection and Analysis in Airborne Lidar Point Clouds for Solar Potential Assessment. *Sensors* **2009**, *9*, 5241–5262. [CrossRef]
8. Pohle-Fröhlich, R.; Bohm, A.; Korb, M.; Goebbels, S. Roof Segmentation based on Deep Neural Networks. In Proceedings of the 14th International Joint Conference on Computer Vision, Imaging and ComputerGraphics Theory and Applications (VISIGRAPP 2019), Prague, Czech Republic, 25–27 February 2019; pp. 326–333. [CrossRef]
9. Wang, X.; Ji, S. Roof Plane Segmentation from LiDAR Point Cloud Data Using Region Expansion Based L0Gradient Minimization and Graph Cut. *IEEE J. Sel. Top. Appl. Earth Obs. Remote Sens.* **2021**, *14*, 10101–10116. [CrossRef]
10. Zhou, Z.; Gong, J. Automated residential building detection from airborne LiDAR data with deep neural networks. *Adv. Eng. Inform.* **2018**, *36*, 229–241. [CrossRef]
11. ISPRS WGII/4. 2D Semantic Labeling—Vaihingen Data, 2013. Available online: https://www2.isprs.org/commissions/comm2/wg4/benchmark/2d-sem-label-vaihingen/ (accessed on 18 March 2021)
12. Maggiori, E.; Tarabalka, Y.; Charpiat, G.; Alliez, P. Can Semantic Labeling Methods Generalize to Any City? The Inria Aerial Image Labeling Benchmark. In Proceedings of the IEEE International Geoscience and Remote Sensing Symposium (IGARSS), Fort Worth, TX, USA, 23–28 July 2017.
13. Roscher, R.; Volpi, M.; Mallet, C.; Drees, L.; Wegner, J.D.; Dirk, J.; Semcity, W.; Roscher, R.; Volpi, M.; Mallet, C.; et al. SemCity Toulouse: A benchmark for building instance segmentation in satellite images. *ISPRS Ann. Photogramm. Remote Sens. Spat. Inf. Sci.* **2020**, *5*, 109–116. [CrossRef]
14. Sirko, W.; Kashubin, S.; Ritter, M.; Annkah, A.; Bouchareb, Y.S.E.; Dauphin, Y.; Keysers, D.; Neumann, M.; Cisse, M.; Quinn, J. Continental-Scale Building Detection from High Resolution Satellite Imagery. *arXiv* **2021**, arXiv:2107.12283.
15. Li, W.; He, C.; Fang, J.; Zheng, J.; Fu, H.; Yu, L. Semantic Segmentation-Based Building Footprint Extraction Using Very High-Resolution Satellite Images and Multi-Source GIS Data. *Remote Sens.* **2019**, *11*, 403. [CrossRef]
16. Xia, L.; Zhang, J.; Zhang, X.; Yang, H.; Xu, M.; Yan, Q.; Awrangjeb, M.; Sirmacek, B.; Demir, N. Precise Extraction of Buildings from High-Resolution Remote-Sensing Images Based on Semantic Edges and Segmentation. *Remote Sensing* **2021**, 13, 3083. [CrossRef]
17. He, K.; Gkioxari, G.; Dollár, P.; Girshick, R. Mask R-CNN. *IEEE Trans. Pattern Anal. Mach. Intell.* **2017**, *42*, 386–397.
18. Marmanis, D.; Schindler, K.; Wegner, J.D.; Galliani, S.; Datcu, M.; Stilla, U. Classification with an edge: Improving semantic image segmentation with boundary detection. *ISPRS J. Photogramm. Remote Sens.* **2018**, *135*, 158–172.
19. Wu, G.; Guo, Z.; Shi, X.; Chen, Q.; Xu, Y.; Shibasaki, R.; Shao, X. A Boundary Regulated Network for Accurate Roof Segmentation and Outline Extraction. *Remote Sens.* **2018**, *10*, 1195. [CrossRef]
20. Diakogiannis, F.I.; Waldner, F.; Caccetta, P.; Wu, C. ResUNet-a: A deep learning framework for semantic segmentation of remotely sensed data. *ISPRS J. Photogramm. Remote Sens.* **2020**, *162*, 94–114.
21. Hosseinpour, H.; Samadzadegan, F.; Javan, F.D. A Novel Boundary Loss Function in Deep Convolutional Networks to Improve the Buildings Extraction From High-Resolution Remote Sensing Images. *IEEE J. Sel. Top. Appl. Earth Obs. Remote Sens.* **2022**, *15*, 4437–4454. [CrossRef]
22. Li, Q.; Mou, L.; Hua, Y.; Sun, Y.; Jin, P.; Shi, Y.; Zhu, X.X. Instance segmentation of buildings using keypoints. In Proceedings of the International Geoscience and Remote Sensing Symposium (IGARSS), Waikoloa, HI, USA, 26 September–2 October 2020; pp. 1452–1455.
23. Li, Z.; Xin, Q.; Sun, Y.; Cao, M. A deep learning-based framework for automated extraction of building footprint polygons from very high-resolution aerial imagery. *Remote Sens.* **2021**, *13*, 3630. [CrossRef]
24. Chen, Q.; Wang, L.; Waslander, S.L.; Liu, X. An end-to-end shape modeling framework for vectorized building outline generation from aerial images. *ISPRS J. Photogramm. Remote Sens.* **2020**, *170*, 114–126. [CrossRef]
25. Poelmans, L.; Janssen, L.; Hambsch, L. *Landgebruik en Ruimtebeslag in Vlaanderen, Toestand 2019, Uitgevoerd in Opdracht van het Vlaams Planbureau voor Omgeving*; Vlaams Planbureau voor Omgeving: Brussel, Belgium, 2021.
26. Ronneberger, O.; Fischer, P.; Brox, T. U-Net: Convolutional Networks for Biomedical Image Segmentation. *IEEE Access* **2015**, *9*, 16591–16603.

27. Tan, M.; Le, Q.V. EfficientNet: Rethinking Model Scaling for Convolutional Neural Networks. In Proceedings of the 36th International Conference on Machine Learning, ICML 2019, Long Beach, CA, USA, 9–15 June 2019; pp. 10691–10700.
28. Fei-Fei, L.; Deng, J.; Li, K. ImageNet: Constructing a large-scale image database. *J. Vis.* **2010**, *9*, 1037. [CrossRef]
29. Roy, A.G.; Navab, N.; Wachinger, C. Recalibrating Fully Convolutional Networks with Spatial and Channel 'Squeeze & Excitation' Blocks. *IEEE Trans. Med. Imaging* **2018**, *38*, 540–549.
30. Yakubovskiy, P. Segmentation Models Pytorch. 2020. Available online: https://github.com/qubvel/segmentation_models.pytorch (accessed on 12 January 2022).
31. Paszke, A.; Gross, S.; Massa, F.; Lerer, A.; Bradbury, J.; Chanan, G.; Killeen, T.; Lin, Z.; Gimelshein, N.; Antiga, L.; et al. PyTorch: An Imperative Style, High-Performance Deep Learning Library. *NeurIPS* **2019**. [CrossRef]
32. Chen, Y.; Carlinet, E.; Chazalon, J.; Mallet, C.; Dumenieu, B.; Perret, J. Vectorization of historical maps using deep edge filtering and closed shape extraction. In Proceedings of the 16th International Conference on Document Analysis and Recognition (ICDAR'21), Lausanne, Switzerland, 5–10 September 2021; pp. 510–525. [CrossRef]
33. van der Walt, S.; Schönberger, J.L.; Nunez-Iglesias, J.; Boulogne, F.; Warner, J.D.; Yager, N.; Gouillart, E.; Yu, T. scikit-image: Image processing in Python. *PeerJ* **2014**, *2*, e453. [CrossRef]
34. Shi, W.; Cheung, C.K. Performance Evaluation of Line Simplification Algorithms for Vector Generalization. *Cartogr. J.* **2006**, *43*, 27–44. [CrossRef]
35. Kirillov, A.; He, K.; Girshick, R.; Rother, C.; Dollar, P. Panoptic Segmentation. In Proceedings of the IEEE Computer Society Conference on Computer Vision and Pattern Recognition, Long Beach, CA, USA, 15–20 June 2019; pp. 9396–9405. [CrossRef]
36. Informatie Vlaanderen. *Large-Scale Reference Database (LRD)*; 2021. Available online: https://overheid.vlaanderen.be/en/producten-diensten/large-scale-reference-database-lrd (accessed on 16 March 2022).
37. Chen, L.C.; Zhu, Y.; Papandreou, G.; Schroff, F.; Adam, H. Encoder-decoder with atrous separable convolution for semantic image segmentation. *Proc. Eur. Conf. Comput. Vis. (ECCV)* **2018**, 801–818. [CrossRef]
38. Lin, T.Y.; Dollar, P.; Girshick, R.; He, K.; Hariharan, B.; Belongie, S. Feature Pyramid Networks for Object Detection. In Proceedings of the 2017 IEEE Conference on Computer Vision and Pattern Recognition (CVPR), Honolulu, HI, USA, 21–26 July 2017; pp. 936–944. [CrossRef]
39. Fan, T.; Wang, G.; Li, Y.; Wang, H. Ma-net: A multi-scale attention network for liver and tumor segmentation. *IEEE Access* **2020**, *8*, 179656–179665. [CrossRef]
40. Li, H.; Xiong, P.; An, J.; Wang, L. Pyramid Attention Network for Semantic Segmentation. In Proceedings of the British Machine Vision Conference 2018, BMVC 2018, Newcastle, UK, 3–6 September 2018.
41. Zhao, H.; Shi, J.; Qi, X.; Wang, X.; Jia, J. Pyramid scene parsing network. In Proceedings of the 30th IEEE Conference on Computer Vision and Pattern Recognition, CVPR 2017, Honolulu, HI, USA, 21–26 July 2017; pp. 6230–6239.
42. Zhou, Z.; Siddiquee, M.M.R.; Tajbakhsh, N.; Liang, J. UNet++: Redesigning Skip Connections to Exploit Multiscale Features in Image Segmentation. *IEEE Trans. Med. Imaging* **2020**, *39*, 1856–1867.

remote sensing

MDPI

Article

Research on Self-Supervised Building Information Extraction with High-Resolution Remote Sensing Images for Photovoltaic Potential Evaluation

De-Yue Chen [1,2], Ling Peng [1,2,*], Wen-Yue Zhang [1,2], Yin-Da Wang [1,3] and Li-Na Yang [1,2]

1 Aerospace Information Research Institute, Chinese Academy of Sciences, Beijing 100094, China
2 College of Resources and Environment (CRE), University of Chinese Academy of Sciences, Beijing 100049, China
3 School of Electronic, Electrical and Communication Engineering (EECE), University of Chinese Academy of Sciences, Beijing 100049, China
* Correspondence: pengling@aircas.ac.cn

Abstract: With the rapid development of the energy industry and the growth of the global energy demand in recent years, the development of the photovoltaic industry has become increasingly significant. However, the development of the PV industry is constrained by high land costs, and land in central cities and industrial areas is often very expensive and unsuitable for the installation of PV equipment in large areas. With this background knowledge, the key to evaluating the PV potential is by counting the rooftop information of buildings, and an ideal solution for extracting building rooftop information is from remote sensing satellite images using the deep learning method; however, the deep learning method often requires large-scale labeled samples, and the labeling of remote sensing images is often time-consuming and expensive. To reduce the burden of data labeling, models trained on large datasets can be used as pre-trained models (e.g., ImageNet) to provide prior knowledge for training. However, most of the existing pre-trained model parameters are not suitable for direct transfer to remote sensing tasks. In this paper, we design a pseudo-label-guided self-supervised learning (PGSSL) semantic segmentation network structure based on high-resolution remote sensing images to extract building information. The pseudo-label-guided learning method allows the feature results extracted by the pretext task to be more applicable to the target task and ultimately improves segmentation accuracy. Our proposed method achieves better results than current contrastive learning methods in most experiments and uses only about 20–50% of the labeled data to achieve comparable performance with random initialization. In addition, a more accurate statistical method for building density distribution is designed based on the semantic segmentation results. This method addresses the last step of the extraction results oriented to the PV potential assessment, and this paper is validated in Beijing, China, to demonstrate the effectiveness of the proposed method.

Keywords: remote sensing building extraction; building photovoltaic; self-supervised learning; semantic segmentation

Citation: Chen, D.-Y.; Peng, L.; Zhang, W.-Y.; Wang, Y.-D.; Yang, L.-N. Research on Self-Supervised Building Information Extraction with High-Resolution Remote Sensing Images for Photovoltaic Potential Evaluation. *Remote Sens.* **2022**, *14*, 5350. https://doi.org/10.3390/rs14215350

Academic Editors: Mohammad Awrangjeb, Qin Yan, Beril Sirmacek, Jiaojiao Tian and Nusret Demir

Received: 9 September 2022
Accepted: 20 October 2022
Published: 25 October 2022

1. Introduction

With developments in society in recent years, global energy demand is gradually increasing every day and most consumption still relies on fossil fuels [1]. The use of fossil energy on a global scale has caused a series of problems such as melting glaciers and global warming. It is very important to change the energy structure and gradually replace current fossil energy with renewable energy [2]. It is also important to vigorously develop renewable energy for promoting green and low-carbon development and accelerating the construction of the ecological civilization. Among the many renewable energy sources, solar power generation is one of the most effective [3]. Solar power generation has the advantages of being inexhaustible, environmentally friendly, and not limited by geographical conditions [2]. The International Energy Agency predicts that solar photovoltaic (PV) power generation will become the main energy source and will account for 20–50% of global power generation by 2050 [4].

Solar power generation is mainly realized through crystalline silicon photovoltaic panels, which is also called photovoltaic power generation. Because it can be closely integrated with buildings and close to where electricity is used, it can reduce the transmission pressure on the power grid. However, in our investigation, we found that solar power generation depends on the intensity of sunlight and is greatly affected by extreme weather, which results in instability in both the power generation and supply time. This situation makes it very hard to transmit power to the grid. Therefore, it is necessary to evaluate the potential of PV system distribution and realize the unified dispatch of PV power and deploy PV equipment in areas with high electricity consumption to achieve the effect of "self-generation and self-consumption". However, in areas with high power consumption, the deployment of PV systems is often accompanied by huge land costs. Distributed PV systems, which are necessary, are mainly installed on the roof of a building and are connected directly to a low-voltage distribution network. On the one hand, they have the advantages of proximity to the customer side, on-site consumption, and reduced transportation costs [5]. The roof of the building should have a strong load-bearing capacity and the generated electricity can be used directly in the building, reducing the costs of equipment and transportation dispatch. On the other hand, the building should have a long service life and PV equipment installed on the roof of a building can facilitate the recovery of installation costs; in addition, the roof of a building is the best platform to carry PV equipment [5]. Therefore, effective access to information about a building's roof is the key to assessing the PV potential of buildings [6].

Conventional building information acquisition methods mainly rely on land surveys, which are difficult to acquire and have poor timeliness. Remote sensing satellite technology, which provides extensive access to ground information, has become a popular way to assess a building's rooftop PV potential. In 2014, Flavio Borfecchia et al. used LiDAR technology and gis modeling tools to estimate urban roof levels under the three-dimensional view angle; this method was expensive but achieved significant evaluation results [7]. In 2016, Wong, MS et al. used remote sensing technology and geographic information system (GIS) technology to estimate the potential of photovoltaic power generation for the city of Hong Kong [8]. In 2018, Xiaoyang Song et al. conducted an assessment of building rooftop potential based on Google Images and global DEM data and calculated the annual rooftop photovoltaic power generation of buildings in Chaoyang District, Beijing, China. Specifically, they classified buildings into five types and further considered the tilt angle of PV panel installations for the PV potential assessment analysis, which had quite good inspirational and applied implications [9]. In 2019, Arti Tiwari et al. used orthophoto and LiDAR data with an object-oriented method to realize the evaluation of the solar energy yield in terms of solar irradiance in pixels in a specific period [10]. In 2020, Blazquez, J and Vittorio, M. used nighttime satellite imagery to assess the residential solar rooftop potential in Saudi Arabia [11]. In 2022, Huang, Xiaoxun et al. estimated the rooftop solar power generation potential in western Aichi Prefecture, Japan, based on the use of LiDAR data and AW3D technology [12]. It can be seen that a large number of researchers have chosen to use LiDAR data or high-resolution remote sensing data to extract building rooftop data. LiDAR data provides the possibility to obtain the building height and slope information for a fine-grained assessment of the PV potential. However, the cost of using LiDAR data is too high considering the large-scale statistical analysis. In contrast, using high-resolution remote sensing data is a cost-effective solution. However, when faced with a large amount of high-resolution remote sensing image data, it is time-consuming and laborious to use manual extraction of building information, and the value of remote sensing images is not fully exploited.

In recent years, the rapid development of deep learning algorithms has allowed for the extraction of spatial and spectral features at the same time so they are also widely used for the extraction of building data. Based on the adversarial neural network, Li Xiang et al. jointly trained a deep convolutional neural network (generator) and an adversarial discriminant network for the robust segmentation of building rooftops in remote sensing images and successfully solved the spatial inconsistency problem in classification [13]. Tian,Tian proposed an urban area target detection algorithm based on DCNNs, which still achieved good extraction results while maintaining the detection speed. Specifically, it used visual words based on DCNNs to extract feature information, thus realizing the extraction of data on urban areas without labeling samples; however, it was not accurate for specific buildings and its practical application accuracy was not good [14]. Zeng, Yifu et al. conducted experiments based on GF2 data and successfully realized the rapid extraction of building information. Specifically, their proposed BR-Net model used multi-task learning for segmentation and contour extraction to overcome limitations such as the unavailability of edge information [15]. Its effectiveness also illustrated the potential of multi-task learning. Based on the multi-task learning algorithm, Hui and Jian conducted building extraction experiments on the Massachusetts dataset and achieved good experimental results. The highlight of their article was the merging of distance representation into a multi-tasking framework as an auxiliary task, forcing the shared encoder to implicitly capture the features of the building structure [16]. However, the above methods were all supervised classification, and the effectiveness of the methods depended largely on the huge training samples. When the building information is extracted in a large area, due to the limitation of the samples, the extraction effect and accuracy of the model will be greatly affected. In recent years, many famous datasets have been proposed in remote sensing building information extraction, such as WHU [17], DOTA [18], Massachusetts [19], etc. These have largely advanced the rapid development of remote sensing information extraction technology. However, the applicability of the models obtained from the training of labeled data is often unsatisfactory due to the differences in spatial resolution, acquisition date, image location, and other elements.

Self-supervised learning methods, which have developed rapidly in recent years, have been evolving by automating the task of learning features to achieve the effective utilization of unlabeled samples. The latest comparative learning methods have achieved good results in some tasks and are very suitable for applications in building information extraction. Studying the application of self-supervised methods in remote sensing images is expected to provide a use for the huge amount of data that cannot be used in remote sensing, thus improving the generalization performance of deep learning models in the field of remote sensing information extraction. Among the many self-supervised learning methods, there are two main methods commonly used to train visual representations; one is a self-supervised learning method based on a reconstructed loss function and is often called the representational learning method [20,21], and the other is a self-supervised learning method that measures the contrastive loss of images and is often called the contrastive learning method. Among them, contrastive learning is the most state-of-the-art learning method in most cases [22–25]. Because of the good generalization of contrastive learning, it can also be developed more rapidly compared to representational learning methods. However, good representational learning tasks are more closely related to the target task so many breakthroughs in self-supervised learning have resulted from the discovery of representational learning tasks, although some researchers tend to use pretext tasks to extract features. In order to learn good representations, people have explored a variety of pretext tasks. A pretext task is meant to be a network task that provides pre-trained parameters, which can generally generate samples automatically without human intervention. Examples include colorization [26], contextual autoencoders [27], inpainting [21], spatial puzzles [28], and discriminative orientation [29]. Today, these self-supervised learning methods are collectively known as representational learning (as distinguished from contrast learning) methods and they can achieve very good results under specific tasks. With the development of self-supervised algorithms, the application of self-supervised learning in remote sensing image information extraction is also gradually emerging. Guo and Qing proposed a method for the automatic extraction of road centerlines from high-resolution remote sensing images based on a self-supervised learning framework. Good extraction results were achieved without the manual selection of training samples or optimization steps such as removing non-road areas [30]. Dong, Huihui et al. proposed a new self-supervised approach representing a time-based learning approach to predict remote sensing image change detection. The main idea of the algorithm was to convert two satellite images into a more consistent feature representation through the self-supervision mechanism without any additional computation of semantic supervision [31], thus reducing the propagation error of the final detection results. Li, Wenyuan et al. [32] designed three different pretext tasks to learn a multi-layer network structure simultaneously. The network was trained with a large amount of unlabeled data, fine-tuned with a small number of labeled segmentation datasets, and only used 10–50% of the labeled samples to achieve the original segmentation effect. In summary, it can be seen that self-supervised learning techniques have developed rapidly in recent years and have achieved excellent results in various tasks through the use of unlabeled data. Their success can be attributed to two aspects: the efficient use of unsupervised samples and the proper selection of the pretext task. Its essence involves the fusion of low-level features with high-level features of the image, and numerous experiments have shown that the fusion of these features can achieve even better segmentation results.

However, compared to general natural images (images taken by cameras, mobile cameras, surveillance cameras, and other ground equipment), satellite-derived remote sensing images have random views, more complex backgrounds, richer spectral features, and texture details. In addition, the above self-supervised methods designed for natural images (photos taken by cameras, mobile cameras, surveillance cameras, and other ground equipment) do not fully consider the characteristics of remote sensing images. This migration of features obtained through pretext task learning of remote sensing targets may not have the expected effect. In addition, the evaluation of building rooftop PV power potential involves a large amount of building information extraction, and traditional deep learning methods require a large number of building samples to ensure that the model can have good generalization ability. The extraction structures between deep learning tasks can be used mutually so that the self-supervised model can use pre-training to obtain a priori information and its extraction effect will be better than random initialization. However, existing self-supervised learning methods have a gap between the pretext task and the target task, which is usually more generalized for the pretext task and more aggregated for the features required by the target task; by using practical self-supervised methods for pre-training, it is often difficult to obtain good experimental results [33].

To address the above issues, we propose a pseudo-label-guided self-supervised learning method (called the PGSSL method), which utilizes pseudo-label learning to guide the pretext tasks. In detail, feature layer sharing is used to achieve the mutual utilization of the feature extraction part for the task interaction between different tasks and the utilization of unlabeled data. The effectiveness of our proposed structure is demonstrated by comparison experiments with different sample proportions and ablation experiments with different structures. The main contributions of this paper are as follows:

- In this paper, a self-supervised learning framework for semantic segmentation is proposed considering the characteristics of remote sensing images, and it is demonstrated that a large number of unlabeled remote sensing images can be effectively used to train the network. For the self-supervised learning task, this paper designs a self-supervised structural method for multi-task learning called the PGSSL method. It improves the performance of the semantic segmentation task by guiding feature extraction with a pseudo-labeling task.
- The proposed method is validated on a public dataset (EA Dataset) and an independently constructed Beijing dataset (BJ Dataset), comparing the performance of algorithms under different sample conditions and verifying the good performance of

the self-supervised learning method with a limited sample size. Finally, our method achieves better results than the ImageNet pre-training in the experiments.
- In this paper, we further analyze the distribution of buildings based on the semantic segmentation of the buildings to obtain a more accurate picture of the suitability of building rooftops for the installation of PV equipment.

The full text is organized as follows. Section 2 introduces the self-supervised learning strategy based on pseudo-label guidance and the method of using the semantic segmentation results for the PV potential assessment. Among them, the self-supervised approach is introduced and includes three modules: a pseudo-label learning module, an image inpainting module, and a comparative learning module. Section 3 presents the datasets used for the experiments and the formulas used for the accuracy assessment. The dataset section presents detailed information on the Beijing dataset and the public dataset. It includes the study area, data sources, and the method of obtaining unlabeled data. Section 4 presents three comparative experiments of the proposed method in this paper, including the overall algorithm effect comparison, the sample proportion experiment, and the ablation experiment. Section 5 summarizes and discusses the paper, discusses and analyzes the phenomena that were observed during the experiments, and provides an outlook on some directions that can be pursued.

2. Methods

In this paper, we design a multi-task learning model based on pseudo-label learning. This paper designs contrastive learning and image inpainting tasks to extract features from unlabeled data, where the image restoration task helps the network to learn low-level features and the contrastive learning task enables the network to learn high-level image features. In addition, this paper designs pseudo-label learning to ensure that the learned features can be adapted to the final target task extraction. The relevant code can be found at https://github.com/Chendeyue/pytorch-ssl-building_extract, accessed on 7 August 2022.

The overall structure of the work is shown in Figure 1. It can be seen that the main body consists of three parts: pseudo-label training, image inpainting, and contrastive learning, which share a common feature extraction layer. In the two parts of the pseudo-label training and image restoration, the UNet network is chosen as the structure of the intermediate implementation, and the skip-connection structure of the UNet framework is used to achieve the full utilization of the features. In the final classification layer, pseudo-label learning classifies the targets into two classes corresponding to the probabilities of buildings and non-buildings, whereas the image inpainting task outputs three classes corresponding to the three bands of the newly generated images. The decoding parts of both are independently constructed networks that do not share network parameters. The contrastive learning task part mainly adopts the idea proposed in BYOL to design a twin network structure with a momentum update. For a set of input data-enhanced images, two feature vectors are generated after passing through the twin network separately. The two vectors generated by the twin network are cross-compared separately and a similarity loss function is constructed, as shown in Figure 1, to achieve comparative learning of the network features. Finally, the three unsupervised task loss functions are combined to form a multi-task learning structure to accomplish a self-supervised learning task. Its loss function is shown in Equation (1).

$$L_{final} = \lambda_1 * L_{pseudo} + \lambda_2 * L_{contrast} + \lambda_3 * L_{inpainting} \tag{1}$$

where λ is a hyperparameter used to balance the magnitude difference between three types of loss and $\lambda_1 = \lambda_2 = \lambda_3 = \frac{1}{3}$. L_{pseudo}, $L_{contrast}$, and $L_{inpainting}$ denote the loss of several parts; details can be found in Equations (4), (6), and (8). Since the three loss functions have a similar status and approximate value range, we take the mean value to balance the three tasks.

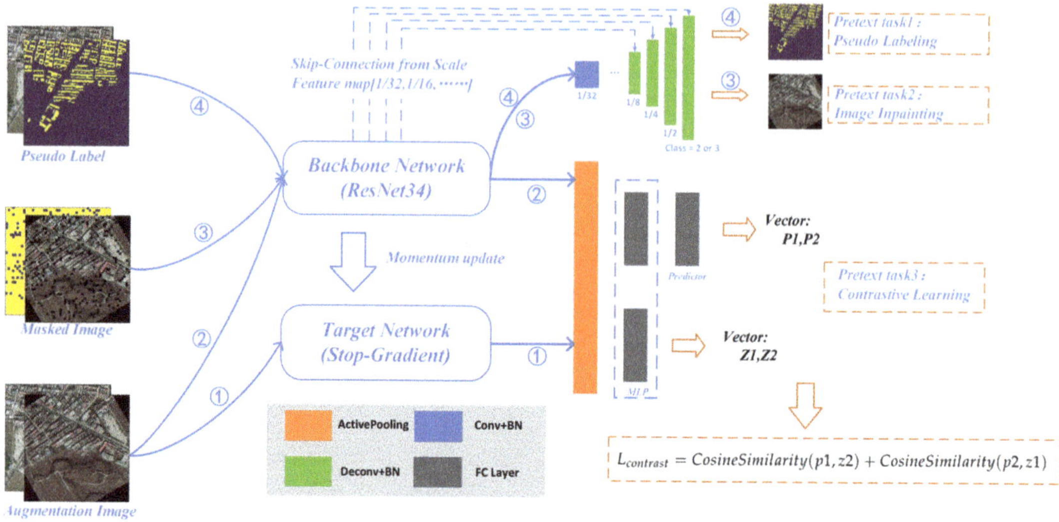

Figure 1. Overview of the proposed pseudo-label-guided self-supervised learning method.

2.1. Weighted Training on Pseudo-Labeled Data

Pseudo-label learning is a semi-supervised learning method that has emerged in recent years. The core idea is to build a model using existing knowledge and use it to predict unlabeled samples to obtain the pseudo label. The pseudo label is trained using prior knowledge to learn information from unlabeled data. Finally, the results with relatively high confidence in prediction are used as labels to ensure the accuracy of the labels.

In this paper, the pre-training trained network is called the teacher network and the pseudo-label learning channel to be trained is called the student network. These networks are structurally similar but are independent network models and their parameters are not shared. Both the teacher network and the student network have unlabeled augmented images as input, where the teacher network is the trained model and the student network relies on the labeled training generated by the teacher network. In addition, the parameters of the teacher network are not updated when training the student network. The training process is as follows: first, we train a teacher network using the enhanced labeled dataset, then, the teacher network predicts the input image as the pseudo-label, named q_j, and finally, the result with high confidence in q_j is compared with the result $\hat{q}_j$ of our student network to be trained. This completes the learning of the pseudo-label part, which is done in parallel with the other two tasks, and with the guidance of pseudo-label learning, the extracted feature layer parameters can be more suitable for migration to the target task. The final loss function is shown in Equation (4):

$$H(\hat{q}_j, q_j) = -\sum_{b=1}^{C} \hat{q}_j(b) \log q_j(b) \tag{2}$$

$$L_x = \frac{1}{N} \sum_{j=1}^{N} (L(max(q_j) \geq \tau) * H(\hat{q}_j, q_j)) \tag{3}$$

$$L_{pseudo} = \frac{1}{2}(L_{x1} + L_{x2}) \tag{4}$$

Here, $q_j = P_t(y|x_j)$ and $\hat{q}_j = argmax(P_s(y|x_j))$, which denote the prediction results of the output of the teacher network that has been fitted and the student network that is being trained, respectively, where τ is the threshold hyperparameter used to filter unsupervised samples. We keep the corresponding pseudo labels only when the maximum predicted probability is higher than the threshold, which is usually taken as 0.5. C is the number of categories, N is the number of samples, and H is the entropy value function. $L(max(q_j) \geq \tau)$ represents the probability that the prediction threshold exceeds a certain value, usually equal to 1 or 0. The loss function, called L_x, is calculated as the result of inputting image x into the teacher network and the student network, respectively, whereas L_{x1} and L_{x2} denote the loss obtained by two different data enhancement methods.

2.2. Contrastive Learning Task

The contrastive learning task achieves the convergence of the network by comparing the similarity of images. The operation is as follows: first, the images are transformed with data augmentation and then, the similarity between the transformed image results is compared to construct a loss function. Theoretically, the higher the similarity of the features in an image, the smaller the loss. This allows all similar objects to be located in adjacent positions in the feature space, whereas dissimilar objects are located in non-adjacent regions.

In recent years, researchers have proposed effective comparative learning methods such as SimCLR [24], MOCO [22], MOCOv2 [34], BYOL [25], etc. In most comparative methods, we must compare each sample to many other negative samples. However, it makes training very unstable and increases the systematic bias of the dataset. The proposal of the BYOL method [25] provides a proper solution to this problem. The BYOL method does not rely on negative samples but only uses similar sample representation types to construct a loss function. The final loss function is shown in Equation (6):

$$CosineSimilarity(p,z) = \frac{\sum_{i=1}^{B} p_i z_i}{\sqrt{\sum_{i=1}^{B} p_i \sum_{j=1}^{B} z_j}} \tag{5}$$

$$L_{contrast} = CosineSimilarity(p1, z2) + CosineSimilarity(p2, z1) \tag{6}$$

Here, $L_{contrast}$ represents the contrastive loss function and B is the dimension of the vector. CosineSimilarity represents the cosine similarity between two vectors. The outputs of p_i and z_i are shown in Figure 1, representing the outputs of the online network and target network, respectively, whereas $p1$ and $p2$ represent the global feature vectors inputted by two different data augmentation methods and $p1$ and $z2$ and $p2$ and $z1$. This alternate combination method for the similarity verification brings greater flexibility to the model, and the main feature extraction network used in the comparative learning process is shared with the other two tasks to ensure that the extracted features are closer to the direction of the target task.

2.3. Inpainting Task

The image inpainting pretext task itself is used to restore the missing parts of an image based on the existing information in the image. However, in the process of image inpainting, because the conventional loss function is generally adjusted globally for the image, this will make the inpainting task only supplement the image information that is similar to the global image and ignore the use of local texture information. To solve this problem, this paper adopts SSIM [35] (structural similarity index) to construct the inpainting loss function, but the global variance of the image fluctuates greatly in the actual operation so the SSIM value is only calculated in one window and then the mean value is taken globally. The specific form is shown in Equation (8):

$$SSIM(x,y) = \frac{(2\mu_x \mu_y + C_1)(2\sigma_x y + C_2)}{(\mu_x^2 + \mu_y^2 + C_1)(\sigma_x^2 + \sigma_y^2 + C_2)} \tag{7}$$

$$L_{inpainting} = MSSIM(x,y) = \frac{1}{N} \sum_{j=1}^{N} SSIM(x_j, y_j) \tag{8}$$

Here, μ_x and μ_y represent the mean values of the input image and the output image; σ_{xy},σ_x, and σ_y represent the covariance and variance of the two images; and C_1 and C_2 are constants. x and y represent the images before and after restoration, respectively, and MSSIM is a better final loss function for the image inpainting task.

2.4. Analysis of Photovoltaic Potential Area Based on Building Semantic Segmentation

The rooftop area of a building is the most important influencing factor for the PV potential assessment, and accurate rooftop extraction results are the basis for the PV potential analysis. However, in practice, the supporting facilities for PV equipment installation, as well as its scale, can also be a constraint to the building's potential. Even if an area has a large building area but the rooftop area of each building is small and the buildings are scattered, the high cost of the supporting facilities would mean that the buildings in the area would not be considered a distributed PV installation area. The semantic segmentation results can only evaluate the installed area of the building and not the distribution. As shown in Figure 2, for the semantic segmentation of the connected domain analysis results, where there is a relatively large degree of separation between buildings, such as in region A and region C, the suitability of PV installation on buildings can be better assessed; however, in the case of the buildings in region B, the actual installation for the patches of private houses is more complex, but it may be identified as an area with better PV potential due to its larger area.

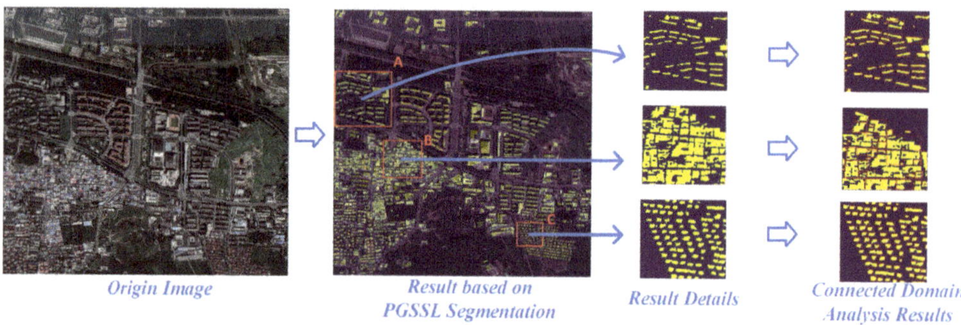

Figure 2. Overview of Regional PV Potential Analysis.

Therefore, in practical applications, when using the semantic segmentation results to evaluate the distribution of buildings, it is necessary to further segment the connected buildings and remove some noise blocks. In this paper, the morphological opening operation is used to exclude some noise spots from the extraction results and the confidence characteristics of the deep learning prediction are used to separate the connected building patches. The specific operation is shown in Figure 3. First, starting from the original image, by controlling the confidence level of the prediction of the original image, the conventional confidence level results for building area extraction are more accurate and the high confidence level results for the separation degree between buildings are obtained.

Then, we perform a morphological opening operation on the conventional results to obtain the building distribution results with the noise patches removed, directly extract the center point of the patch for the high-confidence results, and perform the superposition analysis of the extracted center point results and the opening operation results. After removing the center point of a small part of the noise, a relatively accurate building distribution is obtained.

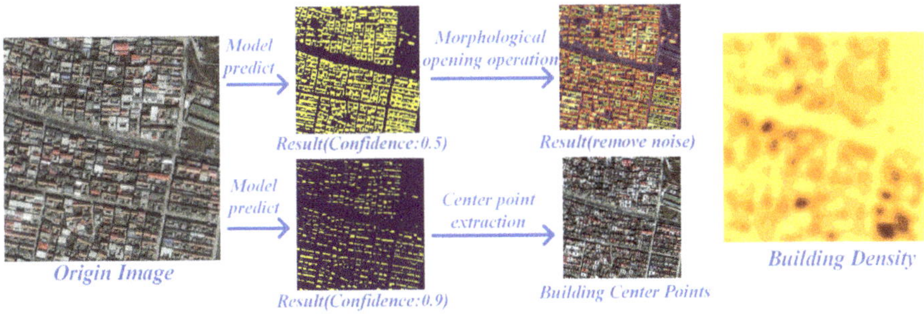

Figure 3. This group of images represents the process diagram for post-processing of building semantic segmentation for photovoltaic potential assessment.

3. Dataset and Evaluation Metrics

3.1. Dataset

In order to fully illustrate the generalization performance of the proposed method in experiments in various regions, this paper independently produced urban and rural datasets in Beijing for the experiments called the BJ Dataset (Beijing Buildings Segmentation Dataset). The remote sensing data used in the experiments include Google data and SV-1 data. Located in Haidian District, Xiangshan District, and Yajishan District of Pinggu District, Beijing, the average resolution was about 1 m. The locations of the training areas and the basic conditions of the images are shown in Figure 4A–D. After using ArcGIS to label all the buildings in the target area, the training area images were cut into a 384 × 384 size and divided into the training set and test set, according to a ratio of 4:1, and finally, 797 training sets and 202 validation sets were obtained.

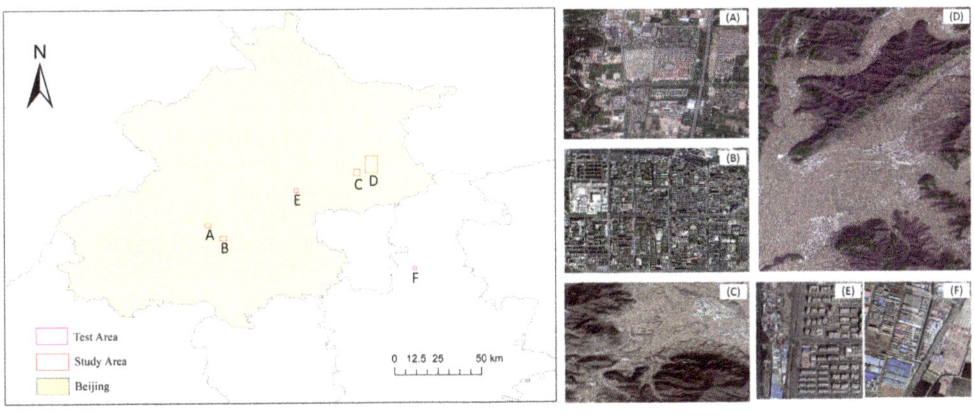

Figure 4. The locations of the Beijing buildings training dataset and test dataset. (**A–F**) represents the training and testing area used in the construction of BJ dataset.

This paper also collected some regional data in Beijing and Tianjin for experimental testing. The data used were all Google receipts and their locations and images are shown in Figure 4E,F. After the images were manually annotated, they were also cropped to a 384 × 384 size, resulting in 16 slice results in the test area in Beijing and 34 slice results in the test area in Tianjin. The schematic diagram of one of the slice results in the Tianjin area used for verification is shown in Figure 5.

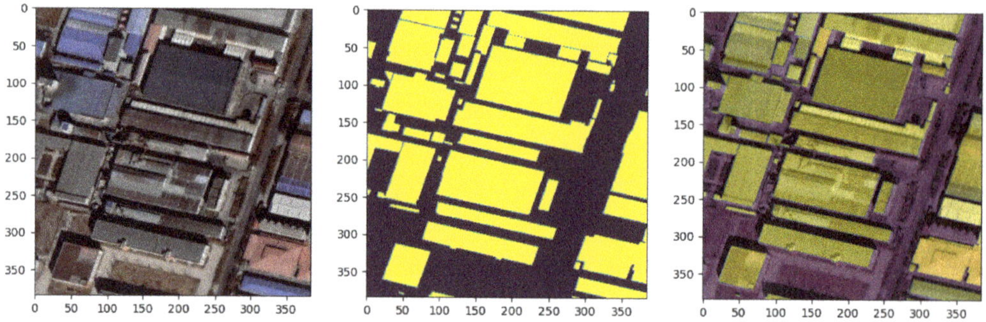

Figure 5. The slice results of Tianjin validation set.

The construction of the dataset also included the acquisition of unlabeled samples because in a whole remote sensing image, if the specific features of buildings are very small when using all the tiles of the region for training, the construction of unsupervised samples of buildings needs to further eliminate the tile images that do not contain buildings. Therefore, this paper first collected the Beijing area 1 m-resolution Google images and cropped them to 384 × 384-size tiles to obtain a total of 21,417 unlabeled building images. Subsequently, a building extraction model was trained to predict the collected image tiles using the above-labeled samples and the tiles without buildings in the prediction results were excluded. Finally, we excluded a few images with relatively poor quality by manual inspection.

In addition, in order to fully verify the actual effect of this model, a public dataset was used for the experiments, which came from Wuhan University [17] (http://study.rsgis.whu.edu.cn/pages/download/building_dataset.html, building Dataset, accessed on 9 September 2022) and is referred to here as the EA (East Asia) dataset. The EA dataset consisted of 6 adjacent satellite images covering 860 square kilometers of East Asia with a ground resolution of 0.45 m. The architectural styles were quite different, and the generalization ability of the deep learning methods on the different data sources was fully evaluated and developed. The vector building map was also drawn manually by ArcGIS software and contained 34,085 buildings. The entire image was seamlessly cropped into 17,388 blocks of 512 × 512 as a result. Excluding the dicing results that did not contain buildings, the remaining training set contained 25,749 buildings (3135 slice results) and the test set contained 8358 buildings (903 slice results). For the convenience of the experimental comparison, the training set was randomly divided into 2508 training sets and 627 validation sets according to a ratio of 4:1, and the result of one slice is shown in Figure 6.

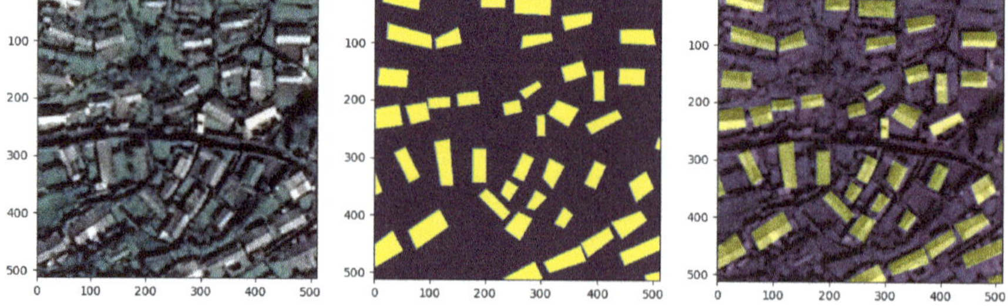

Figure 6. An example image of the EA dataset.

Among them, the unlabeled data used for the self-supervised training for the independently produced the Beijing area sample set were from Google data for the whole of the Beijing area. For the EA dataset, the unlabeled data were from the results obtained by merging all data. A brief introduction to the two datasets is shown in Table 1.

Table 1. The basic information of the datasets

DataSet	Unlabeled	Split: Train/Val/Test	Location	Resolution
BJ Dataset	21,417	797/202/34(16)	Beijing, China	1 m
EA DataSet	4038	2508/903/627	East Asia	0.45 m

3.2. Evaluation

In this article, we used the *F*1-score to evaluate the results. In order to evaluate the effectiveness of the image pixel-level prediction task, we compared the prediction results with the corresponding ground truth and divided each pixel into true positive (*TP*), false positive (*FP*), false negative (*FN*), and true negative (*TN*). The evaluation metrics used to measure the effectiveness of our method were calculated based on these four indicators. The *F*1 score is the reconciled average value of the recall rate and accuracy rate according to specific formulas, as shown in Equations (9)–(12):

$$Precision = \frac{TP}{TP + FP} \tag{9}$$

$$Recall = \frac{TP}{TP + FN} \tag{10}$$

$$IOU = \frac{TP}{TP + FP + FN} \tag{11}$$

$$F1\_score = \frac{2 \times Precision \times Recall}{Precision + Recall} \tag{12}$$

where *TP* denotes true positive, *FP* denotes false positive, and *FN* denotes false negative. These indicators were calculated using each tiled pixel-based confusion matrix or cumulative confusion matrix and are explained in the following section.

4. Results and Discussion

4.1. Experimental Setup

To validate the effectiveness of the proposed method for building information extraction, we conducted experiments on two datasets and compared our model with state-of-the-art methods. For semantic segmentation methods with conventional supervised learning, this paper compared three semantic segmentation frameworks, Unet, PSPNet, and Deeplabv3+. Secondly, based on the Unet structure, we validated the basic experimental results of several popular comparative learning methods and the proposed PGSSL method in this paper on two datasets, as well as the extraction effects under different proportional sample inputs. Finally, this paper conducted ablation experiments on the PGSSL structure under two datasets to verify the effectiveness of each link design. The parameters used for each part of the network are shown in Table 2.

For the selection of the underlying network structure and data augmentation, to ensure the consistency of the experimental process, the experiments in this paper all used ResNet34 as the basic network feature architecture. Firstly, a 7 × 7 convolutional kernel with pooling was used to expand the feature dimension to 64 and the image size was downsampled to 1/4 of the original size. Subsequently, after several convolutional poolings, the final feature map size became 1/32 of the original size and the feature channel was expanded to 512 dimensions. To make full use of the sample features, we randomly augmented the data before training and the data augmentation methods we used included random color transformation, random flip, random rotation, random crop, and resampling in five steps, the specific parameters of which are shown in Table 2 Data Augmentation. Data augmentation was not applied in any of the data testing.

The network training parameters of this paper included two parts, the pretext task network training and the semantic segmentation network training. For the training part of the pretext task network, the input image batch was 8, the total number of iterations was 80,000, the learning rate was set to 3×10^{-4}, and after every 10,000 iterations, it was reduced to 95% of the original, and it took about 11 hours to complete the pretext task network. For the semantic segmentation part of the training, the loss function used the cross-entropy function, the input image batch was also set to 8, and the initial learning rate was set to 0.005. We calculated the accuracy of the validation set every 150 iterations and saved the model with the highest accuracy and stopped training after 80,000 iterations.

Table 2. Network parameter adjustment diagram.

Hyperparameters	Setting Details
Basic Backbone Encoder (ResNet34[Default])	*7 × 7, conv, stride = (2, 2), padding = (3, 3), 64, 3 × 3, maxpool* *[[3 × 3conv, 64] × 2], concat, 1 × 1 conv, 64] × 3,* *3 × 3conv, stride = (2, 2), 128, 3 × 3 conv, 128* *[[3 × 3 conv, 128] × 2], concat, 1 × 1 conv, 128] × 3* *3 × 3 conv, stride = (2, 2), 256, 3 × 3 conv, 256* *[[3 × 3 conv, 256] × 2], concat, 1 × 1 conv, 256] × 5* *3 × 3 conv, stride = (2, 2), 512, 3 × 3 conv, 512* *[[3 × 3 conv, 512] × 2], concat, 1 × 1 conv, 512] × 3*
contrastive Learning	*Q-encoder, Basic Backbone* *K-encoder, Basic Backbone* *Q-mlp, [1 × 1, avgpool, flatten, Liner(512, 128)]* *K-mlp, [1 × 1, avgpool, flatten, Liner(512, 128)]* *projector, [Liner(128, 512), BatchNorm(512), Liner(512, 128)]*
Data Augmentation	*RandomHSV(20, 20, 20), Flip(0.5), Rotate(20), Scale(1), Clip(350, Rescale(384)[BJ Dataset]* *RandomHSV(20, 20, 20), Flip(0.5), Rotate(20), Scale(1), Clip(500), Rescale(512)[EA Dataset]* *ColorJitter(0.4, 0.4, 0.4, 0.1), Flip(0.5), Rotate(20), Scale(1), RandomClip(256), Rescale(224)[Contrastive learning]*
Loss Function Adjustment	*CrossEntropLoss[Default]* *CosineSimilarity[Contrastive]*
Other Hyperparameters	*Batchsize, 4* *iter, 80,000* *Base Learning Rate,* 3×10^{-4}

Data Augmentation: Data augmentation is slightly different in comparative learning. It is necessary to perform data augmentation on the original image twice, and then, respectively, use them as input.

4.2. Comparison of Different Methods

Following the above methods and parameter settings, this paper first conducted experiments using complete training samples on two datasets. For conventional semantic segmentation methods, such as PSPnet [36], Deeplabv3+ [37], and UNet [38], the labeled samples were directly used for training. We kept the best-trained model in the validation set obtained during the training process and finally calculated the accuracy obtained by testing on the test set, as shown in Table 3. The main experiments were divided into two main parts, basic framework training and self-supervised learning training. All experiments in this paper were repeated three times and the best results were used to explore the upper limit of the model approach. In the supervised learning task, the basic network modules were initialized with the ImageNet parameters, except for the special annotation of UNet(Random). After comparison, the Unet structure maintained the best test results among the three traditional semantic segmentation networks.

For the self-supervised methods, such as SimCLR [24], BYOL [25], and PGSSL, in this paper, we first trained the network in unlabeled samples, then trained the network under the self-supervised framework until the loss was minimized, and finally, transferred the parameters of the feature extraction layer to the UNet network framework. In the subsequent training, only the parameters of this part were fine-tuned and the learning rate was set to 0.1 of the regular learning rate. After collecting the best models on the ensemble, the accuracy obtained by testing on the test set was used as the final prediction accuracy. The training of PGSSL consisted of three steps due to pseudo-label learning. First, it provided a basic model for regular training for pseudo-label training and then self-supervised learning training was performed on top of that. This process generated pseudo-label channel prediction results with prediction accuracies as shown in Table 3 PGSSL, and finally, the basic feature structure generated by the above model was used for initial learning and training to obtain the final model test results, as shown in Table 3 PGSSL*. For the BJ Dataset, this paper mainly tested the final results of the model on areas E and F in Figure 4. To illustrate the generalization of the model, the F1-score of the test for the E area was 83.3% and the F1-score for the F area was 77.0%. The accuracy comparison of the subsequent methods was based on the test results of the F area. The overall differences between the methods are shown in Table 3.

Table 3. Overall experimental effect comparison of all methods.

Dataset	BJ Dataset			EA Dataset			
	Precision	Recall	*F*1-Score	Precision	Recall	*F*1-Score	IOU
PSPNet	0.811	0.598	0.688	0.708	0.854	0.774	0.642
DeepLabv3+	0.842	0.550	0.665	0.798	0.839	0.818	0.692
UNet(Random)	0.578	0.836	0.684	0.682	0.850	0.757	0.610
UNet(ImageNet)	0.632	0.839	0.720	0.794	0.852	0.822	0.698
SimCLR	0.706	0.798	0.749	0.805	0.846	0.825	0.702
BYOL	0.823	0.682	0.746	0.790	0.858	0.822	0.704
PGSSL	0.871	0.666	0.755	0.818	0.816	0.817	0.690
PGSSL*	0.853	0.702	0.770	0.796	0.856	0.825	0.706

PGSSL: Test accuracy of the output in the pseudo-label channel. PGSSL* represents the final effect of the PGSSL method after pre-training.

From the comparison in Table 3, it can be seen that the underlying results showed that the Unet structure had the highest accuracy among several semantic segmentation frameworks on both building datasets. Second, initializing the network structure using ImageNet significantly improved the model testing accuracy relative to random initialization, with a 3.6% increase in the *F*1-score in the BJ dataset and a 6.5% increase in the *F*1-score in the public data. Finally, self-supervised learning significantly improved the model results, which was even more significant in the BJ dataset, where the *F*1-scores of the commonly used SimCLR and BYOL methods improved the accuracy by 2.6–2.9% over the original ImageNet, and the final PGSSL method improved it by 5%. The segmentation effects in the two datasets are shown in Figure 3. The effects on the images are shown in Figure 7, where Figure 7a–e show the prediction results for the BJ dataset and Figure 7f–j show the prediction results for the EA dataset.

However, it can also be seen in Table 3 that the improvement effect of the relevant self-supervised methods on the public dataset was much lower than that on the BJ Dataset. After analysis, we believe that this was mainly related to the number of labeled samples and the quality of unlabeled data in the self-supervision. Relatively speaking, the public dataset was relatively rich in labeled data when the learning no longer relied on the prior knowledge provided by the pre-training network and only needed to provide the basic pre-training structure to achieve better segmentation results. In addition, the public data did not additionally collect local unlabeled data. The images used in the self-supervised learning process were from the images needed for the subsequent semantic segmentation and the additional prior knowledge that was provided was relatively limited.

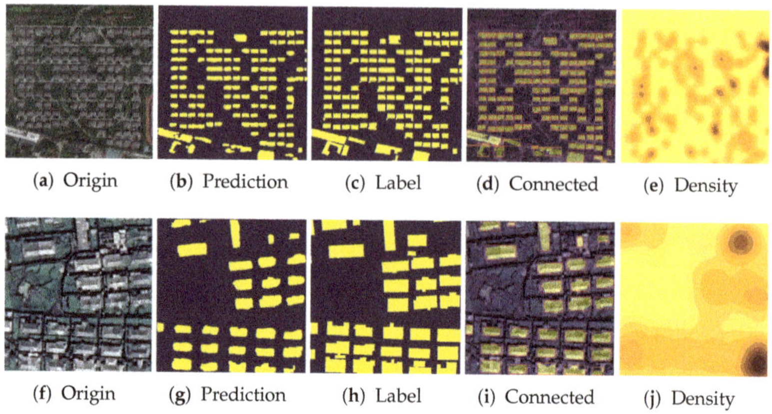

Figure 7. The overall building information extraction effect in both datasets.

4.3. Experiments with Different Sample Ratios

In order to illustrate the effectiveness of the self-supervised method, this paper conducted experiments on the optimization effect of self-supervised learning with a small number of samples. We randomly selected a certain percentage of samples from the training part of the two datasets for the experiments, including 1%, 5%, 10%, 20%, 50%, 80%, and 100%. In this paper, we compared the experimental results of two datasets with different methods at different sample proportions. Among all the methods, the PSPNet, Deeplabv3+, and UNet networks were trained using ImageNet pre-training network initialization, whereas the other self-supervised learning methods were trained on the corresponding structures and migrated to Unet structures. In this paper, we compared our approach with two state-of-the-art self-supervised representation learning methods following the experimental setup in Section 4.1, firstly for the BJ dataset, where PGSSL represented the model output in the pseudo-label channel and PGSSL* represented the further pre-training results; all the experimental results are shown in Table 4.

Table 4. Experimental effects of various methods on BJ dataset with different proportions.

	1%	5%	10%	20%	50%	80%	100%
PSPNet	0.221	0.570	0.588	0.611	0.650	0.702	0.688
Deeplabv3+	0.509	0.548	0.573	0.620	0.723	0.704	0.665
UNet	0.636	0.621	0.598	0.656	0.700	0.703	0.720
SimCLR	0.649	0.624	0.490	0.610	0.754	0.723	0.749
BYOL	0.530	0.572	0.622	0.554	0.714	0.700	0.746
PGSSL	0.050	0.616	0.602	0.575	0.658	0.746	0.755
PGSSL*	0.498	0.500	0.634	0.671	0.719	0.744	0.770

PGSSL: Test accuracy of the output in the pseudo-label channel. PGSSL*: Represents the final effect of the PGSSL method after pre-training.

In the BJ dataset experiments, the accuracy showed a significant upward trend with the increase in the sample proportion. Secondly, when the sample proportion was small, the basic UNet semantic segmentation network structure showed better stability at this time. At 1% of the samples, only about eight images were randomly selected from the original training data center for training and achieved a 63.6% segmentation accuracy, which was second only to the results initialized with SimCLR. The overall change can be seen in Figure 8a. In the EA dataset experiments, the method proposed in our paper always maintained the leading accuracy. However, as the sample size increased, the improvement in accuracy was no longer significant. In the BJ dataset, the training sample size did not reach the limit of potential for this method. As predicted by the current results, it is worth exploring the effect of further supplemental samples to experiment with self-supervised learning. The changing trend is shown in Figure 8b.

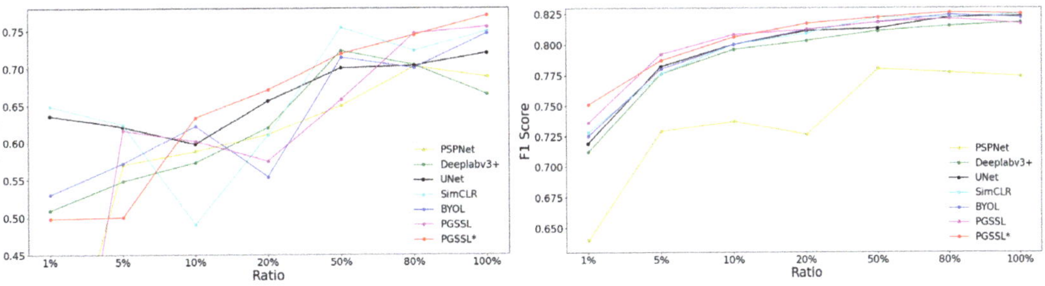

(**a**) (**b**)

Figure 8. The effect diagram of the accuracy changes of various methods under different proportions of training data. (**a**) Effects on the BJ dataset. (**b**) Effects on the EA dataset.

Comparing the UNet network structure with the final segmentation results of PGSSL in this paper, it can be seen that when the sample size was greater than 10%, PGSSL outperformed the former and other self-supervised learning methods. However, when the sample size was very small (the proportion was less than 10%), the results showed a significant decrease. We think this is because pseudo-label learning was used in the structural design to guide the process of self-supervised learning, and when the labeled samples were too small, the actual generalization ability of the model used for guidance was poor, thus providing incorrect guidance to the whole model. Furthermore, it can be seen that the SimCLR self-supervised learning method surpassed the results using all samples by using only 50% of the samples. This was a special case among all the experiments in this paper, but it also shows that the self-supervised learning method has a high upper limit and the prior knowledge from pre-training provides better possibilities for the model.

However, there are some differences between the experiments on the EA dataset and the experimental results of the previous paper. As shown in Table 5, the self-supervised learning method performed better when the sample size was less than 20%, especially when the sample size was 1%; the method proposed in this paper resulted in a 3.2% accuracy optimization, and the other self-supervised learning methods and the output of the pseudo-label channel also improved in this process. After the sample size was increased, it can be seen that the accuracy of the self-supervised learning method was almost the same as the results of conventional ImageNet pre-training. This paper suggests that this reflects a boundary effect that network pre-training can have. When the sample size reaches a certain level, the effect of prior knowledge provided by the pre-training of self-supervised learning decreases.

Table 5. Experimental effects of various methods on EA dataset with different proportions.

	1%	5%	10%	20%	50%	80%	100%
PSPNet	0.640	0.729	0.737	0.726	0.780	0.777	0.774
Deeplabv3+	0.712	0.776	0.796	0.803	0.811	0.815	0.818
Unet	0.719	0.782	0.800	0.811	0.813	0.822	0.823
SimCLR	0.728	0.776	0.800	0.809	0.822	0.822	0.825
BYOL	0.725	0.780	0.800	0.812	0.818	0.824	0.822
PGSSL	0.736	0.792	0.808	0.812	0.818	0.821	0.817
PGSSL*	0.751	0.787	0.806	0.817	0.822	0.826	0.825

PGSSL: Test accuracy of the output in the pseudo-label channel. PGSSL*: Represents the final effect of the PGSSL method after pre-training.

4.4. Ablation Experiment

Finally, in order to verify the effectiveness of the proposed structure, this paper also conducted ablation experiments on several selected proxy tasks to compare the effects of these different proxy tasks on building information extraction. Considering that the method proposed in this paper requires a pre-training structure to provide prior knowledge, this paper selected 100% of the samples on the BJ dataset and 1% of the samples on the EA dataset for the ablation experiments. The final experimental effects are shown in Table 6.

Table 6. Ablation experiment.

Pseudo-Sample Learning	Contrastive Learning	Image Inpainting	EA DataSet [1%]	BJ DataSet [100%]
✘	✘	✘	0.720	0.720
✔	✘	✘	0.740	0.732
✘	✔	✘	0.725	0.746
✘	✘	✔	0.731	0.730
✔	✔	✘	0.748	0.746
✔	✔	✔	0.751	0.770

It can be seen that when the three methods of pseudo-label training, contrastive learning (BYOL), and image inpainting were used alone, the experimental results were better than those of the basic UNet network on both datasets. The experimental results showed that the contrastive learning pre-training had the greatest accuracy in extraction from the BJ dataset, whereas the EA data showed that the pseudo-label learning had a better accuracy improvement. In the experiment, the EA dataset improved by 0.8%, whereas there was almost no improvement in the BJ dataset. In this paper, we suggest that this is related to the data used in the self-supervised learning process. Since the data in the EA dataset were more similar in style and the amount of data for self-supervised learning training was smaller, pseudo-labeling achieved better results. In contrast, the unlabeled data in the BJ dataset were more extensive and pseudo-labeling played a limited role as a guide. Finally, it can be seen that after adding the proxy task of image inpainting, both of the methods had further improved effects, and restraining the spatial information provided by image inpainting was of great significance for the segmentation tasks.

4.5. Regional Photovoltaic Potential Assessment

Large-scale building rooftop information extraction provides important data support for PV potential analysis. Remote sensing satellites are always observing the ground to generate remote sensing data and unlabeled data are simple to obtain. Self-supervised learning can effectively utilize the unlabeled data and reduce the number of labeled samples to be prepared when extracting building information on a large scale. In this paper, we used Google data slicing of the whole Beijing territory for the self-supervised learning pre-training to obtain better building extraction results at the current sample labeling level, which supports the evaluation of the building PV potential on a large scale. Finally, this paper considered the limitations of the semantic segmentation results in the analysis of the PV suitability method and further designed the building density statistics method to obtain the results of the building area and building density distribution within Beijing. The calculation results for the building area are shown in Figure 9.

This paper extracted the building area and center point using the method in Section 2.4 and finally obtained a total building area of 588.24 km^2, with 1.746 million buildings in Beijing. Next, the kernel density analysis of the building distribution results was performed. The circle radius was set to 0.5 km, and the results of the analysis are shown in Figure 10, which shows the number of buildings per square kilometer. Some aggregation centers in urban areas and towns in Beijing are objectively reflected in the figure, which can provide a reference for the construction of distribution facilities related to distributed PV rooftops, as well as provide accurate data support for the assessment of the potential of building PV rooftops.

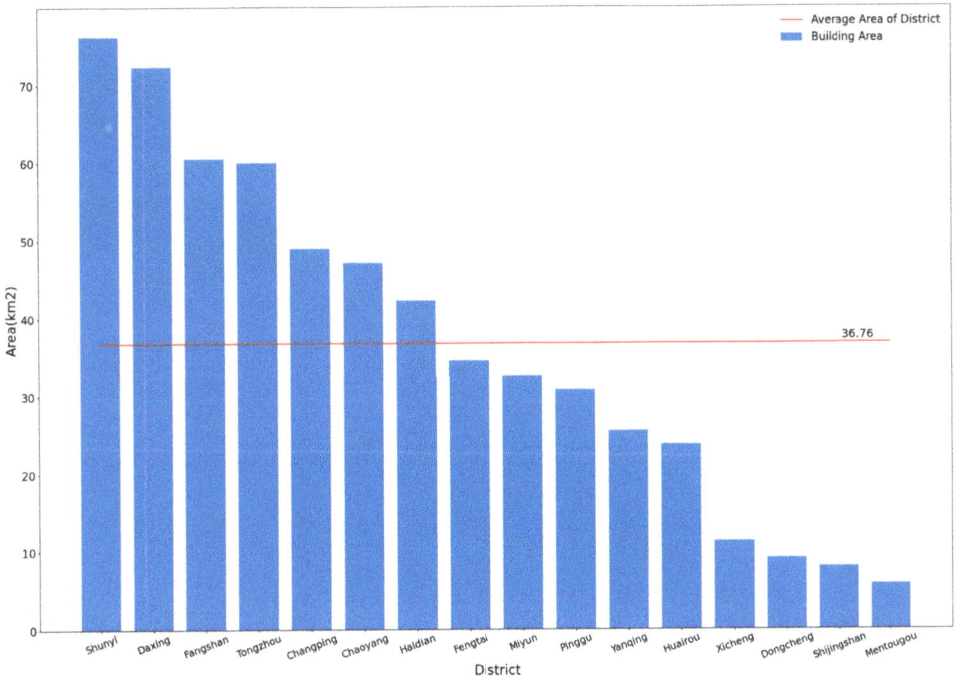

Figure 9. Building area of each district in Beijing.

In addition, this paper compared the building area presented herein with the available literature in a cross-sectional comparison [9,39–41]. In ref. [39], the authors calculated the area based on the NDBI index method using the number of image elements and obtained 560.33 km^2 in 2004. In the statistical yearbook, the built-up area of buildings in Beijing in 2006 was 1254.23 km^2 [40], whereas the built-up area of buildings in Beijing in 2019 was 1469.05 km^2 [41]. In [9], the authors calculated the building rooftop area to be 809.837 m^2 based on Google data in 2018 for a study area of 5 km^2 within Chaoyang District, whereas the total area of Chaoyang District was 470.8 km^2. This paper estimated Chaoyang District according to these proportions. The total building rooftop area was calculated to be about 76.25 km^2, but the actual building area according to the literature is denser and the actual proportions should be smaller. In comparison, our article calculated that the building area of Chaoyang District is 47.13 km^2 based on Google data in 2020, which should be similar to reality, whereas the total building area of Beijing was 588.24 km^2. According to the ratio of the urban built-up area to the building rooftop area, the results of this paper should be close to those mentioned in the literature, and it can be seen that the building extraction method in this paper has a good reference value.

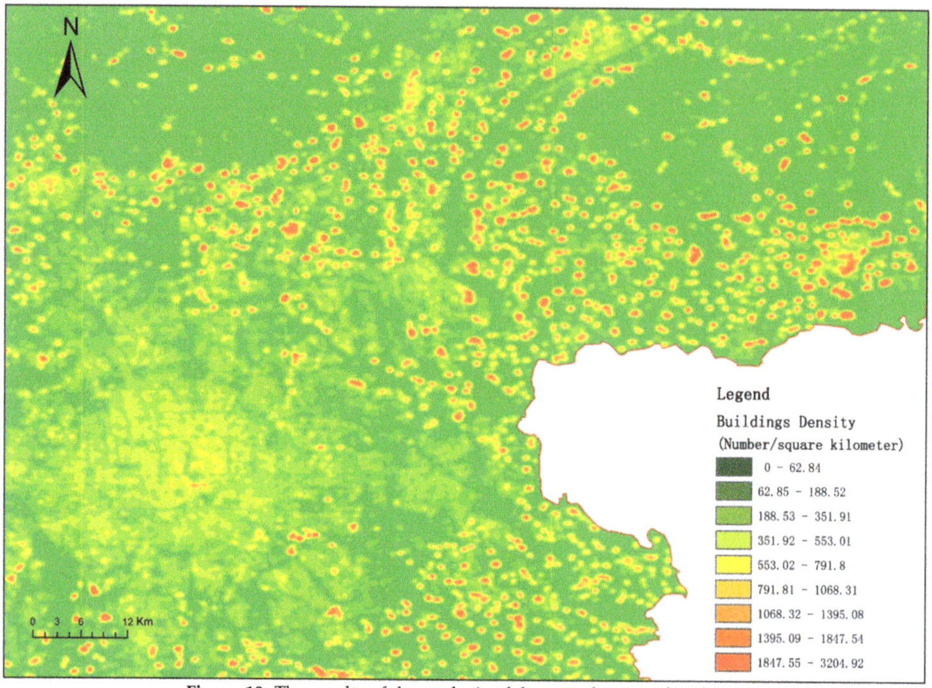

Figure 10. The results of the analysis of the core density of buildings in Beijing.

5. Conclusions

In recent years, the energy industry has developed rapidly and awareness of sustainable development has intensified. The assessment of the potential of distributed PV systems is of great importance for solar energy policy planning and industrial development. It has prompted more attention to be focused on the potential distribution of the PV industry in China. In this paper, a self-supervised learning method for remote sensing building rooftop extraction called the PGSSL method is proposed, which alleviates the problem of the high dependence on samples for deep learning methods and provides the possibility of large-scale building rooftop information extraction. The method uses contrastive learning and image restoration as the base methods to extract the global and local features of buildings and proposes pseudo-label learning to guide these features and drive them to focus on our target. In this paper, experiments are conducted on two datasets independently, and the advantages of the proposed method are demonstrated by comparing it with other deep learning methods. Moreover, in this paper, we conduct comparative experiments on sample size by setting different sample ratios to demonstrate the excellent effect of the method when the labeled samples are few. In addition, this paper also conducts ablation experiments on the proposed method, which proves the rationality and effectiveness of the method design in this paper.

Finally, this paper also proposes a post-processing scheme for the semantic segmentation results in the photovoltaic potential analysis. Based on the Google data of 1 m resolution in 2020, this paper extracts the rooftop area of buildings in Beijing (588.24 km^2 in total) and further analyzes the density of buildings in Beijing based on the results, which can provide a positive reference value for the layout of the distribution network and the scale suitability of building rooftop photovoltaic systems. The application results show that the method proposed in this paper can extract building information in a wide range based on high-resolution remote sensing Images, which provides a very effective method for large-scale building photovoltaic potential assessments and solar energy utilization.

Author Contributions: Conceptualization, D.-Y.C.; Data curation, D.-Y.C. and L.P.; Formal analysis, D.-Y.C.; Funding acquisition, L.P.; Methodology, D.-Y.C.; Resources, L.-N.Y. and L.P.; Validation, W.-Y.Z. and L.-N.Y.; Writing—original draft, D.-Y.C.; Writing—review and editing, L.P. and Y.-D.W. All authors have read and agreed to the published version of the manuscript.

Funding: This work was supported by the Global Energy Internet Group Co., Ltd. Technology Project: Building Photovoltaic Power Generation Potential Evaluation Method and Empirical Research (SGGEIG00JYJS2100032).

Institutional Review Board Statement: Not applicable.

Informed Consent Statement: Not applicable.

Data Availability Statement: Data sharing not applicable.

Acknowledgments: We would like to express our special thanks to Liu Yufei for her outstanding contribution to the compilation of the experimental data.

Conflicts of Interest: The authors declare no conflicts of interest.

References

1. Olejarnik, P. *World Energy Outlook 2013*; International Energy Agency: Paris, France, 2013; pp. 1–7
2. Ramachandra, T.; Shruthi, B. Spatial mapping of renewable energy potential. *Renew. Sustain. Energy Rev.* **2007**, *11*, 1460–1480. [CrossRef]
3. IRENA. *Renewable Capacity Statistics 2019*; International Renewable Energy Agency (IRENA): Masdar, AbuDhabi, 2019; ISBN 978-92-9260-123-2.
4. Chen, Y.; Peng, Y.; He, S.; Hou, Y.; Qin, H. A method for predicting the solar photovoltaic (PV) potential in China. *IOP Conf. Ser. Earth Environ. Sci.* **2020**, *585*, 012012. [CrossRef]
5. Gassar, A.A.A.; Cha, S.H. Review of geographic information systems-based rooftop solar photovoltaic potential estimation approaches at urban scales. *Appl. Energy* **2021**, *291*, 116817. [CrossRef]
6. Lukač, N.; Seme, S.; Žlaus, D.; Štumberger, G.; Žalik, B. Buildings roofs photovoltaic potential assessment based on LiDAR (Light Detection And Ranging) data. *Energy* **2014**, *66*, 598–609. [CrossRef]
7. Borfecchia, F.; Caiaffa, E.; Pollino, M.; De Cecco, L.; Martini, S.; La Porta, L.; Marucci, A. Remote Sensing and GIS in planning photovoltaic potential of urban areas. *Eur. J. Remote Sens.* **2014**, *47*, 195–216. [CrossRef]
8. Wong, M.S.; Zhu, R.; Liu, Z.; Lu, L.; Peng, J.; Tang, Z.; Lo, C.H.; Chan, W.K. Estimation of Hong Kong's solar energy potential using GIS and remote sensing technologies. *Renew. Energy* **2016**, *99*, 325–335. [CrossRef]
9. Song, X.; Huang, Y.; Zhao, C.; Liu, Y.; Lu, Y.; Chang, Y.; Yang, J. An approach for estimating solar photovoltaic potential based on rooftop retrieval from remote sensing images. *Energies* **2018**, *11*, 3172. [CrossRef]
10. Tiwari, A.; Meir, I.A.; Karnieli, A. Object-based image procedures for assessing the solar energy photovoltaic potential of heterogeneous rooftops using airborne LiDAR and orthophoto. *Remote Sens.* **2020**, *12*, 223. [CrossRef]
11. Lopez-Ruiz, H.G.; Blazquez, J.; Vittorio, M. Assessing residential solar rooftop potential in Saudi Arabia using nighttime satellite images: A study for the city of Riyadh. *Energy Policy* **2020**, *140*, 111399. [CrossRef]
12. Huang, X.; Hayashi, K.; Matsumoto, T.; Tao, L.; Huang, Y.; Tomino, Y. Estimation of Rooftop Solar Power Potential by Comparing Solar Radiation Data and Remote Sensing Data—A Case Study in Aichi, Japan. *Remote Sens.* **2022**, *14*, 1742. [CrossRef]
13. Li, X.; Yao, X.; Fang, Y. Building-a-nets: Robust building extraction from high-resolution remote sensing images with adversarial networks. *IEEE J. Sel. Top. Appl. Earth Obs. Remote Sens.* **2018**, *11*, 3680–3687. [CrossRef]
14. Tian, T.; Li, C.; Xu, J.; Ma, J. Urban area detection in very high resolution remote sensing images using deep convolutional neural networks. *Sensors* **2018**, *18*, 904. [CrossRef] [PubMed]
15. Zeng, Y.; Guo, Y.; Li, J. Recognition and extraction of high-resolution satellite remote sensing image buildings based on deep learning. *Neural Comput. Appl.* **2022**, *34*, 2691–2706. [CrossRef]
16. Hui, J.; Du, M.; Ye, X.; Qin, Q.; Sui, J. Effective building extraction from high-resolution remote sensing images with multitask driven deep neural network. *IEEE Geosci. Remote Sens. Lett.* **2018**, *16*, 786–790. [CrossRef]
17. Ji, S.; Wei, S.; Lu, M. Fully convolutional networks for multisource building extraction from an open aerial and satellite imagery data set. *IEEE Trans. Geosci. Remote Sens.* **2018**, *57*, 574–586. [CrossRef]
18. Xia, G.S.; Bai, X.; Ding, J.; Zhu, Z.; Belongie, S.; Luo, J.; Datcu, M.; Pelillo, M.; Zhang, L. DOTA: A Large-Scale Dataset for Object Detection in Aerial Images. In Proceedings of the IEEE Conference on Computer Vision and Pattern Recognition (CVPR), Salt Lake City, UT, USA, 18–22 June 2018.
19. Mnih, V. Machine Learning for Aerial Image Labeling. Ph.D. Thesis, University of Toronto, Toronto, ON, Canada, 2013.
20. Goodfellow, I.; Pouget-Abadie, J.; Mirza, M.; Xu, B.; Warde-Farley, D.; Ozair, S.; Courville, A.; Bengio, Y. Generative adversarial nets. In Proceedings of the 27th International Conference on Neural Information Processing Systems, Montreal, QC, Canada, 8–13 December 2014; pp. 2672–2680.

21. Pathak, D.; Krahenbuhl, P.; Donahue, J.; Darrell, T.; Efros, A.A. Context encoders: Feature learning by inpainting. In Proceedings of the IEEE Conference on Computer Vision and Pattern Recognition, Las Vegas, NV, USA, 16 June–1 July 2016; pp. 2536–2544.
22. He, K.; Fan, H.; Wu, Y.; Xie, S.; Girshick, R. Momentum contrast for unsupervised visual representation learning. In Proceedings of the IEEE/CVF Conference on Computer Vision and Pattern Recognition, Virtual, 14–19 June 2020; pp. 9729–9738.
23. Chaitanya, K.; Erdil, E.; Karani, N.; Konukoglu, E. Contrastive learning of global and local features for medical image segmentation with limited annotations. *Adv. Neural Inf. Process. Syst.* **2020**, *33*, 12546–12558.
24. Chen, T.; Kornblith, S.; Norouzi, M.; Hinton, G. A simple framework for contrastive learning of visual representations. In Proceedings of the International Conference on Machine Learning, PMLR, Virtual, 13–18 July 2020; pp. 1597–1607.
25. Grill, J.B.; Strub, F.; Altché, F.; Tallec, C.; Richemond, P.; Buchatskaya, E.; Doersch, C.; Avila Pires, B.; Guo, Z.; Gheshlaghi Azar, M.; et al. Bootstrap your own latent-a new approach to self-supervised learning. *Adv. Neural Inf. Process. Syst.* **2020**, *33*, 21271–21284.
26. Zhang, R.; Isola, P.; Efros, A.A. Colorful image colorization. In Proceedings of the European Conference on Computer Vision, Amsterdam, The Netherlands, 11–14 October 2016; Springer: Berlin/Heidelberg, Germany, 2016; pp. 649–666.
27. Doersch, C.; Gupta, A.; Efros, A.A. Unsupervised visual representation learning by context prediction. In Proceedings of the IEEE International Conference on Computer Vision, Santiago, Chile, 13–16 December 2015; pp. 1422–1430.
28. Noroozi, M.; Favaro, P. Unsupervised learning of visual representations by solving jigsaw puzzles. *arXiv* **2016**, arXiv:1603.09246.
29. Gidaris, S.; Singh, P.; Komodakis, N. Unsupervised representation learning by predicting image rotations. *arXiv* **2018**, arXiv:1803.07728.
30. Guo, Q.; Wang, Z. A self-supervised learning framework for road centerline extraction from high-resolution remote sensing images. *IEEE J. Sel. Top. Appl. Earth Obs. Remote Sens.* **2020**, *13*, 4451–4461. [CrossRef]
31. Dong, H.; Ma, W.; Wu, Y.; Zhang, J.; Jiao, L. Self-supervised representation learning for remote sensing image change detection based on temporal prediction. *Remote Sens.* **2020**, *12*, 1868. [CrossRef]
32. Li, W.; Chen, H.; Shi, Z. Semantic segmentation of remote sensing images with self-supervised multitask representation learning. *IEEE J. Sel. Top. Appl. Earth Obs. Remote Sens.* **2021**, *14*, 6438–6450. [CrossRef]
33. Kalibhat, N.M.; Narang, K.; Tan, L.; Firooz, H.; Sanjabi, M.; Feizi, S. Understanding Failure Modes of Self-Supervised Learning. *arXiv* **2022**, arXiv:2203.01881.
34. Chen, X.; Fan, H.; Girshick, R.B.; He, K. Improved Baselines with Momentum Contrastive Learning. *arXiv* **2020**, arXiv:2003.04297.
35. Wang, Z.; Bovik, A.C.; Sheikh, H.R.; Simoncelli, E.P. Image quality assessment: From error visibility to structural similarity. *IEEE Trans. Image Process.* **2004**, *13*, 600–612. [CrossRef]
36. Zhao, H.; Shi, J.; Qi, X.; Wang, X.; Jia, J. Pyramid scene parsing network. In Proceedings of the IEEE Conference on Computer Vision and Pattern Recognition, Honolulu, HI, USA, 22–25 July 2017; pp. 2881–2890.
37. Chen, L.C.; Zhu, Y.; Papandreou, G.; Schroff, F.; Adam, H. Encoder-decoder with atrous separable convolution for semantic image segmentation. In Proceedings of the European Conference on Computer Vision (ECCV), Munich, Germany, 8–14 September 2018; pp. 801–818.
38. Ronneberger, O.; Fischer, P.; Brox, T. Convolutional networks for biomedical image segmentation. In Proceedings of the Medical Image Computing and Computer-Assisted Intervention, Munich, Germany, 5–9 October 2015; pp. 234–241.
39. Jia, L. The Remote Sensing Analysis of Urban Sprawl and Environment Change in Beijing City. Master's Thesis, Northeast Normal University, Changchun, China, 2006.
40. Comprehensive Finance Department of the Ministry of Construction, C.F.D. *China Urban-Rural Construction Statistical Yearbook*; China Statistics Press: Beijing, China, 2006.
41. Hu, Z. *China Urban-Rural Construction Statistical Yearbook*; China Statistics Press: Beijing, China, 2019.

remote sensing

MDPI

Review

Review on Active and Passive Remote Sensing Techniques for Road Extraction

Jianxin Jia [1], Haibin Sun [1,2], Changhui Jiang [1], Kirsi Karila [1], Mika Karjalainen [1], Eero Ahokas [1], Ehsan Khoramshahi [1], Peilun Hu [1,3], Chen Chen [1,4], Tianru Xue [1,2], Tinghuai Wang [5], Yuwei Chen [1,*] and Juha Hyyppä [1]

1 Department of Remote Sensing and Photogrammetry, Finnish Geospatial Research Institute, 02430 Kirkkonummi, Finland; jianxin.jia@nls.fi (J.J.); sunhaibin007@gmail.com (H.S.); changhui.jiang@nls.fi (C.J.); kirsi.karila@nls.fi (K.K.); mika.karjalainen@nls.fi (M.K.); eero.ahokas@nls.fi (E.A.); ehsan.khoramshahi@nls.fi (E.K.); peilun.hu@helsinki.fi (P.H.); chenchen115039@njust.edu.cn (C.C.); xuetianru@mail.sitp.ac.cn (T.X.); juha.hyyppa@nls.fi (J.H.)
2 Key Laboratory of Intelligent Infrared Perception, Shanghai Institute of Technical Physics, Chinese Academy of Sciences, Shanghai 200083, China
3 Department of Forest Science, University of Helsinki, 00100 Helsinki, Finland
4 School of Automation, Nanjing University of Science and Technology, Nanjing 210094, China
5 Huawei Helsinki Research Centre, 00180 Helsinki, Finland; tinghuaiwang@huawei.com
* Correspondence: yuwei.chen@nls.fi

Abstract: Digital maps of road networks are a vital part of digital cities and intelligent transportation. In this paper, we provide a comprehensive review on road extraction based on various remote sensing data sources, including high-resolution images, hyperspectral images, synthetic aperture radar images, and light detection and ranging. This review is divided into three parts. Part 1 provides an overview of the existing data acquisition techniques for road extraction, including data acquisition methods, typical sensors, application status, and prospects. Part 2 underlines the main road extraction methods based on four data sources. In this section, road extraction methods based on different data sources are described and analysed in detail. Part 3 presents the combined application of multisource data for road extraction. Evidently, different data acquisition techniques have unique advantages, and the combination of multiple sources can improve the accuracy of road extraction. The main aim of this review is to provide a comprehensive reference for research on existing road extraction technologies.

Keywords: road extraction; high-resolution image; hyperspectral image; synthetic aperture radar (SAR); light detection and ranging (LiDAR)

Citation: Jia, J.; Sun, H.; Jiang, C.; Karila, K.; Karjalainen, M.; Ahokas, E.; Khoramshahi, E.; Hu, P.; Chen, C.; Xue, T.; et al. Review on Active and Passive Remote Sensing Techniques for Road Extraction. *Remote Sens.* **2021**, *13*, 4235. https://doi.org/10.3390/rs13214235

Academic Editors: Mohammad Awrangjeb, Qin Yan, Beril Sirmacek, Jiaojiao Tian and Nusret Demir

Received: 1 September 2021
Accepted: 18 October 2021
Published: 21 October 2021

1. Introduction

Digital mapping of road networks is necessary for various industrial applications such as land use and land cover mapping [1], geographic information system updates [2,3] and natural disaster warning [4]. Moreover, it is a critical requirement for digital cities and intelligent transportation [5]. Traditional cartographic techniques are time-consuming and labour-intensive [6,7]. In comparison, remote sensing techniques changed the mapping community fundamentally without relying entirely on surveyed ground measurements [6]. Remote sensing data used for road extraction include ground moving target indicator (GMTI) tracking, smart phones global positioning system (GPS) data, street view images, synthetic aperture radar (SAR) images, light detection and ranging (LiDAR) data, high-resolution images, and hyperspectral images. GMIT radar has been used for extracting road map information due to the advantages of all-weather, real-time capabilities, and wide-area [8,9]. Smart phone GPS data were used for extracting road centrelines and monitoring road and traffic conditions [10,11]. Street view images obtained by Google of USA and Baidu of China companies have been used to detect, classify, and map traffic signs and road crack information extraction [12,13]. In this paper, based on the different data sources, the existing road extraction technology can be roughly divided into four methods: high-resolution imaging-based, hyperspectral imaging-based, SAR imaging-based and LiDAR-based methods. SAR and LiDAR are active information acquisition methods. In contrast, high-resolution and hyperspectral imaging are passive optical imaging approaches. Each road extraction method, based on a different data source, has unique characteristics. For instance, the high-resolution imaging technology can be used to obtain images with centimetre-level accuracy and detailed target information [14]; hyperspectral remote sensing images are used for conventional road extraction; they also demonstrate excellent potential for road condition detection owing to a large number of bands (generally more than 100 bands) and continuous spectrum coverage [15]; SAR and LiDAR datasets are not easily affected by environmental factors such as changes in environmental illumination conditions or weather [16].

Road extraction is a popular research topic and has attracted the interest of numerous researchers. More than 2,770,000 results can be found in Google Scholar using the keyword 'road extraction', including some state-of-the-art reviews published in recent years. Wang et al. [17] summarised the main road extraction methods from 1984 to 2014 based on high-resolution images. In this study, road information extraction methods were divided into knowledge-based [18,19], classification-based [20–26], active contour-based [27–29], mathematical morphology-based [30,31], and dynamic programming-based grouping methods [32,33]. However, these methods were chiefly heuristic, and deep learning-based approaches were not presented [34]. This scenario has undergone a drastic change in recent years with the rapid development of patch-based convolutional neural networks (CNNs) [35–40], full convolutional network (FCN)-based [41–46], deconvolutional net-based [47–64], generative adversarial network (GAN)-based [41,65,66], and graph-based deep learning methods [67–70] for road extraction. Road extraction methods based on deep learning are collectively referred to as data-driven approaches in [34,71]. Abdollahi et al. [34] and Lian et al. [71] presented and compared the deep learning-based state-of-the-art road extraction methods using publicly available high-resolution image datasets. Sun et al. [72] reviewed the SAR image-based road extraction method. This review initially introduces the road characteristics and basic strategies and then presents a summary of the main road extraction techniques based on SAR images. Similarly, Sun et al. [73] analysed and summarised the SAR image-based road segmentation methods. They introduced the traditional edge detection and deep learning-based road segmentation methods and predicted that new segmentation methods of deep neural networks based on the self-attention mechanism and capsule paradigm as the future development trends. Wang and Weng [74] summarised the road extraction techniques based on LiDAR. In this report, road clusters were defined using the classification framework and algorithms for LiDAR point data-based road identification. Furthermore, techniques for generating road networks, including road classification refinement and centreline extraction, were summarised as well. Several other similar reviews [75–77] that provide scientific references for road extraction have been reported in recent years.

To the best of our understanding, the existing reviews on road extraction methods are commonly based on only a single data source; hence, they fail to provide a comprehensive view that can be derived using different data sources. However, a comprehensive road extraction review based on high-resolution imaging, hyperspectral imaging, SAR imaging, and LiDAR technologies is crucial to bridge the gap between potential applications and available technologies of road extraction. Thus, the aim of this study was to achieve this goal by combining the road extraction techniques based on diverse data sources, including high-resolution images, hyperspectral images, SAR images, and LiDAR data. In Section 2, we provide an overview of the four techniques and summarise the typically used sensors. In Section 3, we introduce and analyse the main road extraction methods using various data sources as well as summarise the road extraction status and prospects for different data sources. Finally, different combinations of the road extraction techniques are presented in Section 4. To the best of our knowledge, in this paper, we present the first comprehensive review of road extraction, including high-resolution images, hyperspectral images, SAR images and LiDAR data sources.

2. Overview of the Existing Data Acquisition Techniques for Road Extraction

2.1. High-Resolution Imaging Technology

In this review, high-resolution images refer to high-spatial-resolution images (resolution of less than 10 m) that were mainly acquired using airborne or spaceborne sensors. The spatial resolution of the images refers to the size of a single pixel. High-resolution images are usually divided into two categories: panchromatic and multispectral images [78].

2.1.1. Data Acquisition Methods and Characteristics

High-resolution images are primarily recorded using spaceborne and airborne sensors. Spaceborne high-resolution imaging techniques have a broad area coverage and stable revisit periods; however, the cost of the satellite is high, and the images are easily affected by the atmosphere [79]. Compared to the spaceborne high-resolution images, the airborne high-resolution images possess higher resolutions and are less affected by the atmosphere. However, the working efficiency of airborne cameras is lower than that of the spaceborne instruments because of the lower flight altitude and smaller coverage [80]. Airborne high-resolution images can be obtained using manned aircrafts and unmanned aerial vehicles (UAVs). In recent years, several imaging systems with high-resolution cameras mounted on UAVs have been rapidly developed; these systems can achieve centimetre-level spatial resolution [81].

2.1.2. Typical Sensors

Spaceborne high-resolution imaging is still the main technical approach for earth observation. As shown in Table 1, an increasing number of high-resolution satellites have been developed; some of these satellites have been developed in series and are constantly being upgraded [82–86]. It can be seen from Table 1 that most high-resolution satellites were developed and launched by the USA, while the number of high-resolution satellites in China has increased in recent years. With the development of related technologies, the performance of spaceborne cameras continues to improve, and images with a spatial resolution better than 1 m can now be obtained.

Table 1. Main parameters of typical high-resolution satellites.

Satellite	Launch (Year)	Swath (km)	PAN (m)	R (m)	G (m)	B (m)	NIR (m)
Gaofen 1 (CN)	2013 [87]		2	8	8	8	8
Gaofen 2 (CN)	2014 [88]	70	0.8	3.2	3.2	3.2	3.2
Gaofen 6 (CN)	2015 [89]		2	8	8	8	8
SuperView (CN)	2016 [90]	12	0.5	2	2	2	2
GeoEye 1 (US)	2008 [91]	15.2	0.41	1.65	1.65	1.65	1.65
IKONOS (US)	1999 [85]	11.3	1	4	4	4	4
PlanetScope (US)	2018 [92]	24.6	/	3	3	3	3
QuickBirds (US)	2001 [93]	16.5	0.6	2.4	2.4	2.4	2.4
WorldView 1 (US)	2007 [94]	17	0.5	/	/	/	/
WorldView 2 (US)	2009 [91]	17	0.5	2	2	2	2
WorldView 3 (US)	2014 [95]	13.1	0.31	1.24	1.24	1.24	1.24
WorldView 4 (US)	2016 [96]	13.1	0.31	1.24	1.24	1.24	1.24
OrbView 3 (US)	2003 [97]	8	1	4	4	4	4
RapidEye (DE)	2008 [98]	77	/	6.5	6.5	6.5	6.5
KOMPSAT 2 (KR)	2006 [99]	15	1	4	4	4	4
KOMPSAT 3 (KR)	2012 [100]	16	0.7	2.8	2.8	2.8	2.8
KOMPSAT 3A (KR)	2015 [101]	12	0.55	2.2	2.2	2.2	2.2
Pléiades 1A (FR)	2011 [102]	20	0.7	2.8	2.8	2.8	2.8
Pléiades 1B (FR)	2012 [103]	20	0.7	2.8	2.8	2.8	2.8
SPOT 6 (FR)	2012 [104]	60	1.5	6	6	6	6
SPOT 7 (FR)	2014 [105]	60	1.5	6	6	6	6
DubaiSat 1 (AE)	2009 [106]	12	2.5	5	5	5	5
DubaiSat 2 (AE)	2013 [107]	12	1	4	4	4	4

PAN: panchromatic. R: red. G: green. B: blue. NIR: near-infrared. WorldView 2 [91] and WorldView 3 [95] have four other multispectral bands (red edge, coastal, yellow and NIR), and RapidEye has another multispectral band (red edge) [98].

2.1.3. Application Status and Prospects

High-resolution images usually contain feature-rich information such as spectral characteristics, geometric features, and texture features, and hence, a significant amount of useful information can be extracted from such images. High-resolution imaging has been widely used in forest management [108], urban mapping [109], farmland management [110], disaster and security mapping, public information service and environmental monitoring. Numerous high-resolution satellites have been developed and launched in recent years. These satellites can form satellite networks to obtain image data with a wide coverage. However, the huge amounts of data also bring new challenges to data transmission and processing [80]. Such high-resolution images have been extensively used for road extraction [34,71]. Moreover, several commercial products such as Google Maps based on high-resolution images have been successfully developed and applied in many fields in recent years.

2.2. Hyperspectral Imaging Technology

Hyperspectral imaging technology is another commonly used technique for obtaining the spectra of a target [111,112]. The hyperspectral image containing two-dimensional (2D) spatial and 1D spectral information comprises a 3D data cube [113]. Notably, different objects exhibit different spectra, which can be used for the identification and detection of such objects. In the 1980s, Goetz et al. [114] began a revolution in remote sensing by developing an airborne visible infrared imaging spectrometer (AVIRIS) [113], which initiated the development of hyperspectral imagers. The number of bands in a multispectral image is usually less than five, while that in a hyperspectral image is more than 100; moreover, continuous spectral information is obtained from a hyperspectral image.

2.2.1. Data Acquisition Methods and Characteristics

Hyperspectral images are obtained by an imaging spectrometer, which is a complex and sophisticated optical system that includes several subsystems and components. The main components of the sensor are a scan mirror, fore-optics, spectrometers, detectors, onboard calibrators and electronic units. The fore-optics of the system receive light, which is dispersed by a spectrometer and converted from photons to electrons by the detector to yield an electronic signal. This electronic signal is then amplified, digitised and recorded by the electronic unit. The instrument performance and preprocessing data results are the main factors that aid in acquiring high-accuracy surface reflectance data. Such an instrument is characterised by its field-of-view (FOV), spectral range, spatial and spectral resolution and sensitivity. Data preprocessing includes geometric rectification, calibration and atmospheric correction [80].

2.2.2. Typical Sensors

According to the installed platforms, hyperspectral sensors can be divided into spaceborne, airborne, UAV [115–117], car-borne [118], and ground-based [119] imaging systems. Airborne hyperspectral imaging systems were the first hyperspectral imagers to be developed and used for verifying the design of later spaceborne instruments. Jia et al. [80] presented a comprehensive review of airborne hyperspectral imagers, including key design technologies, preprocessing and new applications. Based on this review, we added spaceborne sensors in this paper and summarised the typical hyperspectral imagers in Table 2. Evidently, most hyperspectral sensors, especially the spaceborne hyperspectral imagers, were developed in the USA. In addition, there are more airborne hyperspectral imagers are than spaceborne hyperspectral imagers due to the hardware investment. The future spaceborne program includes the HyspIRI hyperspectral satellite of America [120] and the ENMAP hyperspectral satellite of Germany [121]. Most hyperspectral imagers can acquire data in the visible to near-infrared spectral range owing to the availability of silicon detectors with wide spectral detection ranges. Additionally, shortwave infrared and longwave infrared hyperspectral imagers have emerged in recent years [80].

Table 2. Typical airborne and spaceborne hyperspectral sensors.

Name	References	Platform	Wavelength Range (μm)	Channel	Spectral Resolution (nm)	IFOV (mrad)	FOV/Swath
AISA-FENIX 1K	[122], 2018	Airborne	0.38–0.97, 0.97–2.5	348, 246	≤4.5, ≤12	0.68	40°
APEX	[123], 2015	Airborne	0.372–1.015 0.94–2.54	114, 198	0.45–0.75, 5–10	0.489	28.1°
AVIRIS-NG	[124,125], 2016, 2017	Airborne	0.38–2.52	430	5	1	34°
CASI-1500 SASI-1000A TASI-600A	[126], 2014	Airborne	0.38–1.05, 0.95–2.45, 8–11.5	288, 100,32	2.3, 15, 110	0.49, 1.22, 1.19	40°
AMMIS	[127,128], 2019, 2020	Airborne	0.4–0.95, 0.95–2.5, 8–12.5	256, 512, 128	2.34, 3, 32	0.25, 0.5, 1	40°
SYSIPHE	[129], 2016	Airborne	0.4–1, 0.95–2.5, 3–5.4, 8.1–11.8	560 (total)	5, 6.1, 11 cm^{-1}, 5 cm^{-1}	0.25	15°
HSI	[130], 1996	LEWIS Satellite	0.4–1, 1–2.5	128, 256	5, 5.8	0.057	7.68 km
Hyperion	[131], 2003	EO-1 Satellite	0.4–1, 0.9–2.5	242 (total)	10	0.043	7.7 km
CHRIS	[132], 2004	PROBA-1 Satellite	0.4–1.05	18/62	1.25–11	0.03	18.6 km
CRISM	[133], 2007	MRO Satellite	0.362–1.053, 1.002–3.92	544 (total)	6.55	0.061	>7.5 km
AHSI	[134], 2019	Gaofen-5 Satellite	0.39–2.51	330 (total)	5, 10	0.043	60 km

IFOV: instantaneous field of view.

2.2.3. Application Status and Prospects

Hyperspectral imaging—a quantitative remote sensing approach—has been widely applied in environmental monitoring, vegetation analysis, geologic mapping, atmospheric characterisation, biological detection, camouflage detection and disaster assessment [135–141]. However, the application requirements for hyperspectral sensors underwent significant variations with the development of advanced sensors. First, a wide spectrum covering range is required to enhance the monitoring and detection capabilities of this system in various applications. This can be achieved by combining multiple sensors with different wavelength detection ranges [142] or by using an integrated system with a wide spectral range [80]. Second, the system sensitivity and preprocessing accuracies are equally important along with the spatial and spectral resolutions. For example, the AVIRIS next generation system [143] has been applied to detect methane, owing to its high signal-to-noise ratio and high data preprocessing accuracy. Finally, this technology facilitates the advantages of characterisation and quantification of targets; for example, hyperspectral imagers have been used for road network extraction.

2.3. SAR Imaging Technology

2.3.1. Data Acquisition Methods and Characteristics

SAR is an active remote sensing technology that uses microwaves with wavelengths of few centimetres as opposed to LiDAR, which functions using optical wavelengths (ultraviolet, visible, near-infrared or shortwave infrared light). Both these sensors measure the distance between the instrument and the target using the time delay of the echoes.

One of the main benefits of SAR is the acquisition of fine and detailed images through the clouds; moreover, this sensor can even work at night. In SAR, initially, short-pulsed microwave radiation is emitted and backscattered by the target; this backscattered signal from the illuminated area is then recorded. A SAR system with a long virtual antenna can generate fine spatial resolution. Among the currently available SAR systems, the spaceborne SAR sensors can provide sub-metre spatial resolutions. Notably, the long antenna aperture is realised in the cross-range direction and that the range resolution is given by the pulse width (in general, the bandwidth of the signal). Another particularity of the SAR compared with optical sensors is the low correlation between range and spatial resolution.

SAR uses a side-looking imaging geometry to generate 2D image data of the target area. In target areas with uneven topography, the side-looking imaging geometry distorts the SAR images because of various factors such as foreshortening, layover and radar shadows [144]. These distortions are challenging for mapping applications, especially in urban areas with high-rise buildings.

SAR sensors collect data in a complex domain, and the acquired data can be converted into intensity and phase information. SAR data are usually presented as 2D intensity images that provide information on the amount of backscattered signal. Since the backscattered signal is a combination of signals from multiple scatterers, the images have granular noise called speckle. Sensors transmit and receive horizontally and vertically polarised signals and provide either single, dual or quad polarisation data [145–147]. Phase information is used in SAR polarimetry and interferometry. Interferometric coherence, that is, the complex correlation coefficient between two SAR images, can provide information on the changes in the target changes and can be used in target classification as well [148–151].

2.3.2. Typical Sensors

A list of typical SAR satellite systems was recently presented in [152]. With the spotlight imaging mode, less than 1 m spatial resolutions can be achieved (e.g., RCM, Cosmo-Skymed, Terrasar-X, ICEYE). However, the area covered by the image is limited. In general, the swath width varies from 5 to 500 km, and for a wide swath, the resolution is of the order of tens of meters. Spaceborne systems use the X-, C-, S-, L- or P-band sensors. X-band sensors provide the highest resolution; however, their penetration into the vegetation canopy is limited [153]. Furthermore, the polarimetric capabilities of satellites vary significantly; for example, Alos-2, Radarsat-2 and Terrasar-X are fully polarimetric, providing HH, VV, VH and HV data (where HH is horizontal transmit, horizontal receive; VV is vertical transmit, vertical receive; VH is vertical transmit, horizontal receive and HV is horizontal transmit, vertical receive); Cosmo-Skymed and Sentinel-1 provide dual polarimetric HH and VV data and ICEYE provides single polarisation (VV) data. The incidence angle in most satellites can be adjusted between 10° and 60°.

The data recorded by Sentinel-1 of the European Copernicus system are openly and freely accessible. The Sentinel-1 data is similar to those of the previous European Envisat SAR; however, the data availability has increased because of the use of multiple satellites.

When satellites fly in a constellation, short revisit times are possible. Cosmo-Skymed (four satellites) and SAR-Lupe (five satellites) constellations were designed to provide intelligence information. New commercial microsatellite constellations such as ICEYE (10/18 satellites launched), Spacety (18/56 launched) and Capella XSAR (18/36 launched), can provide good temporal coverage. A particular constellation is also formed by TerraSAR-X and Tandem-X, allowing single-pass interferometry. Based on the data collected by these satellite constellations, a global digital elevation model (DEM) has been developed [154].

Airborne SAR systems that enable data acquisition at different wavelengths, higher resolutions and single-pass interferometry are more versatile than the spaceborne systems. Most of these airborne systems are used for research purposes; for instance, the German Aerospace Centre (DLR) has an F-SAR [155] system operating on a Dornier 228 aircraft, providing fully polarimetric data in the X-, C-, S-, L- and P-bands (maximum four bands simultaneously); this system enables single-pass interferometry in the X- and S-bands. In addition, several UAV SAR sensors are also available in the market; for example, SAR Aero offers 1.8 kg SAR sensors in the X-or L-band with 0.3 to 3 m for a range of up to 10 km.

2.3.3. Application Status and Prospects

The principal benefit of SAR is its all-day and all-weather imaging capability, enabling rapid mapping [156]. Therefore, the main application areas are related to emergency and security-related services, where the timeliness and availability of data are critical; for example, SAR data are operationally used in sea-ice mapping, which cannot be easily extracted using other remote sensing techniques, especially in cloudy winters. In addition, considerable scientific research has been conducted on agricultural monitoring, forest mapping and topographic mapping. Continuous environmental monitoring is possible using spaceborne SAR datasets. Moreover, some previously reported studies utilised SAR images for road extraction [72,157,158].

2.4. Airborne Laser Scanning (ALS)

2.4.1. Data Acquisition Methods and Characteristics

In airborne laser scanning (ALS), a LiDAR sensor is mounted on an aircraft, along with an inertial measurement unit and a global navigation satellite system (GNSS) receiver. The LiDAR sensor transmits narrow laser pulses towards the ground and generates a scanning pattern over the target area. ALS systems, typically based on an oscillating mirror and scanning patterns, receive the return signal, measure the time of signal travel and associate each return pulse with the GNSS time and scan angle at which the pulse was transmitted. The travel time can be converted to distance and then to height. The ALS technique can produce georeferenced 3D point clouds in the target area [75,159,160].

The operational ALS systems are mostly based on single-wavelength single-pulse linear-mode LiDAR. The new emerging multispectral ALS systems use a combination of LiDARs at different wavelengths. These sensors provide intensity data that can be used to derive colour images, such as optical imagery. The chief advantage of this technique is that the acquired data are independent of illumination conditions and are without shadows. Therefore, multispectral ALS systems have great potential for increasing the automation level in mapping. Geiger-mode LiDAR and single-photon LiDAR (SPL) are new ALS techniques that are sensitive to a single photon and can provide dense point clouds from higher flight altitudes, owing to their higher system sensitivity.

2.4.2. Typical Sensors

The biggest ALS manufacturers include Leica Geosystems (Switzerland), Teledyne Optech (Canada) and RIEGL (Austria). Examples of the current system are listed in Table 3. It can be seen from Table 3 that ALS can collect one to several million points per second, which secure the usability of the collected data for most surveyed and mapping cases. Meanwhile, the lidar systems can be used both the low-altitude platforms (UAV and helicopter) and high-altitude ones (fixed-wing aircraft). In addition, most ALSs do not operate in eye-safety wavelength: typical operating wavelengths for ALS systems are 532 (green), 1064 (near-infrared) and 1550 nm (shortwave infrared). The point density and accuracy depend on the flying height, reaching a maximum of 60 points/m^2. In addition, its accuracy depends on the range measurement accuracy combining with attitude measurement accuracy. In addition, small sensors are available for UAVs.

Table 3. Examples of currently available commercial ALS systems.

	Special Characteristics	WaveLength	Horizontal and Elevation Accuracy	Altitude	Pulse Repetition Frequency	Point Density
Leica Hyperion2+ [161], 2021	Multiple pulses in the air measured	1064 nm	<13 cm, <5 cm	300–5500 m	−2000 kHz	2 pts/m^2/4000 m, 40 pts/m^2/600 m
Leica SPL [162], 2021	Single photon	532 nm	<15 cm, <10 cm	2000–4500 m	20–60 kHz	6 million points per second, 20 pts/m^2 (4000 m AGL)
Optech Galaxy Prime [163], 2020	Wide-area mapping	1064 nm	1/10,000 × altitude, <0.03–0.25 m	150–6000 m	10–1000 kHz	1 million point per s, 60 pts/m^2 (500 m AGL), 2 pts/m^2 (3000 m)
Optech Titan [164], 2015	3 wavelength	1550 nm, 1064 nm, 532 nm	1/7500 × altitude, <5–10 cm	300–2000 m	3 × 50–300 kHz	45 pts/m^2 (400 AGL)
Riegl VQ-1560i-DW [165], 2019	Dual-wavelength, multiple pulses in the air measured.	532 nm, 1064 nm	/	900–2500 m	2 × 700–1000 kHz	2 × 666,000 pts/s, 20 pts/m^2 (1000 m AGL)
Riegl Vux-240 [166], 2021	UAV	1550 nm	<0.05 m <0.1 m	250–1400 m	150–1800 kHz	60 pts/m^2 (300 m)

2.4.3. Application Status and Prospects

ALS produces accurate 3D models of the target areas. The main application areas of ALS are in topographic mapping, particularly DEM production, city modelling and forestry. Even though the area covered by the ALS in a single scan is limited compared to that covered by the spaceborne sensors, nationwide datasets are available, especially for the Nordic countries. In addition, ALS is currently used to detect human activity in archaeology [167].

3. Road Extraction Based on Different Data Sources

High-resolution images, hyperspectral images, SAR images and LiDAR data are primarily used for road extraction. To date, various road extraction methods have been presented in previously reported studies, and the observed differences are due to the use of different data sources. In this section, we summarise the methods, application status and prospects of road extraction based on four data sources.

3.1. *Road Extraction Based on High-Spatial Resolution Images*

Extracting road information from high-resolution images requires clarification of the road features, including the radiation features, geometric features, topological features and texture features [71]. First, information on the various elements such as features, textures and edges are extracted from the image by analysing the road information. Then, the extracted image information is comprehensively analysed, selected and reorganised and combined with the road features. Finally, it is fused with the structural relationship, model and road-related rules of the road elements to identify the road.

3.1.1. Main Methods

Numerous road extraction algorithms based on high-resolution images have been developed over the past few decades, making it difficult to classify them. Traditional methods include automatic and semiautomatic extraction. Some methods based on deep learning have emerged and attracted considerable attention, owing to their high precision, in recent years. In this paper, we summarised the heuristic and data-driven road extraction methods based on two state-of-the-art reviews [34,71]. A comparison between the different data-driven methods applied on the Massachusetts road dataset is shown in Table 4. It can be seen from Table 4 that most data-driven methods can obtain a high precision (better than 0.8). Meanwhile, different methods have advantages and disadvantages.

Table 4. Comparison of different data-driven methods on Massachusetts road dataset.

Method	Advantages	Disadvantages	References	Precision
Patch-based DCNN	Weight sharing, less parameter	Inefficiency, large-scale training samples	[168], 2016 [38], 2017	0.905 0.917
FCN-based	Arbitrary image size, end to end training	Low fitness, low position accuracy, lack of spatial consistency	[36], 2016	0.710
DeconvNet-based	Arbitrary image size, end to end training, better fitness	High cost of computing and storage	[49], 2017 [51], 2018	0.858 0.919
GAN-based	More consistent	Non-convergence, gradient vanishing, and model collapse	[65], 2017 [66], 2017	0.841 0.883
Graph-based	High connectivity	Complex graph reconstruction and optimisation	[169], 2018 [170], 2020	0.835 0.823

The heuristic extraction methods can be subdivided into automatic and semiautomatic methods according to the degree of automation facilitated by the method. The semiautomatic extraction algorithms require initial seeds, and the user needs to check the results frequently. In these methods, the seeds and directions should be provided manually, and the algorithms can recognise and roll back the previous results. Several classic semiautomatic methods exist, such as the active contour model, dynamic programming, geodesic path and template matching. The active contour model, proposed by Kass et al. [171] and also known as the balloon snakes or ribbon snakes model, can extract road information through contour deformations from the labelled lines or points. The geodesic path method can produce a road-probability map through the extracted road edges. The dynamic programming method requires road parameters and mainly focuses on solving optimisation problems [172]. The template matching method can extract the road features through constructed windows and then matching the extracted points. All these semiautomatic methods include automatic processes.

Automatic road extraction uses information such as road topology and context features to extract and recognise roads through methods such as pattern recognition, computer vision, artificial intelligence, and image understanding [173–177]. Even though there are no fully automatic algorithms for all types of high-resolution images, several methods [71] such as segmentation-based, edge analysis, map-based, swarm intelligence-based, object-based and multispectral segmentation methods are much more efficient than the semiautomatic methods. Methods based on segmentation can identify regions through numerous algorithms [71] such as support vector machines (SVMs), artificial neural networks, Bayesian classifiers, mean shifts, watershed algorithms, super pixel segmentation, Gaussian mixture models, graph-based segmentation and conditional random field (CRF) models, which are usually used in combination to improve the classification accuracy. The edge analysis method is realised via edge detection, which is suitable for extracting the main roads. The map-based methods, especially OpenStreetMap, focus on urban roads. Swarm intelligence-based methods such as ant colony optimisation, artificial bee colony and firefly algorithms use discretisation and networking to simulate real biological behaviours [178]. Object-based methods are used to classify objects through object segmentation and feature characterisation. Multispectral segmentation methods usually require the support of multispectral or hyperspectral images [2].

With the application of deep learning techniques in road extraction, many data-driven methods such as patch-based CNN models, FCNs, deconvolution networks, GAN models and graph-based methods have been proposed for road extraction [34,71]. The patch-based CNN method can exploit a large image and predict a small patch of labels through the CNN models based on structured and refined CNNs; however, this method is time-consuming and computationally inefficient [34]. The FCN method predicts images by replacing connected layers with output labels and classifies images at the pixel level using the FCN-32 and U-shaped FCN models, thereby improving the road extraction accuracy [44,45]. A variant of FCN—DenseNets—including SegNet, DeepLab, RCFs, Y-net and U-Net decoder [49,51,52,54] can extract hierarchical features from images, especially high-resolution images. The GAN method includes a generator and discriminator to segment images and distinguishes between forged and real images, respectively [179]. The graph-based method realises the vector representation of road maps through iterative road tracking and polygon detection strategies.

3.1.2. Status and Prospects

Each algorithm has its advantages and disadvantages. For example, Lian et al. [71] compared the heuristic and data-driven road extraction methods based on four public datasets. Their results indicated that data-driven methods could achieve more than 10% accuracy compared to that achieved using heuristic methods. Similar conclusions were reported in [34]. However, data-driven methods have limitations such as the requirement of large training samples, long processing time and high-speed computing (most algorithms require graphics processing units). In addition, the training parameters used in one dataset may not be able to achieve high accuracy when used in another dataset. In contrast, some heuristic methods require less time and can meet real-time demands in some applications.

Road extraction using the currently available high-spatial-resolution images has some challenges. The key step in road extraction from high-resolution images is to describe the road features. Describing the linear or narrow bright band of a road, which provides good detection results, is the main focus of most existing methods. However, with the improvement in image resolution, we can obtain more noise interference (shadows, buildings and road obstructions) and more detailed road features; therefore, the road objects can also be described more precisely in this case [34,173]. Furthermore, road objects include many complex phenomena such as occlusion or shadows, discontinuities, sharp bends and near-parallel boundaries with constant widths. Incorporating all these factors and modelling them into a single model is almost impossible. Therefore, it is essential to establish a multimode approach to extract roads from images with high spatial resolutions.

3.2. *Road Extraction Based on Hyperspectral Images*

Multispectral remote sensing images have been used in road extraction because of their high spatial resolution and multiple spectral features. The commonly used data sources are the satellite multispectral images including QuickBird, IKONOS [180,181], Worldview 2 satellite [182], Landsat satellites [183] and Gaofen 1 and Gaofen 2 satellites. In addition, because of the large number of bands (generally more than 100 bands) and continuous spectrum bands, hyperspectral images are used for conventional road extraction as well as show great potential for road condition detection, road material identification, road pothole detection and crack detection.

3.2.1. Main Methods

Most road extraction methods using multispectral images are based on high-spatial-resolution images, and these methods can also be divided into heuristic and data-driven methods. Similar to the road extraction from high-resolution images, the heuristic methods in this case can also be divided into semiautomatic and automatic categories; some applications of these heuristic methods are reported in [184–189]. However, few data-driven methods are exclusively used for road extraction based on multispectral images, owing to the lack of public datasets. There are fewer road extraction methods based on hyperspectral images than on high-resolution images, and some of them are included in hyperspectral classifications. In this paper, we introduce road extraction methods using hyperspectral images based on different platforms.

The spaceborne hyperspectral imagers can realise a larger swath and provide a better stability platform than do the airborne instruments; hence, the spaceborne hyperspectral imagers are suitable for large-area work. However, they are mainly used to identify and extract arterial roads such as highways because of their low spatial resolution. The Hyperion hyperspectral imager of the US—EO-1 satellite—is the currently available spaceborne hyperspectral sensor for road extraction. Sun [190] used Hyperion hyperspectral data to complete the road network extraction from an image using three steps: road searching, road tracking and road connecting. In road searching, the spectral information in the image is used to find road features, and several different road feature extraction methods are qualitatively compared; however, no quantitative extraction accuracy results are provided in Sun's report.

Airborne hyperspectral imagers can achieve higher spatial and spectral resolutions than do the spaceborne platforms; however, their operating efficiency is lower than that of the spaceborne instruments. They are mainly used for the extraction of urban roads and detection of road conditions. Airborne hyperspectral instruments currently used for road extraction mainly include AVIRIS, CASI, HyMap, HYDICE and AsiaFenix [122]. In 2001, Gardner et al. [191] used the hyperspectral dataset of AVIRIS in Santa Barbara, USA, to map the different types of typical urban surface features through multiterminal spectral analysis. Then, the Q-tree filter was used to distinguish between the roofs and the roads constructed using the same materials and exhibiting similar spectra. The visible results showed that the main roads in the image can be preliminarily extracted; however, several shortcomings were observed in the extraction of roads blocked by vegetation and connectivity of the road network. Noronha et al. [192] used the urban hyperspectral dataset of AVIRIS and a spectral database, based on the surface materials of the main urban features collected in the field, to extract the road centreline and observe the road surface conditions. Furthermore, the optimal parameters for designing a multispectral instrument to extract urban land-use types was proposed based on these reported results. The overall classification accuracy and kappa coefficient were 73.5% and 72.5%, respectively [192]. Based on the airborne hyperspectral image data of HYDICE and HyMap, Huang and Zhang [193] used an adaptive mean-shift method to accurately classify six major urban features in the image, including roads, houses, and grass. The overall classification accuracy was above 97%, and the road classification accuracy was above 95%. Resende et al. [194] used CASI-1500 airborne hyperspectral data to study the extraction of asphalt roads in cities. The results based on ISODATA unsupervised classification and maximum likelihood supervised classification methods qualitatively showed that the hyperspectral image could be used to extract the main asphalt roads in the city; however, they did not report the extraction accuracy. In 2012, Mohammadi [195] used HyMap airborne hyperspectral image data to study the classification of materials used in urban roads and the state of asphalt road conditions. He mainly distinguished asphalt roads, cement roads, and gravel roads, and based on this result, reported three road conditions: good, medium and poor asphalt roads. However, the experimental results were limited by the spatial resolution of the dataset, and a large number of unclassified pixels were not analysed by the method. Therefore, further studies are required to improve the methods used for reducing the number of unclassified pixels.

Currently, some publicly available airborne hyperspectral datasets such as Pavia Centre and University area hyperspectral datasets and Indian Pine hyperspectral datasets [196,197] are also used for road classification and recognition. In 2012, Liao et al. [196] proposed a directional morphology and semi supervised feature extraction to classify three hyperspectral datasets. The classification accuracy of the roads was the highest, reaching more than 97%; however, this classification accuracy was closely related to the number of training samples and extracted features. Miao et al. [197] studied the extraction of road centrelines from high-resolution images based on shape features and multiple adaptive regression splines. This method combined the shape features and spectral information to extract road segments from high-resolution images and then used multivariate adaptive regression spline functions to extract the road centrelines. The method was applied to the Pavia Centre hyperspectral dataset, and an extraction accuracy of 99% was obtained for the road centreline extraction. This method was based on uniform surface properties; hence, it was suitable only for high-resolution images and not for low-resolution images. In addition, the main limitation of this method was that the threshold in the method must be determined manually.

In addition to spaceborne and airborne hyperspectral imagers, UAV hyperspectral imaging systems that have gradually emerged in recent years have garnered increasing attention owing to their low cost and high spatial resolutions. These systems are mainly used for road condition detection and road material identification in specific areas [15,198]. However, their operating efficiency is lower than that of the spaceborne and airborne platforms, because of their low flight altitude. A summary of road extraction using hyperspectral images is shown in Table 5, which indicates that hyperspectral imaging systems of different platforms have different characteristics for road extraction due to spatial resolution. Spaceborne hyperspectral imagers are primarily used to extract main roads, while airborne hyperspectral imagers can be used for road quality assessment and road condition monitoring.

Table 5. Summary of road extraction using hyperspectral images.

Method	Platform	Characteristic	References
Traditional process includes the spectral information	Spaceborne	Extract the main roads	[190], 2003
Spectral mixture and Q-tree filter	Airborne	Assess road quality	[191], 2001
Pixel to pixel classification	Airborne	Extract asphalted urban roads	[194], 2008
Spectral angle mapper	Airborne	Road classification and condition determination	[196], 2012
Computing the angle from spectral response	UAV	Detect pavement roads	[198], 2019

3.2.2. Status and Prospects

Only a few reports on extracting road information from spaceborne hyperspectral images are available. This is because the spatial resolution is insufficient to extract road information accurately (e.g., the spatial resolution of Hyperion and Gaofen-5 hyperspectral imagers is 30 m), especially for urban roads and narrow roads. In addition, spaceborne hyperspectral images are difficult to acquire, and public datasets for road extraction are not available. Airborne hyperspectral images are still the main data source for the study of road extraction, because of their higher spatial resolutions and lower costs compared to those of the spaceborne data.

Hyperspectral images have shown considerable potential for road condition detection, road material identification, road pothole detection and crack detection. However, the preprocessing accuracy of the hyperspectral images should be improved to promote their applications. Hyperspectral data with geometric correction and relative radiometric calibration can meet this requirement for qualitative applications such as road detection. However, for quantitative applications such as pavement material recognition, more steps are required, such as high-precision absolute radiometric calibration, spectral calibration and atmospheric correction.

3.3. *Road Extraction Based on SAR Images*

3.3.1. Main Methods

In general, roads in SAR images appear as dark linear features. However, the differential orientation of roads and antennas influences the ability of SAR to identify roads. In traditional heuristic methods, road segments are often extracted using an edge/line detector. Then, a graph is generated from the segments and optimised, and the segments are connected to develop a coherent road network. Recently, data-driven deep learning-based semantic segmentation methods have been reported. Moreover, road junctions and bridges are important parts of road networks and have been the topic of some previous studies.

Road network extraction from SAR images has been reported in various studies [199–201]. In [200], two local line detectors were used, and the results were fused to find candidates for road segments. The road segments were connected using a Markov random field, and an active contour model (snake) was used for the postprocessing. In [200], this method was applied on dense urban areas, and very-high-resolution data and different flight directions were combined to improve the results. In [201], constant false alarm rate detection, morphological filtering, segmentation and Hough transformation were integrated to recognise roads in high-resolution polarimetric airborne SAR images. The strong backscattering from fences was used to detect bridges, and then the roads were recognised using a Hough transformation

The fusion of different SAR datasets and algorithms in early studies improved road extraction accuracy. In [202], different preprocessing algorithms, road extractors and different images of the same area were fused, and a multiscale approach was used for road extraction. For SAR, a combination of different view angles is also proposed. Lisini [203] extracted road networks by combining the line extractor results with two classification results. Hedman [204] detected rural and urban areas and then fused a road extractor for rural areas to develop an extractor designed for urban areas.

The latest reported studies used new high-resolution data, usually from the TerraSAR-X or GaoFen-3 satellites. In [205], multiscale geometric analysis was performed on vectorised detector responses for road network grouping using two TerraSAR-X datasets. In [206], a method suitable for analysing SAR images of different resolutions was proposed. A weighted ratio line detector was developed to extract the road ratio and direction information. The road network was constructed using a region-growing method and tested using four SAR datasets obtained from different study areas. Saati [207] extracted road networks based on a network snake model and three TerraSAR-X images. Xu [208] introduced an algorithm in which road segment extraction and network optimisation were performed simultaneously using a Bayesian framework, multiscale feature extractor and CRF; for this analysis, Xu used the TerraSAR-X and airborne SAR data. Xiong [209] proposed a method based on vector Radon transformation, and promising results were presented for six SAR images of different resolutions, bands and polarisations; these images were obtained from airborne SAR, GaoFen-3 and TerraSAR-X. In general, for road extraction from new very-high-spatial-resolution (VHR) data, completeness of 74–93% and correctness of 70–94% have been reported, depending mostly on the study site.

Interferometric information has been used in road extraction by Jiang et al. [210], and the best results were obtained by the fusion of intensity and coherence information. Roads were considered as distributed scatterers and were separated from permanent and temporally variable scatterers. A constant false-alarm-rate line detector based on Wilks' test statistics has been proposed by Jin et al. [211] for polarimetric SAR images.

Deep learning-based methods have been the topic of a few recent studies; they can be used to extract initial road information with good quality for network construction. Zhang [158] compared U-Net (FCN) and CNN to machine learning methods using dual-polarisation Sentinel-1 data; he reported that VV polarisation was better than VH, and dual-polarisation data was better than the single-polarisation data for road extraction. F1 score of 94% was achieved (the same area for training and testing but different shuffled samples were used) using the U-Net and dual-polarisation data. This F1 score was better than that of the best machine learning method (random forest) by 5%. Henry [46] enhanced the fully CNN (FCNN) sensitivity for thin objects and compared the FCN-8s, U-Net and Deeplabv3+ methods for road segmentation. The sensitivity was increased by addressing class imbalance in training and using spatial tolerance rules. A summary of road extraction using SAR images is shown in Table 6. It can be seen that the precision in diverse situations is quite different. Appropriate methods or a combination of multiple methods should be considered according to the scene's characteristics.

Table 6. Summary of road extraction using SAR images.

Method	Category	Characteristic	References	Precision
Multiple Detectors	Heuristic	Fusion of different pre-processing algorithms, road extractors	[202], 2003	0.580 correctness
Line based on vector Radon transform	Heuristic	Suitable for different platform SAR images	[209], 2019	0.700–0.940 correctness
Multitemporal InSAR covariance and information fusion	Heuristic	Use interferometric information	[210], 2017	0.816 correctness
FCN-based	Data-driven	Automatic road extraction	[158], 2019	0.921
FCN-8s	Data-driven	Lack efficiency	[46], 2018	0.717

3.3.2. Status and Prospects

Automatic road extraction from SAR imagery remains solely experimental to date. For operational tasks, semiautomatic methods [172] with human intervention or manual postprocessing are required. Large-scale tests have not been conducted, and for the reported methods, remarkable variations in the results between different test sites have been observed. The studied road types include forest roads, city streets, highways, desert roads, gravel roads, paved roads, icy roads and bridges. However, different or mixed road classes have been evaluated in only a few studies. Different methods are required for different road types and study areas. In general, the developed methods are complex and involve many steps.

Owing to the side-looking sensor [212], SAR produces many linear features, causing many false alarms in road extraction. In addition, road detection varies with the looking direction, especially in urban areas (radar shadows). The speckle and speckle filters also affect the road extraction accuracy. In addition, the roads may appear bright because of the surrounding structures, instead of appearing as dark lines. Therefore, the road appearance differs depending on the SAR image resolution. A highly complex image is obtained with high-resolution data, showing strong geometric effects in urban areas. Therefore, the use of multiple datasets is often required. Conversely, road detection from SAR is independent of road surface material; this feature is less pronounced in the optical detection method. Thus, the fusion of SAR and optical signals can be beneficial.

3.4. Road Extraction Based on LiDAR Data

3.4.1. Main Methods

Ground points and nonground points can be easily distinguished based on pulse and elevation information in airborne LiDAR data. To classify ground and road/nonroad areas, intensity information is needed. Roads, such as water surfaces, have low intensity. In addition to the surface reflectance, the intensity values are affected by atmospheric attenuation, transmitted power, detection range and incidence angle. Therefore, calibration is required to apply the methods over large areas. Road detection using intensity is usually based on the surface homogeneity and consistency of roads. The LiDAR-based methods for road extraction either use point cloud processing and classification or are based on a digital surface model (DSM), digital terrain model (DTM) and intensity raster produced from the point clouds. In some reported studies, the main focus was on data classification, while in others, a complete road network was extracted.

Point cloud-based automatic road extraction using LiDAR height and intensity was first proposed by Clode [213]; in this study, a hierarchical classification method was used to classify point clouds progressively into roads and nonroads. Individual points were selected based on the height difference to generate a DTM, and the intensity value was obtained via filtering based on point density and morphological filtering of a binary image. The method was enhanced in [214], where a phase-coded disk algorithm was introduced to vectorise the binary road network image. Hu [215] proposed a method for road centreline extraction using the salient linear features observed in the images of complex urban areas; in this method, tensor voting was used to eliminate the nonroad areas.

Intensity variations in the road network were considered in [216], where road points were selected from ground points based on a local intensity distribution histogram and filtered by roughness and area. Hui [217] extracted road centrelines using three steps; a skewness-balancing algorithm was proposed to obtain the intensity threshold. A rotating neighbourhood algorithm was proposed to extract the main roads (by removing the narrow roads), and a hierarchical fusion and optimisation algorithm was proposed to extract the road network. For the three test sites, correctness values of 43–92% and completeness of 36–91% were obtained.

A raster-based road extraction method for grid-structured urban areas was proposed by Zhao [218]; in this study, the ground objects were classified, road centrelines were extracted using a total least square line fitting approach, and a voting-based road direction was used to evaluate each road segment's reliability by removing areas such as parking lots from the road segments. For a complete road network extraction, different parts of the road network such as junctions and bridges need to be considered as well. Chen [219] proposed an automatic method to detect and delineate road junctions from rasterised ALS intensity and normalised DSM in three steps: roughness-enhanced Gabor filters for key point extraction, a higher-order tensor voting algorithm to find the junction candidates, and a geometric template matching to identify the road junction positions and road branch directions. A bridge detection algorithm was proposed by Sithole [220]; in this method, DTM cross-sections, that is, profiles, were used to identify bridges. Moreover, forest road detection is possible using detailed ALS DSMs.

Road details and complex structures can be extracted from very dense point clouds as well and modelled. In [221], road points were accurately labelled from a dense point cloud of an urban area, using an approximate 2D road network map as the input. They combined both large-scale (snake smoothness) and small-scale (curb detector) cues to extract roads. The method worked on all types of roads, including tunnels, bridges and multilevel intersections. In [222], UAV data was used to perform fine-scale road mapping. Soilan [223] studied the automatic extraction of road features (sidewalks, pavement areas and road markings) from high-density ALS point clouds.

3D modelling is required for the most complex parts of the road network such as overpasses and multilevel intersections. The use of LiDAR point clouds to derive 3D city models was reviewed in [77]. Cheng [160] studied the detailed 3D reconstruction of multilayer interchange bridges, and satisfactory results were obtained for very complex bridges.

In several studies, road detection has been carried out as a part of land cover mapping. Urban land cover mapping by integrating rasterised LiDAR height and intensity data at the object level was proposed by Zhou [224]; for this method, an accuracy similar to that of multispectral optical imagery accompanied by ALS DSM was obtained. Matkan [225] classified the point cloud into five land cover classes using SVM; during the postprocessing, gaps in the road network were located and filled using a method based on Radon transformation and spline interpolation.

Land cover classification using multispectral ALS (MS-ALS) has been the topic of several recently reported studies. Even though a complete road network was not extracted, the road classification results were promising [164,226]. In [226], the point-based classification completeness for roads was 86% and 92%. In the raster-based classification, accuracies of 92% and 86% were obtained. Karila [227] used rasterised MS-ALS data for a typical road surface (that is, asphalt and gravel) and classified the road types from highways to cycle ways. Due to the lack of shadows, more complete roads (80.5%) were retrieved using this method than using the optical aerial images (71.6%). Ekhtari [228] classified multispectral point clouds into 10 land cover classes using an SVM. Three types of asphalt and concrete classes were included. In general, slightly better results were obtained when the classification was carried out in the point cloud domain; however, the computational costs increased significantly.

Deep learning methods for MS-ALS have been explored in a few recent studies. Pan [229] used deep learning-based high-level feature presentation (deep Boltzmann machine) and machine learning methods for land cover classification. Pan [230] proposed a CNN-based classification approach for MS-ALS data. The classification accuracy and computational performance of the constructed CNN model were superior to those of the classical CNN models. Yu [231] proposed a hybrid capsule network using MS-ALS data. The data were rasterised based on the elevation, the number of returns and intensity of the three channels, and an accuracy of 94% was obtained for the road classification. Dense point clouds from SPL were used for land cover classification in [232]. Due to the rough appearance of the intensity images created from the SPL data, small features such as narrow roads were often difficult to distinguish in the intensity images. A summary of road extraction using LiDAR data is shown in Table 7. Notably, ML-ALC has become the main approach for road extraction in recent years.

Table 7. Summary of road extraction using LiDAR data.

Method	Category	Characteristic	References	Correctness
Hierarchical fusion and optimisation	ALS	Extract road centreline	[217], 2015	0.914
Point-based classification Raster-based classification	MS-ALS	Land cover classification	[226], 2017	0.920 0.860
Object-based image analysis and random forest	MS-ALS	road detection and road surface classification	[227], 2017	0.805
Support vector machine	MS-ALS	Three types of asphalt and a concrete class	[228], 2018	0.947 (Overall accuracy)
Hybrid capsule network	MS-ALS	Land cover classification	[231], 2020	0.979 (Overall accuracy)

3.4.2. Status and Prospects

Airborne LiDAR data are used in the 3D modelling of city road networks [77]. ALS provides direct 3D information for road extraction and is less affected by occlusions and shadows than do optical data. In addition, ALS provides 3D road information with elevation; this is especially useful in complex interchange areas. However, the area covered during the flight considerably limits the automatic road extraction process using ALS. Extensive investigations are still required to address the underlying issues, i.e., the development of fully automatic algorithms suitable for various landscape and road types, application of intensity data over large areas, reducing the number of false positives (car parks, squares, playgrounds, etc.) and identifying as well as connecting road segments shadowed by occlusions (e.g., trees and vehicles). Road markings must be taken into consideration in VHR. The national ALS datasets provide a good basis for mapping; however, these datasets are seldom updated. LiDAR sensors mounted on mobile mapping systems, UAVs or VHR satellite imaging instrument can be used for map updating. In addition, the new MS-ALS data [227,228] and new dense point clouds created by collecting single photons enable the automatic detection of roads with higher accuracy.

4. Combination of Multisource Data for Road Extraction

4.1. Combination of High-Resolution Images with Other Data for Road Extraction

High-resolution images reveal very fine details of the earth's surface and geometry; however, high-resolution imaging also increases the geometric noise and only provides spectral and spatial information on the surface. LiDAR data, as a special data source, can provide 3D information about objects. Tiwari et al. [233] proposed automatic road extraction methods through an integrated approach involving ALS altimetry and high-resolution imaging. The method was used to extract road information without background objects, and showed an increased road extraction accuracy of 90% when applied to Amsterdam data. Hu et al. [234] proposed a grid-structured urban road network extraction method using LiDAR data and high-resolution imagery. A significant improvement in the road extraction accuracy was obtained using this method than using high-resolution imagery or LiDAR data. Zhang et al. [235] proposed a method to improve the accuracy of road centreline extraction using high-resolution images and LiDAR data. The method adopted the minimum area bounding rectangle-based interference-filling approach, multistep approach and Harris corner detection. The experimental results based on the datasets of Vaihingen, New York, and Guangzhou showed that the proposed method was efficient in identifying complex scenes.

4.2. Combination of Hyperspectral Images with Other Data for Road Extraction

Feng et al. [236] fused hyperspectral images and LiDAR data to map urban land use using a state-of-the-art CNN. To improve the speed of the network design, the same structure was used in both the hyperspectral and LiDAR branches. Each branch used a residual block to extract multiscale, parallel, and hierarchical features. The experimental results underlined the efficient road extraction performance of the proposed method. When only hyperspectral images and LiDAR data were used, the highway classification accuracies were found to be 65.35% and 42.08%, respectively. However, this highway classification accuracy increased to 80.89% when fused data was used. Elaksher et al. [237] combined the LiDAR-based hyperspectral images obtained from the AVIRIS and DEM. First, a vector layer of polygons was constructed using the DEM data. Second, the buildings in the hyperspectral images were removed, and then the road and water were classified using a supervised classifier. The experimental results demonstrated that the performance of this classification process could be improved by using LiDAR data to remove the buildings from the hyperspectral image before the classification. The detection rate of the road was 91.3%, and the false alarm rate was 0. Two examples of multiple remote sensing data fusion are presented in [238]. One is the fusion of hyperspectral images with SAR images; this fused data can improve the detection accuracy of the target. The other is the fusion of hyperspectral images with high-resolution images. The spatial–spectral information of the target was fully analysed using the two combined data sources, and the identification accuracy was improved considerably.

4.3. Combination of SAR Images with Other Data for Road Extraction

Several studies have been published on the fusion of optical imagery and SAR data. Cao [239] proposed road extraction via the fusion of infrared and SAR images. Lin et al. [240] compared multiple remote sensing datasets (Spot5, IKONOS, QuickBird, DMC and airborne SAR datasets) and algorithms. Road trackers were designed for five different road types: national highways, interstate highways, railroads, avenues and lanes. Evidently, fused multiple remote sensing data were more efficient than a single data source. Perciano et al. [241] fused TerraSAR-X and QuickBird data for road network extraction at two test sites, and the road extraction accuracy for the fused data was 10–30% higher than those obtained for individual datasets. Multitemporal SAR image stacks (TSX and CSK) were also studied. Bartsch et al. [242] studied the arctic settlements using Sentinel-2 optical and Sentinel-1 SAR satellite images. Pixel-based classification using a gradient boosting machine and a deep learning approach based on windowed semantic segmentation using U-Net architecture were compared. Asphalt roads were easily detected than gravel roads, and for arctic mapping, both methods and sensors were recommended. Liu et al. [243] studied urban area mapping using Sentinel-1 SAR and Sentinel-2 optical data and proposed the integration of object-based postclassification refinement and CNNs for land cover mapping. Notably, for road mapping, SAR backscattering provided different physical information on roads (low backscatters) than that provided by optical remote sensing; moreover, the roads were identified with higher accuracy by combining the optical data with that of the SAR. Lin et al. [244] extracted impervious surfaces using optical, SAR and LiDAR DSM data. The non-shadow and shadow classes were trained using the combined optical–SAR–LiDAR data. As a result, the shadow effects in the classification results were reduced.

4.4. Combination of LiDAR with Other Data for Road Extraction

Aerial imaging cameras are often accompanied by LiDAR sensors in airborne mapping systems. Thus, it is common to fuse LiDAR point clouds and optical aerial imagery. In addition, optical satellite data are used as well. In particular, additional colour information is useful for single-channel LiDAR. Segmentation of the input imagery is often performed to enable object-based fusion of the datasets.

Kim [245] proposed a method to improve the classification of urban areas by fusing high-resolution satellite images (WorldView-2) and ALS data. Special attention was paid to the elevated roads, which were first detected in LiDAR ground points. Then, buildings were detected, and supervised SVM classification was performed on areas without elevated roads or buildings. Liu [246] proposed a road extraction framework based on the fusion of ALS point clouds and aerial imagery; in this framework, pseudo scan lines were created from the fused data, and a rule-based edge-clustering algorithm was used to extract the road segments. Mendes classified road regions using an ANN by integrating aerial RGB (where R is red, G is green and B is blue) images, laser intensity and height images. Compared to the use of optical data alone, incorporating the laser intensity data helped to overcome the road obstructions caused by shadows and trees, and the height information helped in separating the aboveground objects from the ground objects. Zhang [235] proposed an object-based method for road centreline extraction from aerial images and ALS DSMs; using this method, a completeness and correctness of over 90% was obtained for two of the test data sets, and a completeness and correctness of over 80% was obtained for a third large site. Further developments were proposed for curved roads.

4.5. Some Scopes of Future Research in Road Extraction

A summary of road extraction based on different data sources is presented in Table 8. Some prospects can be derived based on the status of current road extraction from high-resolution images. First, data-driven road extraction methods exhibit excellent performance and high extraction accuracy [34]. Therefore, more robust data-driven methods should be developed and verified using different datasets. In addition, it is difficult to obtain high detection accuracy using only one algorithm in some cases; therefore, a combination of multiple road extraction methods must be studied [247]. Finally, the combination of data sources (e.g., hyperspectral, SAR and LiDAR) should be evaluated further. Spatial resolution is one of the most important factors affecting the performance of road extraction methods. High-spatial-resolution images can describe fine objects in detail. However, this increases the spectral variability within the class [248]. Wang et al. [249] demonstrated that images with a spatial resolution of 0.5 m had higher accuracy than those with a spatial resolution of 0.1 m. Data-driven methods show higher road extraction accuracy with improved spatial resolution than do the traditional heuristic methods [34,71].

Table 8. Summary of road extraction based on different data sources.

Data	Resolution/ Mapping Unit	Extent	Advantages	Roads Extracted Mostly by
High spatial resolution [71], 2020	0.5–10 m	Local/regional/global	Most tools available, "basic" software	Colour, texture
Hyperspectral [198], 2019	0.25–30 m/ (>100 channels)	Local/regional	Spectral information	Colour, texture and spectral features
SAR [72], 2014	1–10 m	Local/regional/global	See through clouds, rapid mapping	Linear features/edge
ALS [75], 2017	0.25–2 m	Local (nationwide)	Height information	3D geometry (intensity)

It is necessary to continue to study road extraction from airborne hyperspectral images [80]. In addition, new road extraction methods based on hyperspectral images should be developed. In particular, the advantages of a large number of bands and continuous spectrum coverage of hyperspectral images should be further exploited, and new data-driven methods should be proposed [250]. Finally, hyperspectral image datasets with a wide coverage area and road labels should be produced to promote the road extraction application of hyperspectral data.

Recently, the open-access global SAR satellite datasets (Sentinel-1) have enabled the mapping and monitoring of large areas. However, the resolution of this method is limited. Microsatellite constellations can acquire large amounts of very-high-resolution data with higher frequency; however, this aspect has not been studied in detail. For ALS, deep learning methods [251] for image classification are rapidly emerging, followed by methods for earth observation. Publicly available open training datasets acquired by the earth observation satellite are being used for road extraction. The highest accuracies for road classification have been reported using deep neural network algorithms. However, these studies cover limited areas where they perform relatively well, but no large-scale tests have been conducted yet.

5. Conclusions

High-resolution and hyperspectral images have been widely used in digital road network extraction. More satellites with high spatial resolution and short revisit periods are being developed and launched, promoting the development of heuristic and data-driven road extraction methods. However, few of these hyperspectral satellite data can be used for road extraction. Therefore, airborne systems are still the main approach for the acquisition of hyperspectral data [80,252]. Data-driven methods have high accuracy and show significant potential; however, transfer learning needs to be improved. In addition, the combination of high-resolution image data with other data sources such as LiDAR is one feasible approach to solve some challenging issues such as occlusion or shadows.

Large areas can be rapidly mapped using weather-independent spaceborne SAR images; for example, images acquired after sudden changes in the target area. In addition, global datasets enable global mapping. However, the roads are challenging to interpret because of the interaction of the signal with the surrounding areas. ALS provides excellent data for the generation of topographic databases and detailed mapping of limited areas. Multispectral ALS may be the best remote sensing data source for road mapping; however, only small areas are covered at a time in this case. Further studies are required for developing sensors for various landscapes and road types; moreover, fully automated road detection is still in its infancy.

High-resolution imaging, hyperspectral imaging, SAR imaging, and LiDAR are currently the primary techniques for road extraction. As shown in Table 8, different data sources have unique characteristics. For example, high-resolution images have high spatial resolution and contain rich textures, shapes, structures, and neighbourhood relations of ground objects. Hyperspectral images have multiple data dimensions and rich spectral features. In addition, SAR imaging and LiDAR are less affected by external environmental factors such as clouds, fog, and light and can operate in all weather conditions. Combining different remote sensing data to use their respective advantages is a notable approach for developing advanced road extraction methods in the future.

Author Contributions: Bibliographic review, J.J., H.S., C.J., K.K. and M.K.; paper organisation, Y.C. and J.J.; drawing of conclusions, J.J.; writing—original draft preparation, J.J., H.S., C.J., K.K. and M.K.; writing—review and editing, E.A., E.K., P.H., C.C. and T.X.; supervision, Y.C.; project administration, J.H.; funding acquisition, Y.C. and T.W. All authors have read and agreed to the published version of the manuscript.

Funding: This research was financially supported by Academy of Finland projects "Ultrafast Data Production with Broadband Photodetectors for Active Hyperspectral Space Imaging (No. 336145)", Forest-Human-Machine Interplay-Building Resilience, Redefining Value Networks and Enabling Meaningful Experiences (UNITE), (No. 337656) and Strategic Research Council project "Competence-Based Growth Through Integrated Disruptive Technologies of 3D Digitalization, Robotics, Geospatial Information and Image Processing/Computing–Point Cloud Ecosystem (No. 314312). Additionally, the Chinese Academy of Science (No. 181811KYSB20160040 XDA22030202), Shanghai Science and Technology Foundations (No. 18590712600) and Jihua lab (No. X190211TE190) and Huawei (No. 9424877) are acknowledged.

Institutional Review Board Statement: Not applicable.

Informed Consent Statement: Not applicable.

Data Availability Statement: Not applicable.

Acknowledgments: The authors would like to thank the contributions of the editor and reviewers.

Conflicts of Interest: The authors declare no conflict of interest.

References

1. Huang, B.; Zhao, B.; Song, Y. Urban Land-Use Mapping Using a Deep Convolutional Neural Network with High Spatial Resolution Multispectral Remote Sensing Imagery. *Remote Sens. Environ.* **2018**, *214*, 73–86. [CrossRef]
2. Wang, J.; Treitz, P.M.; Howarth, P.J. Road Network Detection from SPOT Imagery for Updating Geographical Information Systems in the Rural–Urban Fringe. *Int. J. Geogr. Inf. Syst.* **1992**, *6*, 141–157. [CrossRef]
3. Mena, J.B. State of the Art on Automatic Road Extraction for GIS Update: A Novel Classification. *Pattern Recognit. Lett.* **2003**, *24*, 3037–3058. [CrossRef]
4. Coulibaly, I.; Spiric, N.; Sghaier, M.O.; Manzo-Vargas, W.; Lepage, R.; St-Jacques, M. Road Extraction from High Resolution Remote Sensing Image Using Multiresolution in Case of Major Disaster. In Proceedings of the 2014 IEEE Geoscience and Remote Sensing Symposium, Quebec City, QC, Canada, 13–18 July 2014; pp. 2712–2715.
5. Cheng, G.; Zhu, F.; Xiang, S.; Pan, C. Road Centerline Extraction via Semisupervised Segmentation and Multidirection Nonmaximum Suppression. IEEE Geosci. *Remote Sens. Lett.* **2016**, *13*, 545–549. [CrossRef]
6. McKeown, D.M. Toward automatic cartographic feature extraction. In *Mapping and Spatial Modelling for Navigation*; Pau, L.F., Ed.; Springer: Berlin, Heidelberg, 1990; pp. 149–180.
7. Robinson, A.H.; Morrison, J.L.; Muehrcke, P.C. Cartography 1950–2000. *Trans. Inst. Br. Geogr.* **1977**, *2*, 3–18. [CrossRef]
8. Ulmke, M.; Koch, W. Road Map Extraction Using GMTI Tracking. In Proceedings of the 2006 9th International Conference on Information Fusion, Florence, Italy, 10–13 July 2006; pp. 1–7.
9. Koch, W.; Koller, J.; Ulmke, M. Ground Target Tracking and Road Map Extraction. *ISPRS J. Photogramm. Remote Sens.* **2006**, *61*, 197–208. [CrossRef]
10. Niu, Z.; Li, S.; Pousaeid, N. Road Extraction Using Smart Phones GPS. In Proceedings of the 2nd International Conference on Computing for Geospatial Research & Applications, Washington, DC, USA, 23–25 May 2011; Association for Computing Machinery: New York, NY, USA, 2011; pp. 1–6.
11. Bhoraskar, R.; Vankadhara, N.; Raman, B.; Kulkarni, P. Wolverine: Traffic and Road Condition Estimation Using Smartphone Sensors. In Proceedings of the 2012 Fourth International Conference on Communication Systems and Networks (COMSNETS 2012), Bangalore, India, 3–7 January 2012; pp. 1–6.
12. Balali, V.; Ashouri Rad, A.; Golparvar-Fard, M. Detection, Classification, and Mapping of U.S. Traffic Signs Using Google Street View Images for Roadway Inventory Management. *Vis. Eng.* **2015**, *3*, 15. [CrossRef]
13. Zhang, M.; Liu, Y.; Luo, S.; Gao, S. Research on Baidu Street View Road Crack Information Extraction Based on Deep Learning Method. *J. Phys. Conf. Ser.* **2020**, *1616*, 012086. [CrossRef]
14. Li, D.; Ke, Y.; Gong, H.; Li, X. Object-Based Urban Tree Species Classification Using Bi-Temporal WorldView-2 and WorldView-3 Images. *Remote Sens.* **2015**, *7*, 16917–16937. [CrossRef]
15. Pan, Y.; Zhang, X.; Cervone, G.; Yang, L. Detection of Asphalt Pavement Potholes and Cracks Based on the Unmanned Aerial Vehicle Multispectral Imagery. *IEEE J. Sel. Top. Appl. Earth Obs. Remote Sens.* **2018**, *11*, 3701–3712. [CrossRef]
16. Irwin, K.; Beaulne, D.; Braun, A.; Fotopoulos, G. Fusion of SAR, Optical Imagery and Airborne LiDAR for Surface Water Detection. *Remote Sens.* **2017**, *9*, 890. [CrossRef]
17. Wang, W.; Yang, N.; Zhang, Y.; Wang, F.; Cao, T.; Eklund, P. A Review of Road Extraction from Remote Sensing Images. *J. Traffic Transp. Eng.* **2016**, *3*, 271–282. [CrossRef]
18. Hu, J.; Razdan, A.; Femiani, J.C.; Cui, M.; Wonka, P. Road Network Extraction and Intersection Detection From Aerial Images by Tracking Road Footprints. *IEEE Trans. Geosci. Remote Sens.* **2007**, *45*, 4144–4157. [CrossRef]
19. Shen, J.; Lin, X.; Shi, Y.; Wong, C. Knowledge-Based Road Extraction from High Resolution Remotely Sensed Imagery. In Proceedings of the 2008 Congress on Image and Signal Processing, Sanya, China, 27–30 May 2008; IEEE: New York, NY, USA, 2008; Volume 4, pp. 608–612.
20. George, J.; Mary, L.; Riyas, K.S. Vehicle Detection and Classification from Acoustic Signal Using ANN and KNN. In Proceedings of the 2013 International Conference on Control Communication and Computing (ICCC), Thiruvananthapuram, India, 13–15 December 2013; IEEE: New York, NY, USA, 2013; pp. 436–439.
21. Li, J.; Chen, M. On-Road Multiple Obstacles Detection in Dynamical Background. In Proceedings of the 2014 Sixth International Conference on Intelligent Human-Machine Systems and Cybernetics, Hangzhou, China, 26–27 August 2014; IEEE: New York, NY, USA, 2014; Volume 1, pp. 102–105.
22. Simler, C. An Improved Road and Building Detector on VHR Images. In Proceedings of the 2011 IEEE International Geoscience and Remote Sensing Symposium, Vancouver, BC, Canada, 24–29 July 2011; IEEE: New York, NY, USA, 2011; pp. 507–510.
23. Zhu, D.-M.; Wen, X.; Ling, C.-L. Road Extraction Based on the Algorithms of MRF and Hybrid Model of SVM and FCM. In Proceedings of the 2011 International Symposium on Image and Data Fusion, Tengchong, China, 9–11 August 2011; IEEE: New York, NY, USA, 2011; pp. 1–4.

24. Zhou, J.; Bischof, W.F.; Caelli, T. Road Tracking in Aerial Images Based on Human–Computer Interaction and Bayesian Filtering. *ISPRS J. Photogramm. Remote Sens.* **2006**, *61*, 108–124. [CrossRef]
25. Miao, Z.; Wang, B.; Shi, W.; Zhang, H. A Semi-Automatic Method for Road Centerline Extraction From VHR Images. *IEEE Geosci. Remote Sens. Lett.* **2014**, *11*, 1856–1860. [CrossRef]
26. Pawar, V.; Zaveri, M. Graph Based K-Nearest Neighbor Minutiae Clustering for Fingerprint Recognition. In Proceedings of the 2014 10th International Conference on Natural Computation (ICNC), Xiamen, China, 19–21 August 2014; IEEE: New York, NY, USA, 2014; pp. 675–680.
27. Anil, P.N.; Natarajan, S. A Novel Approach Using Active Contour Model for Semi-Automatic Road Extraction from High Resolution Satellite Imagery. In Proceedings of the 2010 Second International Conference on Machine Learning and Computing, Bangalore, India, 9–11 February 2010; IEEE: New York, NY, USA, 2010; pp. 263–266.
28. Abraham, L.; Sasikumar, M. A Fuzzy Based Road Network Extraction from Degraded Satellite Images. In Proceedings of the 2013 International Conference on Advances in Computing, Communications and Informatics (ICACCI), Mysore, India, 22–25 August 2013; IEEE: New York, NY, USA, 2013; pp. 2032–2036.
29. Awrangjeb, M. Road Traffic Island Extraction from High Resolution Aerial Imagery Using Active Contours. In Proceedings of the Australian Remote Sensing & Photogrammetry Conference (ARSPC 2010), Alice Springs, Australia, 13–17 September 2010.
30. Valero, S.; Chanussot, J.; Benediktsson, J.A.; Talbot, H.; Waske, B. Advanced Directional Mathematical Morphology for the Detection of the Road Network in Very High Resolution Remote Sensing Images. *Pattern Recognit. Lett.* **2010**, *31*, 1120–1127. [CrossRef]
31. Ma, R.; Wang, W.; Liu, S. Extracting Roads Based on Retinex and Improved Canny Operator with Shape Criteria in Vague and Unevenly Illuminated Aerial Images. *J. Appl. Remote Sens.* **2012**, *6*, 063610.
32. Movaghati, S.; Moghaddamjoo, A.; Tavakoli, A. Road Extraction from Satellite Images Using Particle Filtering and Extended Kalman Filtering. *IEEE Trans. Geosci. Remote Sens.* **2010**, *48*, 2807–2817. [CrossRef]
33. Barzohar, M.; Cooper, D.B. Automatic Finding of Main Roads in Aerial Images by Using Geometric-Stochastic Models and Estimation. *IEEE Trans. Pattern Anal. Mach. Intell.* **1996**, *18*, 707–721. [CrossRef]
34. Abdollahi, A.; Pradhan, B.; Shukla, N.; Chakraborty, S.; Alamri, A. Deep Learning Approaches Applied to Remote Sensing Datasets for Road Extraction: A State-Of-The-Art Review. *Remote Sens.* **2020**, *12*, 1444. [CrossRef]
35. He, K.; Zhang, X.; Ren, S.; Sun, J. *Identity Mappings in Deep Residual Networks, Proceedings of the Computer Vision—ECCV 2016, Amsterdam, The Netherlands, 8–16 October 2016*; Leibe, B., Matas, J., Sebe, N., Welling, M., Eds.; Springer International Publishing: Berlin/Heidelberg, Germany, 2016; pp. 630–645.
36. Zhong, Z.; Li, J.; Cui, W.; Jiang, H. Fully Convolutional Networks for Building and Road Extraction: Preliminary Results. In Proceedings of the 2016 IEEE International Geoscience and Remote Sensing Symposium (IGARSS), Beijing, China, 10–15 July 2016; pp. 1591–1594.
37. Wei, Y.; Wang, Z.; Xu, M. Road Structure Refined CNN for Road Extraction in Aerial Image. *IEEE Geosci. Remote Sens. Lett.* **2017**, *14*, 709–713. [CrossRef]
38. Alshehhi, R.; Marpu, P.R.; Woon, W.L.; Dalla Mura, M. Simultaneous Extraction of Roads and Buildings in Remote Sensing Imagery with Convolutional Neural Networks. *ISPRS J. Photogramm. Remote Sens.* **2017**, *130*, 139–149. [CrossRef]
39. Liu, R.; Miao, Q.; Song, J.; Quan, Y.; Li, Y.; Xu, P.; Dai, J. Multiscale Road Centerlines Extraction from High-Resolution Aerial Imagery. *Neurocomputing* **2019**, *329*, 384–396. [CrossRef]
40. Li, P.; Zang, Y.; Wang, C.; Li, J.; Cheng, M.; Luo, L.; Yu, Y. Road Network Extraction via Deep Learning and Line Integral Convolution. In Proceedings of the 2016 IEEE International Geoscience and Remote Sensing Symposium (IGARSS), Beijing, China, 10–15 July 2016; pp. 1599–1602.
41. Varia, N.; Dokania, A.; Senthilnath, J. DeepExt: A Convolution Neural Network for Road Extraction Using RGB Images Captured by UAV. In Proceedings of the 2018 IEEE Symposium Series on Computational Intelligence (SSCI), Bangalore, India, 18–21 November 2018; pp. 1890–1895.
42. Abdollahi, A.; Pradhan, B.; Shukla, N. Extraction of Road Features from UAV Images Using a Novel Level Set Segmentation Approach. *Int. J. Urban Sci.* **2019**, *23*, 391–405. [CrossRef]
43. Moranduzzo, T.; Melgani, F. Detecting Cars in UAV Images With a Catalog-Based Approach. *IEEE Trans. Geosci. Remote Sens.* **2014**, *52*, 6356–6367. [CrossRef]
44. Yang, B.; Chen, C. Automatic Registration of UAV-Borne Sequent Images and LiDAR Data. *ISPRS J. Photogramm. Remote Sens.* **2015**, *101*, 262–274. [CrossRef]
45. Kestur, R.; Farooq, S.; Abdal, R.; Mehraj, E.; Narasipura, O.S.; Mudigere, M. UFCN: A Fully Convolutional Neural Network for Road Extraction in RGB Imagery Acquired by Remote Sensing from an Unmanned Aerial Vehicle. *J. Appl. Remote Sens.* **2018**, *12*, 016020. [CrossRef]
46. Henry, C.; Azimi, S.M.; Merkle, N. Road Segmentation in SAR Satellite Images With Deep Fully Convolutional Neural Networks. *IEEE Geosci. Remote Sens. Lett.* **2018**, *15*, 1867–1871. [CrossRef]
47. Panboonyuen, T.; Vateekul, P.; Jitkajornwanich, K.; Lawawirojwong, S. *An Enhanced Deep Convolutional Encoder-Decoder Network for Road Segmentation on Aerial Imagery, Proceedings of the Recent Advances in Information and Communication Technology, Bangkok, Thailand, 5–6 July 2017*; Meesad, P., Sodsee, S., Unger, H., Eds.; Springer International Publishing: Berlin/Heidelberg, Germany, 2018; pp. 191–201.

48. Wang, J.; Song, J.; Chen, M.; Yang, Z. Road Network Extraction: A Neural-Dynamic Framework Based on Deep Learning and a Finite State Machine. *Int. J. Remote Sens.* **2015**, *36*, 3144–3169. [CrossRef]
49. Panboonyuen, T.; Jitkajornwanich, K.; Lawawirojwong, S.; Srestasathiern, P.; Vateekul, P. Road Segmentation of Remotely-Sensed Images Using Deep Convolutional Neural Networks with Landscape Metrics and Conditional Random Fields. *Remote Sens.* **2017**, *9*, 680. [CrossRef]
50. Constantin, A.; Ding, J.-J.; Lee, Y.-C. Accurate Road Detection from Satellite Images Using Modified U-Net. In Proceedings of the 2018 IEEE Asia Pacific Conference on Circuits and Systems (APCCAS), Chengdu, China, 26–30 October 2018; pp. 423–426.
51. Zhang, Z.; Liu, Q.; Wang, Y. Road Extraction by Deep Residual U-Net. *IEEE Geosci. Remote Sens. Lett.* **2018**, *15*, 749–753. [CrossRef]
52. Hong, Z.; Ming, D.; Zhou, K.; Guo, Y.; Lu, T. Road Extraction From a High Spatial Resolution Remote Sensing Image Based on Richer Convolutional Features. *IEEE Access* **2018**, *6*, 46988–47000. [CrossRef]
53. Xin, J.; Zhang, X.; Zhang, Z.; Fang, W. Road Extraction of High-Resolution Remote Sensing Images Derived from DenseUNet. *Remote Sens.* **2019**, *11*, 2499. [CrossRef]
54. Li, Y.; Xu, L.; Rao, J.; Guo, L.; Yan, Z.; Jin, S. A Y-Net Deep Learning Method for Road Segmentation Using High-Resolution Visible Remote Sensing Images. *Remote Sens. Lett.* **2019**, *10*, 381–390. [CrossRef]
55. Cheng, G.; Wang, Y.; Xu, S.; Wang, H.; Xiang, S.; Pan, C. Automatic Road Detection and Centerline Extraction via Cascaded End-to-End Convolutional Neural Network. *IEEE Trans. Geosci. Remote Sens.* **2017**, *55*, 3322–3337. [CrossRef]
56. Xu, Y.; Xie, Z.; Feng, Y.; Chen, Z. Road Extraction from High-Resolution Remote Sensing Imagery Using Deep Learning. *Remote Sens.* **2018**, *10*, 1461. [CrossRef]
57. Buslaev, A.; Seferbekov, S.; Iglovikov, V.; Shvets, A. Fully Convolutional Network for Automatic Road Extraction from Satellite Imagery. In Proceedings of the 2018 IEEE/CVF Conference on Computer Vision and Pattern Recognition Workshops (CVPRW), Salt Lake City, UT, USA, 18–22 June 2018; pp. 197–1973.
58. Zhou, L.; Zhang, C.; Wu, M. D-LinkNet: LinkNet with Pretrained Encoder and Dilated Convolution for High Resolution Satellite Imagery Road Extraction. In Proceedings of the 2018 IEEE/CVF Conference on Computer Vision and Pattern Recognition Workshops (CVPRW), Salt Lake City, UT, USA, 18–22 June 2018; pp. 192–1924.
59. Doshi, J. Residual Inception Skip Network for Binary Segmentation. In Proceedings of the 2018 IEEE/CVF Conference on Computer Vision and Pattern Recognition Workshops (CVPRW), Salt Lake City, UT, USA, 18–22 June 2018; pp. 206–2063.
60. Xu, Y.; Feng, Y.; Xie, Z.; Hu, A.; Zhang, X. A Research on Extracting Road Network from High Resolution Remote Sensing Imagery. In Proceedings of the 2018 26th International Conference on Geoinformatics, Kunming, China, 18–30 June 2018; pp. 1–4.
61. He, H.; Yang, D.; Wang, S.; Wang, S.; Liu, X. Road Segmentation of Cross-Modal Remote Sensing Images Using Deep Segmentation Network and Transfer Learning. *Ind. Robot Int. J. Robot. Res. Appl.* **2019**, *46*, 384–390. [CrossRef]
62. Xia, W.; Zhang, Y.-Z.; Liu, J.; Luo, L.; Yang, K. Road Extraction from High Resolution Image with Deep Convolution Network—A Case Study of GF-2 Image. *Proceedings* **2018**, *2*, 325. [CrossRef]
63. Gao, L.; Song, W.; Dai, J.; Chen, Y. Road Extraction from High-Resolution Remote Sensing Imagery Using Refined Deep Residual Convolutional Neural Network. *Remote Sens.* **2019**, *11*, 552. [CrossRef]
64. Xie, Y.; Miao, F.; Zhou, K.; Peng, J. HsgNet: A Road Extraction Network Based on Global Perception of High-Order Spatial Information. *ISPRS Int. J. Geo-Inf.* **2019**, *8*, 571. [CrossRef]
65. Costea, D.; Marcu, A.; Slusanschi, E.; Leordeanu, M. Creating Roadmaps in Aerial Images with Generative Adversarial Networks and Smoothing-Based Optimization. In Proceedings of the IEEE International Conference on Computer Vision Workshops, Venice, Italy, 22–29 October 2017; pp. 2100–2109.
66. Shi, Q.; Liu, X.; Li, X. Road Detection from Remote Sensing Images by Generative Adversarial Networks. *IEEE Access* **2017**, *6*, 25486–25494. [CrossRef]
67. Belli, D.; Kipf, T. Image-Conditioned Graph Generation for Road Network Extraction. *arXiv* **2019**, arXiv:1910.14388, 4388.
68. Castrejon, L.; Kundu, K.; Urtasun, R.; Fidler, S. Annotating Object Instances With a Polygon-RNN. *arXiv* **2017**, arXiv:1704.05548, 5230–5238.
69. Acuna, D.; Ling, H.; Kar, A.; Fidler, S. Efficient Interactive Annotation of Segmentation Datasets With Polygon-RNN++. *arXiv* **2018**, arXiv:1803.09693, 859–868.
70. Li, Z.; Wegner, J.D.; Lucchi, A. Topological Map Extraction From Overhead Images. *arXiv* **2019**, arXiv:1812.01497, 1715–1724.
71. Lian, R.; Wang, W.; Mustafa, N.; Huang, L. Road Extraction Methods in High-Resolution Remote Sensing Images: A Comprehensive Review. *IEEE J. Sel. Top. Appl. Earth Obs. Remote Sens.* **2020**, *13*, 5489–5507. [CrossRef]
72. Sun, N.; Zhang, J.X.; Huang, G.M.; Zhao, Z.; Lu, L.J. Review of Road Extraction Methods from SAR Image. *IOP Conf. Ser. Earth Environ. Sci.* **2014**, *17*, 012245. [CrossRef]
73. Sun, Z.; Geng, H.; Lu, Z.; Scherer, R.; Woźniak, M. Review of Road Segmentation for SAR Images. *Remote Sens.* **2021**, *13*, 1011. [CrossRef]
74. Wang, G.; Weng, Q. *Remote Sensing of Natural Resources*; CRC Press: Boca Raton, FL, USA, 2013. ISBN 978-1-4665-5692-8.
75. Gargoum, S.; El-Basyouny, K. Automated Extraction of Road Features Using LiDAR Data: A Review of LiDAR Applications in Transportation. In Proceedings of the 2017 4th International Conference on Transportation Information and Safety (ICTIS), Banff, AB, Canada, 8–10 August 2017; pp. 563–574.
76. Ma, L.; Li, Y.; Li, J.; Wang, C.; Wang, R.; Chapman, M.A. Mobile Laser Scanned Point-Clouds for Road Object Detection and Extraction: A Review. *Remote Sens.* **2018**, *10*, 1531. [CrossRef]

77. Wang, R.; Peethambaran, J.; Chen, D. LiDAR Point Clouds to 3-D Urban Models: A Review. *IEEE J. Sel. Top. Appl. Earth Obs. Remote Sens.* **2018**, *11*, 606–627. [CrossRef]
78. Alparone, L.; Baronti, S.; Garzelli, A.; Nencini, F. A Global Quality Measurement of Pan-Sharpened Multispectral Imagery. *IEEE Geosci. Remote Sens. Lett.* **2004**, *1*, 313–317. [CrossRef]
79. French, A.N.; Norman, J.M.; Anderson, M.C. A Simple and Fast Atmospheric Correction for Spaceborne Remote Sensing of Surface Temperature. *Remote Sens. Environ.* **2003**, *87*, 326–333. [CrossRef]
80. Jia, J.; Wang, Y.; Chen, J.; Guo, R.; Shu, R.; Wang, J. Status and Application of Advanced Airborne Hyperspectral Imaging Technology: A Review. *Infrared Phys. Technol.* **2020**, *104*, 103115. [CrossRef]
81. Turner, D.; Lucieer, A.; Watson, C. An Automated Technique for Generating Georectified Mosaics from Ultra-High Resolution Unmanned Aerial Vehicle (UAV) Imagery, Based on Structure from Motion (SfM) Point Clouds. *Remote Sens.* **2012**, *4*, 1392–1410. [CrossRef]
82. Ozesmi, S.L.; Bauer, M.E. Satellite Remote Sensing of Wetlands. *Wetl. Ecol. Manag. Vol.* **2002**, *10*, 381–402. [CrossRef]
83. Sato, H.P.; Hasegawa, H.; Fujiwara, S.; Tobita, M.; Koarai, M.; Une, H.; Iwahashi, J. Interpretation of Landslide Distribution Triggered by the 2005 Northern Pakistan Earthquake Using SPOT 5 Imagery. *Landslides* **2007**, *4*, 113–122. [CrossRef]
84. Yadav, S.K.; Singh, S.K.; Gupta, M.; Srivastava, P.K. Morphometric Analysis of Upper Tons Basin from Northern Foreland of Peninsular India Using CARTOSAT Satellite and GIS. *Geocarto Int.* **2014**, *29*, 895–914. [CrossRef]
85. Dial, G.; Bowen, H.; Gerlach, F.; Grodecki, J.; Oleszczuk, R. IKONOS Satellite, Imagery, and Products. *Remote Sens. Environ.* **2003**, *88*, 23–36. [CrossRef]
86. Li, D.; Wang, M.; Jiang, J. China's High-Resolution Optical Remote Sensing Satellites and Their Mapping Applications. *Geo-Spat. Inf. Sci.* **2021**, *24*, 85–94. [CrossRef]
87. Hao, P.; Wang, L.; Niu, Z. Potential of Multitemporal Gaofen-1 Panchromatic/Multispectral Images for Crop Classification: Case Study in Xinjiang Uygur Autonomous Region, China. *J. Appl. Remote Sens.* **2015**, *9*, 096035. [CrossRef]
88. Zheng, Y.; Dai, Q.; Tu, Z.; Wang, L. Guided Image Filtering-Based Pan-Sharpening Method: A Case Study of GaoFen-2 Imagery. *ISPRS Int. J. Geo-Inf.* **2017**, *6*, 404. [CrossRef]
89. Yang, A.; Zhong, B.; Hu, L.; Wu, S.; Xu, Z.; Wu, H.; Wu, J.; Gong, X.; Wang, H.; Liu, Q. Radiometric Cross-Calibration of the Wide Field View Camera Onboard GaoFen-6 in Multispectral Bands. *Remote Sens.* **2020**, *12*, 1037. [CrossRef]
90. Liu, Y.-K.; Liu, Y.-K.; Ma, L.-L.; Ma, L.-L.; Wang, N.; Qian, Y.-G.; Qian, Y.-G.; Zhao, Y.-G.; Qiu, S.; Gao, C.-X.; et al. On-Orbit Radiometric Calibration of the Optical Sensors on-Board SuperView-1 Satellite Using Three Independent Methods. *Opt. Express* **2020**, *28*, 11085–11105. [CrossRef]
91. Aguilar, M.A.; Saldaña, M.M.; Aguilar, F.J. GeoEye-1 and WorldView-2 Pan-Sharpened Imagery for Object-Based Classification in Urban Environments. *Int. J. Remote Sens.* **2013**, *34*, 2583–2606. [CrossRef]
92. Shi, Y.; Huang, W.; Ye, H.; Ruan, C.; Xing, N.; Geng, Y.; Dong, Y.; Peng, D. Partial Least Square Discriminant Analysis Based on Normalized Two-Stage Vegetation Indices for Mapping Damage from Rice Diseases Using PlanetScope Datasets. *Sensors* **2018**, *18*, 1901. [CrossRef] [PubMed]
93. Meusburger, K.; Bänninger, D.; Alewell, C. Estimating Vegetation Parameter for Soil Erosion Assessment in an Alpine Catchment by Means of QuickBird Imagery. *Int. J. Appl. Earth Obs. Geoinf.* **2010**, *12*, 201–207. [CrossRef]
94. Alkan, M.; Buyuksalih, G.; Sefercik, U.G.; Jacobsen, K. Geometric Accuracy and Information Content of WorldView-1 Images. *Opt. Eng.* **2013**, *52*, 026201. [CrossRef]
95. Ye, B.; Tian, S.; Ge, J.; Sun, Y. Assessment of WorldView-3 Data for Lithological Mapping. *Remote Sens.* **2017**, *9*, 1132. [CrossRef]
96. Akumu, C.E.; Amadi, E.O.; Dennis, S. Application of Drone and WorldView-4 Satellite Data in Mapping and Monitoring Grazing Land Cover and Pasture Quality: Pre- and Post-Flooding. *Land* **2021**, *10*, 321. [CrossRef]
97. Mulawa, D. On-Orbit Geometric Calibration of the OrbView-3 High Resolution Imaging Satellite. *Int. Arch. Photogramm. Remote Sens. Spat. Inf. Sci.* **2004**, *35*, 1–6.
98. Tyc, G.; Tulip, J.; Schulten, D.; Krischke, M.; Oxfort, M. The RapidEye Mission Design. *Acta Astronaut.* **2005**, *56*, 213–219. [CrossRef]
99. Oh, K.-Y.; Jung, H.-S. Automated Bias-Compensation Approach for Pushbroom Sensor Modeling Using Digital Elevation Model. *IEEE Trans. Geosci. Remote Sens.* **2016**, *54*, 3400–3409. [CrossRef]
100. Kim, J.; Jin, C.; Choi, C.; Ahn, H. Radiometric Characterization and Validation for the KOMPSAT-3 Sensor. *Remote Sens. Lett.* **2015**, *6*, 529–538. [CrossRef]
101. Seo, D.; Oh, J.; Lee, C.; Lee, D.; Choi, H. Geometric Calibration and Validation of Kompsat-3A AEISS-A Camera. *Sensors* **2016**, *16*, 1776. [CrossRef] [PubMed]
102. Kubik, P.; Lebègue, L.; Fourest, S.; Delvit, J.-M.; de Lussy, F.; Greslou, D.; Blanchet, G. First In-Flight Results of Pleiades 1A Innovative Methods for Optical Calibration. In Proceedings of the International Conference on Space Optics—ICSO 2012; International Society for Optics and Photonics, Ajaccio, France, 9–12 October 2012; Volume 10564, p. 1056407.
103. Panagiotakis, E.; Chrysoulakis, N.; Charalampopoulou, V.; Poursanidis, D. Validation of Pleiades Tri-Stereo DSM in Urban Areas. *ISPRS Int. J. Geo-Inf.* **2018**, *7*, 118. [CrossRef]
104. Yang, G.D.; Zhu, X. Ortho-Rectification of SPOT 6 Satellite Images Based on RPC Models. *Appl. Mech. Mater.* **2013**, *392*, 808–814. [CrossRef]

105. Wilson, K.L.; Skinner, M.A.; Lotze, H.K. Eelgrass (Zostera Marina) and Benthic Habitat Mapping in Atlantic Canada Using High-Resolution SPOT 6/7 Satellite Imagery. *Estuar. Coast. Shelf Sci.* **2019**, *226*, 106292. [CrossRef]
106. Rais, A.A.; Suwaidi, A.A.; Ghedira, H. DubaiSat-1: Mission Overview, Development Status and Future Applications. In Proceedings of the 2009 IEEE International Geoscience and Remote Sensing Symposium, Cape Town, South Africa, 12–17 July 2009; Volume 5, pp. V-196–V-199.
107. Suwaidi, A.A. DubaiSat-2 Mission Overview. In *Sensors, Systems, and Next-Generation Satellites XVI*; International Society for Optics and Photonics: Edinburgh, UK, 2012; Volume 8533, p. 85330W.
108. Immitzer, M.; Böck, S.; Einzmann, K.; Vuolo, F.; Pinnel, N.; Wallner, A.; Atzberger, C. Fractional Cover Mapping of Spruce and Pine at 1ha Resolution Combining Very High and Medium Spatial Resolution Satellite Imagery. *Remote Sens. Environ.* **2018**, *204*, 690–703. [CrossRef]
109. Hamedianfar, A.; Shafri, H.Z.M. Detailed Intra-Urban Mapping through Transferable OBIA Rule Sets Using WorldView-2 Very-High-Resolution Satellite Images. *Int. J. Remote Sens.* **2015**, *36*, 3380–3396. [CrossRef]
110. Diaz-Varela, R.A.; Zarco-Tejada, P.J.; Angileri, V.; Loudjani, P. Automatic Identification of Agricultural Terraces through Object-Oriented Analysis of Very High Resolution DSMs and Multispectral Imagery Obtained from an Unmanned Aerial Vehicle. *J. Environ. Manag.* **2014**, *134*, 117–126. [CrossRef]
111. Goetz, A.F.H.; Vane, G.; Solomon, J.E.; Rock, B.N. Imaging Spectrometry for Earth Remote Sensing. *Science* **1985**, *228*, 1147–1153. [CrossRef] [PubMed]
112. Plaza, A.; Benediktsson, J.A.; Boardman, J.W.; Brazile, J.; Bruzzone, L.; Camps-Valls, G.; Chanussot, J.; Fauvel, M.; Gamba, P.; Gualtieri, A.; et al. Recent Advances in Techniques for Hyperspectral Image Processing. *Remote Sens. Environ.* **2009**, *113*, S110–S122. [CrossRef]
113. Green, R.O.; Eastwood, M.L.; Sarture, C.M.; Chrien, T.G.; Aronsson, M.; Chippendale, B.J.; Faust, J.A.; Pavri, B.E.; Chovit, C.J.; Solis, M.; et al. Imaging Spectroscopy and the Airborne Visible/Infrared Imaging Spectrometer (AVIRIS). *Remote Sens. Environ.* **1998**, *65*, 227–248. [CrossRef]
114. Goetz, A.F.H.; Srivastava, V. Mineralogical Mapping in the Cuprite Mining District, Nevada. In Proceedings of the Airborne Imaging Spectrometer Data Analysis Workshop, JPL Publication 85-41, Jet Propulsion Laboratory. Pasadena, CA, USA, 8–10 April 1985; pp. 22–29.
115. Zarco-Tejada, P.J.; Guillén-Climent, M.L.; Hernández-Clemente, R.; Catalina, A.; González, M.R.; Martín, P. Estimating Leaf Carotenoid Content in Vineyards Using High Resolution Hyperspectral Imagery Acquired from an Unmanned Aerial Vehicle (UAV). *Agric. For. Meteorol.* **2013**, *171–172*, 281–294. [CrossRef]
116. Hruska, R.; Mitchell, J.; Anderson, M.; Glenn, N.F. Radiometric and Geometric Analysis of Hyperspectral Imagery Acquired from an Unmanned Aerial Vehicle. *Remote Sens.* **2012**, *4*, 2736–2752. [CrossRef]
117. Yue, J.; Yang, G.; Li, C.; Li, Z.; Wang, Y.; Feng, H.; Xu, B. Estimation of Winter Wheat Above-Ground Biomass Using Unmanned Aerial Vehicle-Based Snapshot Hyperspectral Sensor and Crop Height Improved Models. *Remote Sens.* **2017**, *9*, 708. [CrossRef]
118. Lu, J.; Liu, H.; Yao, Y.; Tao, S.; Tang, Z.; Lu, J. Hsi Road: A Hyper Spectral Image Dataset For Road Segmentation. In Proceedings of the 2020 IEEE International Conference on Multimedia and Expo (ICME), London, UK, 6–10 July 2020; pp. 1–6.
119. Wendel, A.; Underwood, J. Illumination Compensation in Ground Based Hyperspectral Imaging. ISPRS J. Photogramm. *Remote Sens.* **2017**, *129*, 162–178. [CrossRef]
120. Van der Meer, F.D.; van der Werff, H.M.A.; van Ruitenbeek, F.J.A.; Hecker, C.A.; Bakker, W.H.; Noomen, M.F.; van der Meijde, M.; Carranza, E.J.M.; de Smeth, J.B.; Woldai, T. Multi- and Hyperspectral Geologic Remote Sensing: A Review. *Int. J. Appl. Earth Obs. Geoinf.* **2012**, *14*, 112–128. [CrossRef]
121. Stuffler, T.; Kaufmann, C.; Hofer, S.; Förster, K.P.; Schreier, G.; Mueller, A.; Eckardt, A.; Bach, H.; Penné, B.; Benz, U.; et al. The EnMAP Hyperspectral Imager—An Advanced Optical Payload for Future Applications in Earth Observation Programmes. *Acta Astronaut.* **2007**, *61*, 115–120. [CrossRef]
122. Carmon, N.; Ben-Dor, E. Mapping Asphaltic Roads' Skid Resistance Using Imaging Spectroscopy. *Remote Sens.* **2018**, *10*, 430. [CrossRef]
123. Schaepman, M.E.; Jehle, M.; Hueni, A.; D'Odorico, P.; Damm, A.; Weyermann, J.; Schneider, F.D.; Laurent, V.; Popp, C.; Seidel, F.C.; et al. Advanced Radiometry Measurements and Earth Science Applications with the Airborne Prism Experiment (APEX). *Remote Sens. Environ.* **2015**, *158*, 207–219. [CrossRef]
124. Edberg, S.J.; Evans, D.L.; Graf, J.E.; Hyon, J.J.; Rosen, P.A.; Waliser, D.E. Studying Earth in the New Millennium: NASA Jet Propulsion Laboratory's Contributions to Earth Science and Applications Space Agencies. *IEEE Geosci. Remote Sens. Mag.* **2016**, *4*, 26–39. [CrossRef]
125. Green, R.O.; Team, C. New Measurements of the Earth's Spectroscopic Diversity Acquired during the AVIRIS-NG Campaign to India. In Proceedings of the 2017 IEEE International Geoscience and Remote Sensing Symposium (IGARSS), Fort Worth, TX, USA, 23–28 July 2017; pp. 3066–3069.
126. Jie-lin, Z.; Jun-hu, W.; Mi, Z.; Yan-ju, H.; Ding, W. Aerial Visible-Thermal Infrared Hyperspectral Feature Extraction Technology and Its Application to Object Identification. *IOP Conf. Ser. Earth Environ. Sci.* **2014**, *17*, 012184. [CrossRef]
127. Jia, J.; Wang, Y.; Cheng, X.; Yuan, L.; Zhao, D.; Ye, Q.; Zhuang, X.; Shu, R.; Wang, J. Destriping Algorithms Based on Statistics and Spatial Filtering for Visible-to-Thermal Infrared Pushbroom Hyperspectral Imagery. *IEEE Trans. Geosci. Remote Sens.* **2019**, *57*, 4077–4091. [CrossRef]

128. Jia, J.; Zheng, X.; Guo, S.; Wang, Y.; Chen, J. Removing Stripe Noise Based on Improved Statistics for Hyperspectral Images. *IEEE Geosci. Remote Sens. Lett.* **2020**, 1–5. [CrossRef]
129. Rouvière, L.R.; Sisakoun, I.; Skauli, T.; Coudrain, C.; Ferrec, Y.; Fabre, S.; Poutier, L.; Boucher, Y.; Løke, T.; Blaaberg, S. Sysiphe, an Airborne Hyperspectral System from Visible to Thermal Infrared. In Proceedings of the 2016 IEEE International Geoscience and Remote Sensing Symposium (IGARSS), Beijing, China, 10–15 July 2016; pp. 1947–1949.
130. Marmo, J.; Folkman, M.A.; Kuwahara, C.Y.; Willoughby, C.T. Lewis Hyperspectral Imager Payload Development. *Proc. SPIE* **1996**, *2819*, 80–90.
131. Pearlman, J.S.; Barry, P.S.; Segal, C.C.; Shepanski, J.; Beiso, D.; Carman, S.L. Hyperion, a Space-Based Imaging Spectrometer. *IEEE Trans. Geosci. Remote Sens.* **2003**, *41*, 1160–1173. [CrossRef]
132. Barnsley, M.J.; Settle, J.J.; Cutter, M.A.; Lobb, D.R.; Teston, F. The PROBA/CHRIS Mission: A Low-Cost Smallsat for Hyperspectral Multiangle Observations of the Earth Surface and Atmosphere. *IEEE Trans. Geosci. Remote Sens.* **2004**, *42*, 1512–1520. [CrossRef]
133. Murchie, S.; Arvidson, R.; Bedini, P.; Beisser, K.; Bibring, J.-P.; Bishop, J.; Boldt, J.; Cavender, P.; Choo, T.; Clancy, R.T.; et al. Compact Reconnaissance Imaging Spectrometer for Mars (CRISM) on Mars Reconnaissance Orbiter (MRO). *J. Geophys. Res. Planets* **2007**, 112. [CrossRef]
134. Liu, Y.; Sun, D.; Hu, X.; Ye, X.; Li, Y.; Liu, S.; Cao, K.; Chai, M.; Zhou, W.; Zhang, J.; et al. The Advanced Hyperspectral Imager: Aboard China's GaoFen-5 Satellite. *IEEE Geosci. Remote Sens. Mag.* **2019**, *7*, 23–32. [CrossRef]
135. Kimuli, D.; Wang, W.; Wang, W.; Jiang, H.; Zhao, X.; Chu, X. Application of SWIR Hyperspectral Imaging and Chemometrics for Identification of Aflatoxin B1 Contaminated Maize Kernels. *Infrared Phys. Technol.* **2018**, *89*, 351–362. [CrossRef]
136. Ambrose, A.; Kandpal, L.M.; Kim, M.S.; Lee, W.H.; Cho, B.K. High Speed Measurement of Corn Seed Viability Using Hyperspectral Imaging. *Infrared Phys. Technol.* **2016**, *75*, 173–179. [CrossRef]
137. He, H.; Sun, D. Hyperspectral Imaging Technology for Rapid Detection of Various Microbial Contaminants in Agricultural and Food Products. *Trends Food Sci. Technol.* **2015**, *46*, 99–109. [CrossRef]
138. Randolph, K.; Wilson, J.; Tedesco, L.; Li, L.; Pascual, D.L.; Soyeux, E. Hyperspectral Remote Sensing of Cyanobacteria in Turbid Productive Water Using Optically Active Pigments, Chlorophyll a and Phycocyanin. *Remote Sens. Environ.* **2008**, *112*, 4009–4019. [CrossRef]
139. Huang, H.; Liu, L.; Ngadi, M.O. Recent Developments in Hyperspectral Imaging for Assessment of Food Quality and Safety. *Sensors* **2014**, *14*, 7248–7276. [CrossRef] [PubMed]
140. Brando, V.E.; Dekker, A.G. Satellite Hyperspectral Remote Sensing for Estimating Estuarine and Coastal Water Quality. *IEEE Trans. Geosci. Remote Sens.* **2003**, *41*, 1378–1387. [CrossRef]
141. Shimoni, M.; Haelterman, R.; Perneel, C. Hypersectral Imaging for Military and Security Applications: Combining Myriad Processing and Sensing Techniques. *IEEE Geosci. Remote Sens. Mag.* **2019**, *7*, 101–117. [CrossRef]
142. Kruse, F.A. Comparative Analysis of Airborne Visible/Infrared Imaging Spectrometer (AVIRIS), and Hyperspectral Thermal Emission Spectrometer (HyTES) Longwave Infrared (LWIR) Hyperspectral Data for Geologic Mapping. In Proceedings of the Algorithms and Technologies for Multispectral, Hyperspectral, and Ultraspectral Imagery XXI; International Society for Optics and Photonics, Baltimore, MD, USA, 21 May 2015; Volume 9472, p. 94721F.
143. Duren, R.M.; Thorpe, A.K.; Foster, K.T.; Rafiq, T.; Hopkins, F.M.; Yadav, V.; Bue, B.D.; Thompson, D.R.; Conley, S.; Colombi, N.K.; et al. California's Methane Super-Emitters. *Nature* **2019**, *575*, 180–184. [CrossRef]
144. Gelautz, M.; Frick, H.; Raggam, J.; Burgstaller, J.; Leberl, F. SAR Image Simulation and Analysis of Alpine Terrain. ISPRS J. Photogramm. *Remote Sens.* **1998**, *53*, 17–38. [CrossRef]
145. Haldar, D.; Das, A.; Mohan, S.; Pal, O.; Hooda, R.S.; Chakraborty, M. Assessment of L-Band SAR Data at Different Polarization Combinations for Crop and Other Landuse Classification. *Prog. Electromagn. Res. B* **2012**, *36*, 303–321. [CrossRef]
146. Raney, R.K. Hybrid-Polarity SAR Architecture. *IEEE Trans. Geosci. Remote Sens.* **2007**, *45*, 3397–3404. [CrossRef]
147. McNairn, H.; Brisco, B. The Application of C-Band Polarimetric SAR for Agriculture: A Review. *Can. J. Remote Sens.* **2004**, *30*, 525–542. [CrossRef]
148. Jung, J.; Kim, D.; Lavalle, M.; Yun, S.-H. Coherent Change Detection Using InSAR Temporal Decorrelation Model: A Case Study for Volcanic Ash Detection. *IEEE Trans. Geosci. Remote Sens.* **2016**, *54*, 5765–5775. [CrossRef]
149. Liu, J.G.; Black, A.; Lee, H.; Hanaizumi, H.; Moore, J.M.c.M. Land Surface Change Detection in a Desert Area in Algeria Using Multi-Temporal ERS SAR Coherence Images. *Int. J. Remote Sens.* **2001**, *22*, 2463–2477. [CrossRef]
150. Monti-Guarnieri, A.V.; Brovelli, M.A.; Manzoni, M.; Mariotti d'Alessandro, M.; Molinari, M.E.; Oxoli, D. Coherent Change Detection for Multipass SAR. *IEEE Trans. Geosci. Remote Sens.* **2018**, *56*, 6811–6822. [CrossRef]
151. Wahl, D.E.; Yocky, D.A.; Jakowatz, C.V.; Simonson, K.M. A New Maximum-Likelihood Change Estimator for Two-Pass SAR Coherent Change Detection. *IEEE Trans. Geosci. Remote Sens.* **2016**, *54*, 2460–2469. [CrossRef]
152. Vosselman, G.; Maas, H.G. *Airborne and Terrestrial Laser Scanning*; CRC Press: Boca Raton, FL, USA, 2010. ISBN 978-1-904445-87-6.
153. Garestier, F.; Dubois-Fernandez, P.C.; Papathanassiou, K.P. Pine Forest Height Inversion Using Single-Pass X-Band PolInSAR Data. *IEEE Trans. Geosci. Remote Sens.* **2007**, *46*, 59–68. [CrossRef]
154. Rizzoli, P.; Martone, M.; Gonzalez, C.; Wecklich, C.; Borla Tridon, D.; Bräutigam, B.; Bachmann, M.; Schulze, D.; Fritz, T.; Huber, M.; et al. Generation and Performance Assessment of the Global TanDEM-X Digital Elevation Model. *ISPRS J. Photogramm. Remote Sens.* **2017**, *132*, 119–139. [CrossRef]

155. Horn, R.; Nottensteiner, A.; Reigber, A.; Fischer, J.; Scheiber, R. F-SAR—DLR's New Multifrequency Polarimetric Airborne SAR. In Proceedings of the 2009 IEEE International Geoscience and Remote Sensing Symposium, Cape Town, South Africa, 12–17 July 2009; Volume 2, pp. II-902–II-905.
156. Dell'Acqua, F.; Gamba, P. Rapid Mapping Using Airborne and Satellite SAR Images. In *Radar Remote Sensing of Urban Areas*; Soergel, U., Ed.; Remote Sensing and Digital Image Processing; Springer Netherlands: Dordrecht, The Netherlands, 2010; pp. 49–68. ISBN 978-90-481-3751-0.
157. Xiao, F.; Tong, L.; Luo, S. A Method for Road Network Extraction from High-Resolution SAR Imagery Using Direction Grouping and Curve Fitting. *Remote Sens.* **2019**, *11*, 2733. [CrossRef]
158. Zhang, Q.; Kong, Q.; Zhang, C.; You, S.; Wei, H.; Sun, R.; Li, L. A New Road Extraction Method Using Sentinel-1 SAR Images Based on the Deep Fully Convolutional Neural Network. *Eur. J. Remote Sens.* **2019**, *52*, 572–582. [CrossRef]
159. Harvey, W.A.; McKeown, D.M., Jr. Automatic Compilation of 3D Road Features Using LIDAR and Multi-Spectral Source Data. In Proceedings of the ASPRS Annual Conference, Portland, OR, USA, 28 April–2 May 2008; p. 11.
160. Cheng, L.; Wu, Y.; Wang, Y.; Zhong, L.; Chen, Y.; Li, M. Three-Dimensional Reconstruction of Large Multilayer Interchange Bridge Using Airborne LiDAR Data. *IEEE J. Sel. Top. Appl. Earth Obs. Remote Sens.* **2015**, *8*, 691–708. [CrossRef]
161. How to Plan for a Leica CityMapper-2 Project. Available online: https://blog.hexagongeosystems.com/how-to-plan-for-a-leica-citymapper-2-project/ (accessed on 21 July 2021).
162. Leica SPL100 Single Photon LiDAR Sensor. Available online: https://leica-geosystems.com/products/airborne-systems/topographic-lidar-sensors/leica-spl100 (accessed on 21 July 2021).
163. Communicatie, F.M. ALTM Galaxy PRIME. Available online: https://geo-matching.com/airborne-laser-scanning/altm-galaxy-prime (accessed on 21 July 2021).
164. Wichmann, V.; Bremer, M.; Lindenberger, J.; Rutzinger, M.; Georges, C.; Petrini-Monteferri, F. Evaluating the Potential of Multispectral Airborne LiDAR for Topographic Mapping and Land Cover Classification. *ISPRS Ann. Photogramm. Remote Sens. Spat. Inf. Sci.* **2015**, 2. [CrossRef]
165. Pilarska, M.; Ostrowski, W. Evaluating the Possibility of Tree Species Calssification with Dual-Wavelength ALS Data. Int. Arch. Photogramm. *Remote Sens. Spat. Inf. Sci.* **2019**.
166. RIEGL—RIEGL VUX-240. Available online: http://www.riegl.com/products/unmanned-scanning/riegl-vux-240/ (accessed on 21 July 2021).
167. Magnoni, A.; Stanton, T.W.; Barth, N.; Fernandez-Diaz, J.C.; León, J.F.O.; Ruíz, F.P.; Wheeler, J.A. Detection Thresholds of Archaeological Features in Airborne LiDAR Data from Central Yucatán. *Adv. Archaeol. Pract.* **2016**, *4*, 232–248. [CrossRef]
168. Saito, S.; Yamashita, T.; Aoki, Y. Multiple Object Extraction from Aerial Imagery with Convolutional Neural Networks. *Electron. Imaging* **2016**, *2016*, 1–9. [CrossRef]
169. Ventura, C.; Pont-Tuset, J.; Caelles, S.; Maninis, K.-K.; Van Gool, L. Iterative Deep Learning for Road Topology Extraction. *arXiv* **2018**, arXiv:1808.09814.
170. Lian, R.; Huang, L. DeepWindow: Sliding Window Based on Deep Learning for Road Extraction from Remote Sensing Images. *IEEE J. Sel. Top. Appl. Earth Obs. Remote Sens.* **2020**, *13*, 1905–1916. [CrossRef]
171. Kass, M.; Witkin, A.; Terzopoulos, D. Snakes: Active Contour Models. *Int. J. Comput. Vis.* **1988**, *1*, 321–331. [CrossRef]
172. Gruen, A.; Li, H. Semi-Automatic Linear Feature Extraction by Dynamic Programming and LSB-Snakes. *Photogramm. Eng. Remote Sens.* **1997**, *63*, 985–994.
173. Jagalingam, P.; Vittal, V.H.; Vittal, A. Hegde Review of Quality Metrics for Fused Image. *Aquat. Procedia* 2015.
174. Song, M.; Civco, D. Road Extraction Using SVM and Image Segmentation. *Photogramm. Eng. Remote Sens.* **2004**, *70*, 1365–1371. [CrossRef]
175. Mayer, H. Object Extraction in Photogrammetric Computer Vision. *ISPRS J. Photogramm. Remote Sens.* **2008**, *63*, 213–222. [CrossRef]
176. Kirthika, A.; Mookambiga, A. Automated Road Network Extraction Using Artificial Neural Network. In Proceedings of the 2011 International Conference on Recent Trends in Information Technology (ICRTIT), Chennai, India, 3–5 June 2011; pp. 1061–1065.
177. Li, M.; Zang, S.; Zhang, B.; Li, S.; Wu, C. A Review of Remote Sensing Image Classification Techniques: The Role of Spatio-Contextual Information. *Eur. J. Remote Sens.* **2014**, *47*, 389–411. [CrossRef]
178. Yang, X.-S.; Cui, Z.; Xiao, R.; Gandomi, A.H.; Karamanoglu, M. *Swarm Intelligence and Bio-Inspired Computation: Theory and Applications*; Elsevier: Waltham, MA, USA, 2013. ISBN 0-12-405177-4.
179. Goodfellow, I.J.; Pouget-Abadie, J.; Mirza, M.; Xu, B.; Warde-Farley, D.; Ozair, S.; Courville, A.; Bengio, Y. Generative Adversarial Networks. *arXiv* **2014**, arXiv:1406.2661.
180. Zhang, Q.; Couloigner, I. Benefit of the Angular Texture Signature for the Separation of Parking Lots and Roads on High Resolution Multi-Spectral Imagery. *Pattern Recognit. Lett.* **2006**, *27*, 937–946. [CrossRef]
181. Zhang, Q.; Couloigner, I. Automated Road Network Extraction from High Resolution Multi-Spectral Imagery. In Proceedings of the ASPRS 2006 Annual Conference, Reno, NV, USA, 1–5 May 2006.
182. Manandhar, P.; Marpu, P.R.; Aung, Z. Segmentation Based Traversing-Agent Approach for Road Width Extraction from Satellite Images Using Volunteered Geographic Information. *Appl. Comput. Inform.* **2018**, *17*, 131–152. [CrossRef]
183. Boggess, J.E. *Identification of Roads in Satellite Imagery Using Artificial Neural Networks: A Contextual Approach*; Mississippi State University: Starkville, MS, USA, 1993.

184. Doucette, P.; Agouris, P.; Stefanidis, A.; Musavi, M. Self-Organised Clustering for Road Extraction in Classified Imagery. *ISPRS J. Photogramm. Remote Sens.* **2001**, *55*, 347–358. [CrossRef]
185. Shackelford, A.K.; Davis, C.H. A Hierarchical Fuzzy Classification Approach for High-Resolution Multispectral Data over Urban Areas. *IEEE Trans. Geosci. Remote Sens.* **2003**, *41*, 1920–1932. [CrossRef]
186. Doucette, P.; Agouris, P.; Stefanidis, A. Automated Road Extraction from High Resolution Multispectral Imagery. *Photogramm. Eng. Remote Sens.* **2004**, *70*, 1405–1416. [CrossRef]
187. Jin, X.; Davis, C.H. An Integrated System for Automatic Road Mapping from High-Resolution Multi-Spectral Satellite Imagery by Information Fusion. *Inf. Fusion* **2005**, *6*, 257–273. [CrossRef]
188. Shi, W.; Miao, Z.; Debayle, J. An Integrated Method for Urban Main-Road Centerline Extraction From Optical Remotely Sensed Imagery. *IEEE Trans. Geosci. Remote Sens.* **2014**, *52*, 3359–3372. [CrossRef]
189. Liu, W.; Zhang, Z.; Chen, X.; Li, S.; Zhou, Y. Dictionary Learning-Based Hough Transform for Road Detection in Multispectral Image. *IEEE Geosci. Remote Sens. Lett.* **2017**, *14*, 2330–2334. [CrossRef]
190. Sun, T.-L. A Detection Algorithm for Road Feature Extraction Using EO-1 Hyperspectral Images. In Proceedings of the IEEE 37th Annual 2003 International Carnahan Conference onSecurity Technology, Taipei, Taiwan, 14–16 October 2003; pp. 87–95.
191. Gardner, M.E.; Roberts, D.A.; Funk, C. Road Extraction from AVIRIS Using Spectral Mixture and Q-Tree Filter Techniques. In Proceedings of the AVIRIS Airborne Geoscience Workshop, Santa Barbara, CA, USA, 1 December 2001; Volume 27, p. 6.
192. Noronha, V.; Herold, M.; Roberts, D.; Gardner, M. Spectrometry and Hyperspectral Remote Sensing for Road Centerline Extraction and Evaluation of Pavement Condition. In Proceedings of the Pecora Conference, San Diego, CA, USA, 11–13 March 2002.
193. Huang, X.; Zhang, L. An Adaptive Mean-Shift Analysis Approach for Object Extraction and Classification From Urban Hyperspectral Imagery. *IEEE Trans. Geosci. Remote Sens.* **2008**, *46*, 4173–4185. [CrossRef]
194. Resende, M.; Jorge, S.; Longhitano, G.; Quintanilha, J.A. Use of Hyperspectral and High Spatial Resolution Image Data in an Asphalted Urban Road Extraction. In Proceedings of the IGARSS 2008—2008 IEEE International Geoscience and Remote Sensing Symposium, Boston, MA, USA, 8–11 July 2008; IEEE: New York, NY, USA, 2008; pp. III-1323–III-1325.
195. Mohammadi, M. Road Classification and Condition Determination Using Hyperspectral Imagery. *ISPRS—Int. Arch. Photogramm. Remote Sens. Spat. Inf. Sci.* **2012**, *XXXIX-B7*, 141–146. [CrossRef]
196. Liao, W.; Bellens, R.; Pizurica, A.; Philips, W.; Pi, Y. Classification of Hyperspectral Data Over Urban Areas Using Directional Morphological Profiles and Semi-Supervised Feature Extraction. *IEEE J. Sel. Top. Appl. Earth Obs. Remote Sens.* **2012**, *5*, 1177–1190. [CrossRef]
197. Miao, Z.; Shi, W.; Zhang, H.; Wang, X. Road Centerline Extraction From High-Resolution Imagery Based on Shape Features and Multivariate Adaptive Regression Splines. *IEEE Geosci. Remote Sens. Lett.* **2013**, *10*, 583–587. [CrossRef]
198. Abdellatif, M.; Peel, H.; Cohn, A.G.; Fuentes, R. Hyperspectral Imaging for Autonomous Inspection of Road Pavement Defects. In Proceedings of the 36th International Symposium on Automation and Robotics in Construction (ISARC), Banff, AB, Canada, 24 May 2019.
199. Tupin, F.; Maitre, H.; Mangin, J.-F.; Nicolas, J.-M.; Pechersky, E. Detection of Linear Features in SAR Images: Application to Road Network Extraction. *IEEE Trans. Geosci. Remote Sens.* **1998**, *36*, 434–453. [CrossRef]
200. Tupin, F.; Houshmand, B.; Datcu, M. Road Detection in Dense Urban Areas Using SAR Imagery and the Usefulness of Multiple Views. *IEEE Trans. Geosci. Remote Sens.* **2002**, *40*, 2405–2414. [CrossRef]
201. Wang, Y.; Zheng, Q. Recognition of Roads and Bridges in SAR Images. *Pattern Recognit.* **1998**, *31*, 953–962. [CrossRef]
202. Dell'Acqua, F.; Gamba, P.; Lisini, G. Road Map Extraction by Multiple Detectors in Fine Spatial Resolution SAR Data. *Can. J. Remote Sens.* **2003**, *29*, 481–490. [CrossRef]
203. Lisini, G.; Tison, C.; Tupin, F.; Gamba, P. Feature Fusion to Improve Road Network Extraction in High-Resolution SAR Images. *IEEE Geosci. Remote Sens. Lett.* **2006**, *3*, 217–221. [CrossRef]
204. Hedman, K.; Stilla, U.; Lisini, G.; Gamba, P. Road Network Extraction in VHR SAR Images of Urban and Suburban Areas by Means of Class-Aided Feature-Level Fusion. *IEEE Trans. Geosci. Remote Sens.* **2010**, *48*, 1294–1296. [CrossRef]
205. He, C.; Liao, Z.; Yang, F.; Deng, X.; Liao, M. Road Extraction From SAR Imagery Based on Multiscale Geometric Analysis of Detector Responses. *IEEE J. Sel. Top. Appl. Earth Obs. Remote Sens.* **2012**, *5*, 1373–1382. [CrossRef]
206. Lu, P.; Du, K.; Yu, W.; Wang, R.; Deng, Y.; Balz, T. A New Region Growing-Based Method for Road Network Extraction and Its Application on Different Resolution SAR Images. *IEEE J. Sel. Top. Appl. Earth Obs. Remote Sens.* **2014**, *7*, 4772–4783. [CrossRef]
207. Saati, M.; Amini, J. Road Network Extraction from High-Resolution SAR Imagery Based on the Network Snake Model. *Photogramm. Eng. Remote Sens.* **2017**, *83*, 207–215. [CrossRef]
208. Xu, R.; He, C.; Liu, X.; Chen, D.; Qin, Q. Bayesian Fusion of Multi-Scale Detectors for Road Extraction from SAR Images. *ISPRS Int. J. Geo-Inf.* **2017**, *6*, 26. [CrossRef]
209. Xiong, X.; Jin, G.; Xu, Q.; Zhang, H.; Xu, J. Robust Line Detection of Synthetic Aperture Radar Images Based on Vector Radon Transformation. *IEEE J. Sel. Top. Appl. Earth Obs. Remote Sens.* **2019**, *12*, 5310–5320. [CrossRef]
210. Jiang, M.; Miao, Z.; Gamba, P.; Yong, B. Application of Multitemporal InSAR Covariance and Information Fusion to Robust Road Extraction. *IEEE Trans. Geosci. Remote Sens.* **2017**, *55*, 3611–3622. [CrossRef]
211. Jin, R.; Zhou, W.; Yin, J.; Yang, J. CFAR Line Detector for Polarimetric SAR Images Using Wilks' Test Statistic. *IEEE Geosci. Remote Sens. Lett.* **2016**, *13*, 711–715. [CrossRef]

212. Scharf, D.P. Analytic Yaw–Pitch Steering for Side-Looking SAR With Numerical Roll Algorithm for Incidence Angle. *IEEE Trans. Geosci. Remote Sens.* **2012**, *50*, 3587–3594. [CrossRef]
213. Clode, S.; Kootsookos, P.J.; Rottensteiner, F. *The Automatic Extraction of Roads from LIDAR Data*; ISPRS: Istanbul, Turkey, 2004.
214. Clode, S.; Rottensteiner, F.; Kootsookos, P.; Zelniker, E. Detection and Vectorization of Roads from Lidar Data. *Photogramm. Eng. Remote Sens.* **2007**, *73*, 517–535. [CrossRef]
215. Hu, X.; Li, Y.; Shan, J.; Zhang, J.; Zhang, Y. Road Centerline Extraction in Complex Urban Scenes From LiDAR Data Based on Multiple Features. *IEEE Trans. Geosci. Remote Sens.* **2014**, *52*, 7448–7456. [CrossRef]
216. Li, Y.; Yong, B.; Wu, H.; An, R.; Xu, H. Road Detection from Airborne LiDAR Point Clouds Adaptive for Variability of Intensity Data. *Optik* **2015**, *126*, 4292–4298. [CrossRef]
217. Hui, Z.; Hu, Y.; Jin, S.; Yevenyo, Y.Z. Road Centerline Extraction from Airborne LiDAR Point Cloud Based on Hierarchical Fusion and Optimization. ISPRS J. Photogramm. *Remote Sens.* **2016**, *118*, 22–36. [CrossRef]
218. Zhao, J.; You, S.; Huang, J. Rapid Extraction and Updating of Road Network from Airborne LiDAR Data. In Proceedings of the 2011 IEEE Applied Imagery Pattern Recognition Workshop (AIPR), Washington, DC, USA, 11–13 October 2011; pp. 1–7.
219. Chen, Z.; Liu, C.; Wu, H. A Higher-Order Tensor Voting-Based Approach for Road Junction Detection and Delineation from Airborne LiDAR Data. *ISPRS J. Photogramm. Remote Sens.* **2019**, *150*, 91–114. [CrossRef]
220. Sithole, G.; Vosselman, G. Bridge Detection in Airborne Laser Scanner Data. ISPRS J. Photogramm. *Remote Sens.* **2006**, *61*, 33–46. [CrossRef]
221. Boyko, A.; Funkhouser, T. Extracting Roads from Dense Point Clouds in Large Scale Urban Environment. *ISPRS J. Photogramm. Remote Sens.* **2011**, *66*, S2–S12. [CrossRef]
222. Lin, Y.; Hyyppä, J.; Jaakkola, A. Mini-UAV-Borne LIDAR for Fine-Scale Mapping. *IEEE Geosci. Remote Sens. Lett.* **2011**, *8*, 426–430. [CrossRef]
223. Soilán, M.; Truong-Hong, L.; Riveiro, B.; Laefer, D. Automatic Extraction of Road Features in Urban Environments Using Dense ALS Data. *Int. J. Appl. Earth Obs. Geoinformation* **2018**, *64*, 226–236. [CrossRef]
224. Zhou, W. An Object-Based Approach for Urban Land Cover Classification: Integrating LiDAR Height and Intensity Data. *IEEE Geosci. Remote Sens. Lett.* **2013**, *10*, 928–931. [CrossRef]
225. Matkan, A.A.; Hajeb, M.; Sadeghian, S. Road Extraction from Lidar Data Using Support Vector Machine Classification. *Photogramm. Eng. Remote Sens.* **2014**, *80*, 409–422. [CrossRef]
226. Morsy, S.; Shaker, A.; El-Rabbany, A. Multispectral LiDAR Data for Land Cover Classification of Urban Areas. *Sensors* **2017**, *17*, 958. [CrossRef]
227. Karila, K.; Matikainen, L.; Puttonen, E.; Hyyppä, J. Feasibility of Multispectral Airborne Laser Scanning Data for Road Mapping. *IEEE Geosci. Remote Sens. Lett.* **2017**, *14*, 294–298. [CrossRef]
228. Ekhtari, N.; Glennie, C.; Fernandez-Diaz, J.C. Classification of Airborne Multispectral Lidar Point Clouds for Land Cover Mapping. *IEEE J. Sel. Top. Appl. Earth Obs. Remote Sens.* **2018**, *11*, 2068–2078. [CrossRef]
229. Pan, S.; Guan, H.; Yu, Y.; Li, J.; Peng, D. A Comparative Land-Cover Classification Feature Study of Learning Algorithms: DBM, PCA, and RF Using Multispectral LiDAR Data. *IEEE J. Sel. Top. Appl. Earth Obs. Remote Sens.* **2019**, *12*, 1314–1326. [CrossRef]
230. Pan, S.; Guan, H.; Chen, Y.; Yu, Y.; Nunes Gonçalves, W.; Marcato Junior, J.; Li, J. Land-Cover Classification of Multispectral LiDAR Data Using CNN with Optimized Hyper-Parameters. *ISPRS J. Photogramm. Remote Sens.* **2020**, *166*, 241–254. [CrossRef]
231. Yu, Y.; Guan, H.; Li, D.; Gu, T.; Wang, L.; Ma, L.; Li, J. A Hybrid Capsule Network for Land Cover Classification Using Multispectral LiDAR Data. *IEEE Geosci. Remote Sens. Lett.* **2020**, *17*, 1263–1267. [CrossRef]
232. Matikainen, L.; Karila, K.; Litkey, P.; Ahokas, E.; Hyyppä, J. Combining Single Photon and Multispectral Airborne Laser Scanning for Land Cover Classification. *ISPRS J. Photogramm. Remote Sens.* **2020**, *164*, 200–216. [CrossRef]
233. Tiwari, P.S.; Pande, H.; Pandey, A.K. Automatic Urban Road Extraction Using Airborne Laser Scanning/Altimetry and High Resolution Satellite Data. *J. Indian Soc. Remote Sens.* **2009**, *37*, 223. [CrossRef]
234. Hu, X.; Tao, C.V.; Hu, Y. Automatic Road Extraction from Dense Urban Area by Integrated Processing of High Resolution Imagery and Lidar Data. *Int. Arch. Photogramm. Remote Sens. Spat. Inf. Sci.* **2004**, *35*, 288–292.
235. Zhang, Z.; Zhang, X.; Sun, Y.; Zhang, P. Road Centerline Extraction from Very-High-Resolution Aerial Image and LiDAR Data Based on Road Connectivity. *Remote Sens.* **2018**, *10*, 1284. [CrossRef]
236. Feng, Q.; Zhu, D.; Yang, J.; Li, B. Multisource Hyperspectral and LiDAR Data Fusion for Urban Land-Use Mapping Based on a Modified Two-Branch Convolutional Neural Network. *ISPRS Int. J. Geo-Inf.* **2019**, *8*, 28. [CrossRef]
237. Elaksher, A.F. Fusion of Hyperspectral Images and Lidar-Based Dems for Coastal Mapping. *Opt. Lasers Eng.* **2008**, *46*, 493–498. [CrossRef]
238. Hsu, S.M.; Burke, H. Multisensor fusion with hyperspectral imaging data: Detection and classification. In *Handbook of Pattern Recognition and Computer Vision*; WORLD SCIENTIFIC: Singapore, 2005; pp. 347–364. ISBN 978-981-256-105-3.
239. Cao, G.; Jin, Y.Q. A Hybrid Algorithm of the BP-ANN/GA for Classification of Urban Terrain Surfaces with Fused Data of Landsat ETM+ and ERS-2 SAR. *Int. J. Remote Sens.* **2007**, *28*, 293–305. [CrossRef]
240. Lin, X.; Liu, Z.; Zhang, J.; Shen, J. Combining Multiple Algorithms for Road Network Tracking from Multiple Source Remotely Sensed Imagery: A Practical System and Performance Evaluation. *Sensors* **2009**, *9*, 1237–1258. [CrossRef]
241. Perciano, T.; Tupin, F.; Jr, R.H.; Jr, R.M.C. A Two-Level Markov Random Field for Road Network Extraction and Its Application with Optical, SAR, and Multitemporal Data. *Int. J. Remote Sens.* **2016**, *37*, 3584–3610. [CrossRef]

242. Bartsch, A.; Pointner, G.; Ingeman-Nielsen, T.; Lu, W. Towards Circumpolar Mapping of Arctic Settlements and Infrastructure Based on Sentinel-1 and Sentinel-2. *Remote Sens.* **2020**, *12*, 2368. [CrossRef]
243. Liu, S.; Qi, Z.; Li, X.; Yeh, A.G.-O. Integration of Convolutional Neural Networks and Object-Based Post-Classification Refinement for Land Use and Land Cover Mapping with Optical and SAR Data. *Remote Sens.* **2019**, *11*, 690. [CrossRef]
244. Lin, Y.; Zhang, H.; Li, G.; Wang, T.; Wan, L.; Lin, H. Improving Impervious Surface Extraction With Shadow-Based Sparse Representation From Optical, SAR, and LiDAR Data. *IEEE J. Sel. Top. Appl. Earth Obs. Remote Sens.* **2019**, *12*, 2417–2428. [CrossRef]
245. Kim, Y.; Kim, Y. Improved Classification Accuracy Based on the Output-Level Fusion of High-Resolution Satellite Images and Airborne LiDAR Data in Urban Area. *IEEE Geosci. Remote Sens. Lett.* **2014**, *11*, 636–640. [CrossRef]
246. Liu, L.; Lim, S. A Framework of Road Extraction from Airborne Lidar Data and Aerial Imagery. *J. Spat. Sci.* **2016**, *61*, 263–281. [CrossRef]
247. Chen, Z.; Fan, W.; Zhong, B.; Li, J.; Du, J.; Wang, C. Corse-to-Fine Road Extraction Based on Local Dirichlet Mixture Models and Multiscale-High-Order Deep Learning. *IEEE Trans. Intell. Transp. Syst.* **2019**, *21*, 4283–4293. [CrossRef]
248. Bruzzone, L.; Carlin, L. A Multilevel Context-Based System for Classification of Very High Spatial Resolution Images. *IEEE Trans. Geosci. Remote Sens.* **2006**, *44*, 2587–2600. [CrossRef]
249. Wang, J.; Qin, Q.; Yang, X.; Wang, J.; Ye, X.; Qin, X. Automated Road Extraction from Multi-Resolution Images Using Spectral Information and Texture. In Proceedings of the 2014 IEEE Geoscience and Remote Sensing Symposium, Quebec City, QC, Canada, 13–18 July 2014; pp. 533–536.
250. Zhao, W.; Du, S. Spectral–Spatial Feature Extraction for Hyperspectral Image Classification: A Dimension Reduction and Deep Learning Approach. *IEEE Trans. Geosci. Remote Sens.* **2016**, *54*, 4544–4554. [CrossRef]
251. Hamraz, H.; Jacobs, N.B.; Contreras, M.A.; Clark, C.H. Deep Learning for Conifer/Deciduous Classification of Airborne LiDAR 3D Point Clouds Representing Individual Trees. ISPRS J. Photogramm. *Remote Sens.* **2019**, *158*, 219–230.
252. Jia, J.; Chen, J.; Zheng, X.; Wang, Y.; Guo, S.; Sun, H.; Jiang, C.; Karjalainen, M.; Karila, K.; Duan, Z.; et al. Tradeoffs in the Spatial and Spectral Resolution of Airborne Hyperspectral Imaging Systems: A Crop Identification Case Study. *IEEE Trans. Geosci. Remote Sens.* **2021**, 1–18. [CrossRef]

MDPI
St. Alban-Anlage 66
4052 Basel
Switzerland
Tel. +41 61 683 77 34
Fax +41 61 302 89 18
www.mdpi.com

Remote Sensing Editorial Office
E-mail: remotesensing@mdpi.com
www.mdpi.com/journal/remotesensing

www.ingramcontent.com/pod-product-compliance
Lightning Source LLC
Chambersburg PA
CBHW061832210926
51850CB00006B/868
9783036570648